FOUNDATIONS OF ALGEBRAIC TOPOLOGY

PRINCETON MATHEMATICAL SERIES

Editors: MARSTON MORSE and A. W. TUCKER

1. The Classical Groups, Their Invariants and Representatives. By HERMANN WEYL
2. Topological Groups. By L. PONTRJAGIN. Translated by EMMA LEHMER.
3. An Introduction to Differential Geometry with Use of the Tensor Calculus. By LUTHER PFAHLER EISENHART.
4. Dimension Theory. By WITOLD HUREWICZ and HENRY WALLMAN.
5. The Analytical Foundations of Celestial Mechanics. By AUREL WINTNER.
6. The Laplace Transform. By DAVID VERNON WIDDER.
7. Integration. By EDWARD JAMES McSHANE.
8. Theory of Lie Groups. I. By CLAUDE CHEVALLEY.
9. Mathematical Methods of Statistics. By HARALD CRAMÉR.
10. Several Complex Variables. By SALOMON BOCHNER and WILLIAM TED MARTIN.
11. Introduction to Topology. By SOLOMON LEFSCHETZ.
12. Algebraic Geometry and Topology. Edited by R. H. FOX, D. C. SPENCER, and A. W. TUCKER.
13. Algebraic Curves. By ROBERT J. WALKER.
14. The Topology of Fibre Bundles. By NORMAN STEENROD.
15. Foundations of Algebraic Topology. By SAMUEL EILENBERG and NORMAN STEENROD.
16. Functionals of Finite Riemann Surfaces. By MENAHEM SCHIFFER and DONALD C. SPENCER.
17. Introduction to Mathematical Logic, Vol. I. By ALONZO CHURCH.
18. Algebraic Geometry. By SOLOMON LEFSCHETZ.
19. Homological Algebra. By HENRI CARTAN and SAMUEL EILENBERG.
20. The Convolution Transform. By I. I. HIRSCHMAN and D. V. WIDDER.
21. Geometric Integration Theory. By HASSLER WHITNEY.

FOUNDATIONS OF ALGEBRAIC TOPOLOGY

BY SAMUEL EILENBERG

AND

NORMAN STEENROD

PRINCETON, NEW JERSEY

PRINCETON UNIVERSITY PRESS

1952

L. C. Card 52-5841

LONDON: OXFORD UNIVERSITY PRESS

SECOND PRINTING 1957

PRINTED IN THE UNITED STATES OF AMERICA

TO SOLOMON LEFSCHETZ
IN ADMIRATION AND GRATITUDE

Preface

1. PREAMBLE

The principal contribution of this book is an axiomatic approach to the part of algebraic topology called homology theory It is the oldest and most extensively developed portion of algebraic topology, and may be regarded as the main body of the subject The present axiomatization is the first which has been given. The dual theory of cohomology is likewise axiomatized

It is assumed that the reader is familiar with the basic concepts of algebra and of point set topology No attempt is made to axiomatize these subjects. This has been done extensively in the literature Our achievement is different in kind Homology theory is a transition (or function) from topology to algebra. It is this transition which is axiomatized

Speaking roughly, a homology theory assigns groups to topological spaces and homomorphisms to continuous maps of one space into another. To each array of spaces and maps is assigned an array of groups and homomorphisms In this way, a homology theory is an *algebraic image* of topology. The *domain* of a homology theory is the topologist's field of study Its *range* is the field of study of the algebraist. Topological problems are converted into algebraic problems.

In this respect, homology theory parallels analytic geometry However, unlike analytic geometry, it is not reversible The derived algebraic system represents only an aspect of the given topological system, and is usually much simpler This has the advantage that the geometric problem is stripped of inessential features and replaced by a familiar type of problem which one can hope to solve It has the disadvantage that some essential feature may be lost. In spite of this, the subject has proved its value by a great variety of successful applications.

Our axioms are statements of the fundamental properties of this assignment of an algebraic system to a topological system. The axioms are *categorical* in the sense that two such assignments give isomorphic algebraic systems.

2. THE NEED FOR AXIOMATIZATION

The construction of a homology theory is exceedingly complicated. It is true that the definitions and necessary lemmas can be compressed within ten pages, and the main properties established within a hundred. But this is achieved by disregarding numerous problems raised by the construction, and ignoring the problem of computing illustrative examples. These are serious problems, as is well known to anyone who has taught the subject There is need for a perspective, and a pattern into which the student can fit the numerous parts

Part of the complexity of the subject is that numerous variants of the basic definitions have appeared, e.g. the singular homology groups of Veblen, Alexander, and Lefschetz, the relative homology groups of Lefschetz, the Vietoris homology groups, the Čech homology groups, the Alexander cohomology groups, etc. The objective of each variant was to extend the validity of some basic theorems, and thereby increase their range of applicability.

In spite of this confusion, a picture has gradually evolved of what is and should be a homology theory. Heretofore this has been an imprecise picture which the expert could use in his thinking but not in his exposition. A precise picture is needed. It is at just this stage in the development of other fields of mathematics that an axiomatic treatment appeared and cleared the air.

The discussion will be advanced by a rough outline of the construction of the homology groups of a space. There are four main steps as follows:

(1) space $\to$ complex
(2) complex $\to$ oriented complex
(3) oriented complex $\to$ groups of chains
(4) groups of chains $\to$ homology groups

In the first form of (1) it was necessary to place on a space the structure of a complex by decomposing it into subsets called cells, each cell being a homeomorph of a euclidean cube of some dimension, and any two cells meeting, if at all, in common faces. It was recognized that only certain spaces, called triangulable, admit such a decomposition. There arose the problem of characterizing triangulable spaces by other properties. This is still unsolved. Special classes of spaces (e g differentiable manifolds) have been proved triangulable and these suffice for many applications.

The assumption of triangulability was eliminated in three different ways by the works of Vietoris, Lefschetz, and Čech. In each case, the relation that the complex be a triangulation of the space was replaced by another more complicated relation, and the complex had to be infinite The gain was made at the cost of effective computability.

Step (2) has also been a source of trouble. The problem is to assign integers (incidence numbers) to each pair consisting of a cell and a face (of one lower dimension) so as to satisfy the condition that the boundary of a cell be a cycle This is always possible, but the general proof requires the existence of a homology theory To avoid circularity, it was necessary to restrict the class of complexes to those for which orientability could be proved directly The simplicial complexes form such a class This feature and several others combined to make simplicial complexes the dominant type Their sole defect is that computations which use them are excessively long, so much so that they are impractical for the computation of the homology groups of a space as simple as a torus

Steps (3) and (4) have not caused trouble They are purely algebraic and unique The only difficulty a student faces is the absence of motivation

The final major problem is the proof of the topological invariance of the composite assignment of homology groups to a space Equivalently, one must show that the homology groups are independent of the choices made in steps (1) and (2) Some thirty years were required for the development of a fully satisfactory proof of invariance Several problems arising along the way have not yet been solved, e g Do homeomorphic complexes have isomorphic subdivisions?

The origin of the present axiomatic treatment was an effort, on the part of the authors, to write a textbook on algebraic topology We were faced with the problem of presenting two parallel lines of thought One was the rigorous and abstract development of the homology groups of a space in the manner of Lefschetz or Čech, a procedure which lacks apparent motivation, and is noneffective so far as calculation is concerned The other was the nonrigorous, partly intuitive, and computable method of assigning homology groups which marked the early historical development of the subject In addition the two lines had to be merged eventually so as to justify the various computations These difficulties made clear the need of an axiomatic approach.

The axioms which we use meet this need in every respect. Their statement requires only the concepts of point set topology and of algebra The concepts of complex, orientation, and chain do not appear here However, the axioms lead one to introduce complexes in order to calculate the homology groups of various spaces. Furthermore, each of the steps (2), (3), and (4) is derived from the axioms These derivations are an essential part of the proof of the categorical nature of the axioms

Summarizing, the construction of homology groups is a long and diverse story, with a fairly obscure motivation. In contrast, the axioms, which are given in a few pages, state precisely the ultimate goal, and motivate every step of the construction

No motivation is offered for the axioms themselves. The beginning student is asked to take these on faith until the completion of the first three chapters This should not be difficult, for most of the axioms are quite natural, and their totality possesses sufficient internal beauty to inspire trust in the least credulous.

3. COMPARISON WITH OTHER AXIOMATIC SYSTEMS

The need for an axiomatic treatment has been felt by topologists for many years This has resulted in the axiomatization of certain stages in the construction of homology groups W Mayer isolated the step (4). He defined the abstract and purely algebraic concept of chain complex, and showed that it was adequate for the completion of step (4) He also demonstrated that a number of mixed geometric–algebraic concepts and arguments could be handled with algebra alone.

A W Tucker axiomatized the notion of an abstract cell complex, i.e. the initial point of step (2), and showed that steps (2), (3), and (4) could be carried through starting with such an object This had the effect of relegating the geometry to step (1) alone where, of course, it is essential.

Recently H. Cartan and J. Leray have axiomatized the concept of a grating (carapace) on a space In essence, it replaces the notion of complex in the four-step construction outlined in §2. Their associated invariance theorem has several advantages Most important is the inclusion of the de Rham theorem which relates the exterior differential forms in a manifold to the cohomology groups of the manifold

It is to be noted that these various systems are axiomatizations of *stages* in the construction of homology groups None of them axiomatize a *transition* from one stage to another Thus they differ both in scope and in nature from the axioms we shall give The latter axiomatize the full transition from spaces to homology groups.

4. NEW METHODS

The great gain of an axiomatic treatment lies in the simplification obtained in proofs of theorems Proofs based directly on the axioms are usually simple and conceptual It is no longer necessary for a proof to be burdened with the heavy machinery used to define the homology groups Nor is one faced at the end of a proof by the question, Does the proof still hold if another homology theory replaces the one used? When a homology theory has been shown to satisfy the axioms, the machinery of its construction may be dropped

Successful axiomatizations in the past have led invariably to new techniques of proof and a corresponding new language The present system is no exception The reader will observe the presence of numerous diagrams in the text Each diagram is a network or linear graph in which each vertex represents a group, and each oriented edge represents a homomorphism connecting the groups at its two ends

A directed path in the network represents the homomorphism which is the composition of the homomorphisms assigned to its edges Two paths connecting the same pair of vertices usually give the same homomorphism This is called a *commutativity* relation The combinatorially minded individual can regard it as a homology relation due to the presence of 2-dimensional cells adjoined to the graph

If, at some vertex of the graph, two abutting edges are in line, one oriented toward the vertex and the other away, it is frequently the case that the *image* of the incoming homomorphism coincides with the *kernel* of the outgoing homomorphism. This property is called *exactness* It asserts that the group at the vertex is determined, up to a group extension, by the two neighboring groups, the kernel of the incoming homomorphism, and the image of the outgoing homomorphism Exact *sequences* of groups and homomorphisms occur throughout Their algebraic properties are readily established, and are very convenient

The reader will note that there is a vague analogy between the commutativity-exactness relations in a diagram and the two Kirchhoff laws for an electrical network

Certain diagrams occur repeatedly in whole or as parts of others. Once the abstract properties of such a diagram have been established, they apply each time it recurs.

The diagrams incorporate a large amount of information Their use provides extensive savings in space and in mental effort In the case of many theorems, the setting up of the correct diagram is the major part of the proof We therefore urge that the reader stop at the end of each theorem and attempt to construct for himself the relevant diagram before examining the one which is given in the text Once this is done, the subsequent demonstration can be followed more readily, in fact, the reader can usually supply it himself

5. STRUCTURE OF THE BOOK

Chapter I presents the axioms for a homology theory, and a body of general theorems deducible from them. Simplicial complexes and triangulable spaces are treated in Chapter II. This chapter is entirely geometric. In Chapter III, a homology theory is assumed to be given on triangulable spaces. We then derive from the axioms the classical algorithms for computing the homology groups of a complex. Using these, we show that the axioms are categorical for homology theories on triangulable spaces.

The first three chapters form a closed unit, but one which is based on the assumption that a homology theory exists, i e. the axioms are consistent in a nontrivial manner. In Chapters IV through X, the existence is established in four different ways. The singular homology theory is given in Chapter VII, the Čech homology theory in IX, and two others in X

The intervening chapters are preparatory. As noted above, the construction of a homology theory is complicated. Not only do we have the four steps outlined in §2 for the construction of homology groups, but also corresponding constructions of homomorphisms. Then the axioms must be verified Finally the dual cohomology theory must be given in each case. With a total of eight theories to present, the tendency to repeat constructions and parallel others is nearly irresistible. We avoid most of this by presenting a number of steps on a sufficiently abstract level to make them usable in all cases These are given in Chaps IV, V

Chapter IV presents the ideas and language of *category* and *functor*. These concepts formalize a point of view which has dominated the development of the entire book. We axiomatize here the notion of a homology theory on an abstract category, and formulate a pattern which the subsequent constructions must fulfill. In Chapter V, the step from chain complexes to homology groups is treated. The chapter is entirely algebraic Chapter VI presents the classical homology theory of simplicial complexes. In Chapter VII, the singular homology theory is defined and proved to satisfy the axioms. This chapter is independent of VI except possibly for motivation. A reader interested in the shortest existence proof need only read Chapter IV, the first four articles of V, and then Chapter VII

Chapter VIII treats direct and inverse systems of groups and their limit groups. This is the algebraic machinery needed for the development of the Čech homology theory given in IX. Chapter X presents additional properties of the Čech theory It is shown that the addition of a single new axiom characterizes the Čech theory on compact spaces. Two additional homology theories are constructed which are extensions of the Čech theory on compact spaces. The first is defined on locally compact spaces, and the second on normal spaces. Both are obtained by processes of compactification.

Chapter XI, which completes this volume, gives a number of the classical applications of homology theory such as the Brouwer fixed-point theorem, invariance of domain, and the fundamental theorem of algebra.

Homology theory and cohomology theory are dual to one another. We treat them in parallel. Throughout Chapters I and III, each section which treats of homology is accompanied by a section on cohomology. The latter contains no proofs. It contains just the list of definitions and theorems dual to those given for homology, and are numbered correspondingly, e g Definition 4 1c is the cohomology form of Definition 4.1. The duality between the two theories has only a semiformal status. It is true that, by the use of special "coefficient groups," Pontrjagin has given a strictly formal duality based on his theory of character groups. However, the duality appears to persist without such restrictions. The reader is urged to supply the proofs for the cohomology sections. In addition to constituting useful exercises, such proofs will familiarize the reader with the language of cohomology.

The device of dual sections occurs rarely in later chapters The greater parts of the constructions of homology theories and their corresponding cohomology theories deal with mechanisms in which the two aspects are not differentiated In the remaining parts, the two dual theories are treated in equal detail.

At the end of each chapter is a list of exercises These cover material which might well have been incorporated in the text but was omitted as not essential to the main line of thought

There are no footnotes Instead, comments on the historical development and on the connections with other subjects are gathered together in the form of notes at the ends of various chapters.

A cross reference gives the chapter number first, then the section, and, lastly, the numbered proposition, e g X,2 6 refers to Proposition 6 of Section 2 of Chapter X The chapter number is omitted in the case where it is the one containing the reference. A reference of the form (3) means the displayed formula number 3 of the section at hand.

We acknowledge with pleasure our indebtedness to Professors S. MacLane, T Rado, and P Reichelderfer who read large portions of the manuscript and whose suggestions and criticisms resulted in substantial improvements.

S. EILENBERG AND N. STEENROD

August, 1951
Columbia University
Princeton University

Contents

FOUNDATIONS OF ALGEBRAIC TOPOLOGY

CHAPTER I

Axioms and general theorems

1. TOPOLOGICAL PRELIMINARIES

The axioms for a homology theory are given in §3. In §§1 and 2, we review the language and notation of topology and algebra, and we introduce a number of definitions and conventions which, as will be seen, are virtually enforced by the nature of our axiomatic system

We define a *pair of sets* (X,A) to be a set X and a subset A of X. In case $A = 0$ is the vacuous subset, the symbol $(X,0)$ is usually abbreviated by (X) or, simply, X

A *map* f of (X,A) into (Y,B), in symbols

$$f\colon\ (X,A) \to (Y,B),$$

is a single-valued function from X to Y such that $f(A) \subset B$. If also $g\colon\ (Y,B) \to (Z,C)$, then the composition of the two functions is a map $gf\colon\ (X,A) \to (Z, C)$ given by $(gf)x = g(fx)$

The relation $(X',A') \subset (X,A)$ means $X' \subset X$ and $A' \subset A$. The map $i.\ (X',A') \to (X,A)$ defined by $ix = x$ for each $x \in X'$ is called the *inclusion* map and is denoted by

$$i\colon\ (X',A') \subset (X,A).$$

If $(X',A') = (X,A)$, then the inclusion map i is called the *identity* map of (X,A)

It will be important for us to distinguish a function from those obtained from it by seemingly trivial modifications of the domain or range Let $f.\ (X,A) \to (Y,B)$ be given, and let (X',A'), (Y',B') be pairs such that $X' \subset X$, $Y' \subset Y$, $f(X') \subset Y'$, and $f(A') \subset B'$ Then the unique map $f'\colon\ (X',A') \to (Y',B')$ such that $f'x = fx$ for each $x \in X'$ is called *the map defined by* f, and f is said *to define* f'. For example any inclusion map is defined by the identity map of its range. If $f\colon\ (X,A) \to (Y,B)$, the map of A into B defined by f is denoted by

$$f|A\colon\ A \to B.$$

The *lattice* of a pair (X,A) consists of the pairs

$$\begin{array}{ccccccc} & & & (X,0) & & & \\ & & \nearrow & & \searrow & & \\ (0,0) & \longrightarrow (A,0) & & & & (X,A) \longrightarrow & (X,X) \\ & & \searrow & & \nearrow & & \\ & & & (A,A) & & & \end{array}$$

all their identity maps, the inclusion maps indicated by arrows, and all their compositions. If $f\colon (X,A) \to (Y,B)$, then f defines a map of every pair of the lattice of (X,A) into the corresponding pair of the lattice of (Y,B). In particular $f|A$ is one of these maps

A *topological space* X is a set X together with a family of subsets of X, called *open* sets, subject to the following conditions:

(1) the set X and the empty set 0 are open,

(2) the union of any family of open sets is open,

(3) the intersection of a finite family of open sets is open

The family of open sets of X is called the *topology* of X The complement $X - U$ of an open set U is called *closed*.

If A is a subset of X, then the union of all the open sets contained in A is called the *interior* of A and is denoted by Int A. The intersection of all the closed sets containing A is called the *closure* of A and is denoted by $\bar{A}$. A topology, called the *relative topology*, is defined in A by the family of intersections $A \cap U$ for all open sets U of X. With this topology, A is called a subspace and (X,A) is called a *pair of topological spaces* or, briefly, a *pair*. If X,X' are spaces, the relation $X' \subset X$ means that X' is a subspace of X.

A map $f\colon X \to Y$ of one topological space into another is said to be continuous if, for every open set V of Y, the set $f^{-1}(V)$ is open in X A map of pairs $f\colon (X,A) \to (Y,B)$ is continuous if the map $X \to Y$ defined by f is continuous. The terms map, mapping, and transformation when applied to topological spaces or pairs will always mean continuous maps. Identity and inclusion maps are continuous.

A space X is a Hausdorff space if, for each pair of distinct points $x_1,x_2 \in X$, there exist disjoint open sets U_1,U_2 in X with $x_1 \in U_1$, $x_2 \in U_2$. A space is called *compact* if it is a Hausdorff space, and if each covering of the space by open sets contains a finite covering A pair (X,A) is called compact if X is compact and A is a closed (and therefore compact) subset of X.

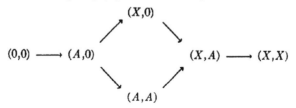

The foregoing is intended as a review of some basic definitions. We shall assume a knowledge of the elementary properties of spaces and

maps such as can be found, for instance, in the book of Alexandroff and Hopf (*Topologie*, J Springer, Berlin 1935) Chapters 1 and 2, or in the book of Lefschetz (*Algebraic Topology*, Colloq. Pub. Amer. Math. Soc. 1942) Chapter 1

DEFINITION A family $\mathfrak{A}$ of pairs of spaces and maps of such pairs which satisfies the conditions (1) to (5) below is called an *admissible category for homology theory*. The pairs and maps of $\mathfrak{A}$ are called *admissible*

(1) If $(X,A) \in \mathfrak{A}$, then all pairs and inclusion maps of the lattice of (X,A) are in $\mathfrak{A}$

(2) If $f: (X,A) \to (Y,B)$ is in $\mathfrak{A}$, then (X,A) and (Y,B) are in $\mathfrak{A}$ together with all maps that f defines of members of the lattice of (X,A) into corresponding members of the lattice of (Y,B).

(3) If f_1 and f_2 are in $\mathfrak{A}$, and their composition f_1f_2 is defined, then $f_1f_2 \in \mathfrak{A}$.

(4) If $I = [0,1]$ is the closed unit interval, and $(X,A) \in \mathfrak{A}$, then the cartesian product

$$(X,A) \times I = (X \times I, A \times I)$$

is in $\mathfrak{A}$ and the maps

$$g_0,g_1: (X,A) \to (X,A) \times I$$

given by

$$g_0(x) = (x,0), \qquad g_1(x) = (x,1)$$

are in $\mathfrak{A}$

(5) There is in $\mathfrak{A}$ a space P_0 consisting of a single point. If X,P are in $\mathfrak{A}$, if $f: P \to X$, and if P is a single point, then $f \in \mathfrak{A}$.

The following are examples of admissible categories for homology theory

$\mathfrak{A}_1$ = the set of all pairs (X,A) and all maps of such pairs. This is the largest admissible category.

$\mathfrak{A}_C$ = the set of all compact pairs and all maps of such pairs

$\mathfrak{A}_{LC}$ = the set of pairs (X,A) where X is a locally compact Hausdorff space, A is closed in X, and all maps of such pairs having the property that the inverse images of compact sets are compact sets

This last example of an admissible category has the property that both X and X' can be admissible, $X' \subset X$, and yet the inclusion map $X' \to X$ may not be admissible This is the case if X is compact and X' is an open but not closed subset of X.

DEFINITION. Two maps f_0, f_1: $(X, A) \to (Y,B)$ in the admissible category $\mathfrak{A}$ are said to be *homotopic* in $\mathfrak{A}$ if there is a map

$$h: (X,A) \times I \to (Y,B)$$

in $\mathfrak{A}$ such that

$$f_0 = hg_0, \qquad f_1 = hg_1$$

or, explicitly,

$$f_0(x) = h(x,0), \qquad f_1(x) = h(x,1).$$

The map h is called a *homotopy*.

2. ALGEBRAIC PRELIMINARIES

Let R be a ring with a unit element. An abelian group G is called an *R-module* if for each $r \in R$ and each $g \in G$ an element $rg \in G$ is defined such that

$$r(g_1 + g_2) = rg_1 + rg_2, \qquad (r_1 + r_2)g = r_1g + r_2g,$$
$$r_1(r_2g) = (r_1r_2)g, \qquad 1g = g.$$

A subgroup H of G such that $rh \in H$ whenever $r \in R$, $h \in H$ is called a *linear subspace* of G. A homomorphism ϕ of G into another R-module G' is called *linear* if $\phi(rg) = r\phi(g)$ holds for all $r \in R$, $g \in G$.

Only two special cases are of importance to us. In the first, $R = F$ is a field. Then G is a vector space over F. In the second, $R = J$ is the ring of integers. In this case G is an ordinary abelian group (without additional structure). The unifying concept of module saves repetition.

In addition to R-modules, we shall also wish to consider compact abelian groups. To avoid a complete duplication of the discussion, we adopt the following convention:

Unless otherwise stated the word "group" will be used to mean either one of the following two objects:

1°. An R-module (over some ring R with a unit element), or

2°. A compact topological abelian group.

Whenever, in a discussion, several "groups" appear, the word "group" is to be interpreted in a fixed manner. In particular, in case 1°, the ring R is the same for all groups. All groups are written additively. If G is a group, $G = 0$ means that G consists of the zero element alone.

The word "subgroup" will mean correspondingly

1°. A linear subspace, or

2°. A closed subgroup.

The word "homomorphism" will mean

1°. A linear map, or

2° A continuous homomorphism

If G and H are groups, the notation

$$\phi: \quad G \to H$$

means that ϕ maps G homomorphically into H. The *kernel* of ϕ is the subgroup of elements of G mapped into the zero element of H. The *image* of ϕ is the subgroup $\phi(G) \subset H$ The statement "ϕ maps G *onto* H" means $\phi(G) = H$ We sometimes abbreviate "ϕ maps G onto H" by "ϕ is onto." The symbolism $\phi = 0$ is used to indicate that the kernel of ϕ is all of G, or equally well that $\phi(G) = 0$. The expression "ϕ has kernel zero" means that the kernel of ϕ contains only the zero element The symbolism

$$\phi: \quad G \approx H$$

means that the map $\phi\colon \ G \to H$ maps G isomorphically onto H, and ϕ is called an *isomorphism* Observe that $\phi\colon G \approx H$ implies $\phi^{-1}\colon H \approx G$ In case 1°, this is obvious In case 2°, it follows from the theorem that the inverse of a continuous 1-1 map between compact spaces is continuous It is precisely the failure of ϕ^{-1} to be continuous in the non-compact case which prevents our unifying the concepts of R-module and compact group by the use of *topological* R-modules

If L is a subgroup of G, G/L denotes the factor (or difference) group, i e the group whose elements are the cosets of L in G The *natural* homomorphism

$$\eta \quad G \to G/L$$

is the function which attaches to each element of G the coset of L which contains it: $\eta(g) = g + L$ In case 1° we define $r(g + L) = rg + L$ so that G/L is an R-module and η is linear In case 2°, we introduce a topology into G/L as follows. a subset U of G/L is open if and only if $\eta^{-1}(U)$ is open in G. It can then be seen clearly that G/L is a compact abelian group and that η is continuous

If ϕ: $G \to G'$ and $L \subset G$, $L' \subset G'$ are subgroups such that $\phi(L) \subset L'$, then the homomorphism $\tilde{\phi}\colon \ G/L \to G'/L'$ *induced* by ϕ attaches to each coset of L in G the coset of L' in G' which contains its image under ϕ. The natural maps η,η' and the homomorphisms $\phi,\tilde{\phi}$ satisfy the commutativity relation

$$\tilde{\phi}\eta = \eta'\phi.$$

It asserts that two homomorphisms of G into G'/L' coincide. As shown in the diagram

$$\begin{array}{ccc} G & \xrightarrow{\phi} & G' \\ \downarrow \eta & & \downarrow \eta' \\ G/L & \xrightarrow{\bar{\phi}} & G'/L' \end{array}$$

the first is obtained by moving down and then over, the second, by moving over and then down.

If $\{G_\alpha\}, \alpha = 1, \cdots, n$, are groups, their (external) *direct sum* $\sum_{\alpha=1}^n G_\alpha$ is defined, in the usual way, as the set of n-tuples $\{g_\alpha\}$, $g_\alpha \in G_\alpha$, with

$$r\{g_\alpha\} = \{rg_\alpha\}, \qquad \{g_\alpha\} + \{g'_\alpha\} = \{g_\alpha + g'_\alpha\}.$$

In case 2°, $\sum G_\alpha$ is given the product topology.

A set of homomorphisms i_α: $G_\alpha \to G$, $\alpha = 1, \cdots, n$, determine a homomorphism i: $\sum_{\alpha=1}^n G_\alpha \to G$ by the rule $i(\{g_\alpha\}) = \sum_{\alpha=1}^n i_\alpha(g_\alpha)$. If i is an isomorphism of $\sum G_\alpha$ onto G, then the set $\{i_\alpha\}$ is called an *injective representation of G as a direct sum*, and each component i_α is called an *injection*. If, in addition, $G_\alpha \subset G$ and each i_α is an inclusion, G is said to *decompose into the (internal) direct sum* $G = \sum G_\alpha$.

A set of homomorphisms j_α: $G \to G_\alpha$, $\alpha = 1, \cdots, n$, determine a homomorphism j: $G \to \sum_{\alpha=1}^n G_\alpha$ by the rule $j(g) = \{j_\alpha g\}$. If j is an isomorphism of G onto $\sum G_\alpha$, then the set $\{j_\alpha\}$ is called a *projective representation of G as a direct sum*, and each component j_α is called a projection

Given an injective representation $\{i_\alpha\}$ as a direct sum, one constructs a projective representation $\{j_\alpha\}$ by defining $j_\alpha g$ to be the α-coordinate of $i^{-1}g$. Similarly a projective representation determines an injective representation The advantage of having the two types is that they are dual and one can state the dual of a direct sum theorem by interchanging injection and projection In Chapter v where we deal with infinite direct sums, the distinction between the two types will be more than formal.

DEFINITION. A *lower sequence* of groups is a collection $\{G_q, \phi_q\}$ where, for each integer q (positive, negative, or zero), G_q is a group, and ϕ_q $G_q \to G_{q-1}$ is a homomorphism. A lower sequence is said to be *exact* if, for each integer q, the image of ϕ_{q+1} in G_q coincides with the kernel of ϕ_q.

DEFINITION. An *upper sequence* of groups is a collection $\{G^q, \phi^q\}$ where, for each integer q (positive, negative, or zero), G^q is a group and

ϕ^q: $G^q \to G^{q+1}$ is a homomorphism. An upper sequence is *exact* if, for each integer q, the image of ϕ^{q-1} in G^q coincides with the kernel of ϕ^q.

If $\{G_q,\phi_q\}$ is a lower sequence, and we set $G^q = G_{-q}$, $\phi^q = \phi_{-q}$ for each q, then $\{G^q,\phi^q\}$ is an upper sequence This transformation sets up a 1-1 correspondence between the set of all lower sequences and the set of all upper sequences In the sequel, definitions are made only for lower sequences The corresponding definitions for upper sequences are to be obtained using this transformation

DEFINITION. A lower sequence $G' = \{G'_q,\phi'_q\}$ is said to be a *subsequence* of the lower sequence $G = \{G_q,\phi_q\}$, written $G' \subset G$, if, for each q, $G'_q \subset G_q$ and $\phi'_q = \phi_q|G'_q$ A subsequence is determined by any set of subgroups $\{G'_q\}$ provided $\phi_q(G'_q) \subset G'_{q-1}$ for each q.

REMARK. The word subsequence is used here in a sense different from the usual one—no terms of the original sequence are discarded It is not to be expected that a subsequence of an exact sequence should be exact. For example, the subsequence K_q = kernel ϕ_q is usually not exact

DEFINITION. If G,G' are two lower sequences, a *homomorphism* ψ: $G \to G'$ is a sequence of homomorphisms $\{\psi_q\}$ such that, for each integer q, ψ_q: $G_q \to G'_q$ and the following commutativity relations hold.

$$\phi'_q\psi_q = \psi_{q-1}\phi_q$$

The subgroups {kernel ψ_q} form a subsequence of G called the *kernel* of ψ, and kernel $\psi = 0$ means kernel $\psi_q = 0$ for each q Likewise image ψ = {image ψ_q} is a subsequence of G'; and when G' = image ψ, we say that ψ is *onto* If each ψ_q is an isomorphism, then ψ is said to be an isomorphism.

The commutativity relations assert that, in the following diagram of homomorphisms

$$\begin{array}{ccc} G_{q-1} & \xleftarrow{\ \phi_q\ } & G_q \\ \downarrow{\scriptstyle \psi_{q-1}} & & \downarrow{\scriptstyle \psi_q} \\ G'_{q-1} & \xleftarrow{\ \phi'_q\ } & G'_q \end{array}$$

the two composite homomorphisms from G_q to G'_{q-1} coincide

DEFINITION. If L is a subsequence of the lower sequence G, the *factor sequence* G/L of G by L is the lower sequence composed of the factor groups G_q/L_q and the homomorphisms $\bar{\phi}_q$: $G_q/L_q \to G_{q-1}/L_{q-1}$ induced by the ϕ_q. Let η_q. $G_q \to G_q/L_q$ be the natural homomorphism.

Since $\tilde{\phi}_q\eta_q = \eta_{q-1}\phi_q$, it follows that $\eta = \{\eta_q\}$: $G \to G/L$. It is called the *natural* homomorphism of G onto G/L.

DEFINITION. If G',G'' are lower sequences, their (external) *direct sum* $G' + G''$ is the lower sequence whose groups are $G_q = G'_q + G''_q$, and whose homomorphisms are defined by $\phi_q(g_q,g'_q) = (\phi'_q(g_q),\phi''_q(g'_q))$.

DEFINITION. If G',G'' are subsequences of the lower sequence G, then G is said to *decompose into the (internal) direct sum* of G' and G'' if, for each q, G_q decomposes into the direct sum of G'_q and G''_q.

The conventions and notations introduced for pairs of sets (X,A) and their maps will also be used for pairs of groups, pairs of sequences, and their homomorphisms.

3. AXIOMS FOR HOMOLOGY

A homology theory H on an admissible category $\mathfrak{A}$ is a collection of three functions as follows: The first is a function $H_q(X,A)$ defined for each pair (X,A) in $\mathfrak{A}$ and each integer q (positive, negative, or zero) The value of the function is an abelian group It is called the *q-dimensional relative homology group of X modulo A*.

The second function is defined for each map

$$f. \quad (X,A) \to (Y,B)$$

in $\mathfrak{A}$ and each integer q, and attaches to such a pair a homomorphism

$$f_{*q}: \ H_q(X,A) \to H_q(Y,B). \tag{1}$$

It is called the homomorphism *induced* by f

The third function $\partial(q,X,A)$ is defined for each (X,A) in $\mathfrak{A}$ and each integer q. Its value is a homomorphism

$$\partial(q,X,A): \ H_q(X,A) \to H_{q-1}(A) \tag{2}$$

called the *boundary operator.*

Since, in (2), the symbol (q,X,A) is redundant, it will be omitted in the future Likewise the index q on f_{*q} in (1) will be omitted

According to the convention of the preceding section, $H_q(X,A)$ is either always an R-module, or always a compact abelian group. The corresponding conventions govern the homomorphisms ∂ and f_*

In addition, the three functions are required to have the following properties:

Axiom 1. *If f = identity, then f_* = identity.*

Explicitly, if f is the identity map of $(X,A) \in \mathfrak{A}$ on itself, then, for each q, f_* is the identity map of $H_q(X,A)$ on itself.

Axiom 2. $(gf)_* = g_* f_*$

Explicitly, if $f\colon (X,A) \to (Y,B)$ and $g\colon (Y,B) \to (Z,C)$ are admissible, then the composition of the induced homomorphisms $f_*\colon H_q(X,A) \to H_q(Y,B)$ and $g_*\colon H_q(Y,B) \to H_q(Z,C)$ is the induced homomorphism $(gf)_*\colon H_q(X,A) \to H_q(Z,C)$.

Axiom 3. $\partial f_* = (f|A)_* \partial$.

Explicitly, if $f\colon (X,A) \to (Y,B)$ is admissible and $f|A\colon A \to B$ is the map defined by f, then there are two ways of mapping $H_q(X,A)$ into $H_{q-1}(B)$. As shown in the diagram,

$$\begin{array}{ccc} H_q(X,A) & \xrightarrow{f_*} & H_q(Y,B) \\ \downarrow{\scriptstyle\partial} & & \downarrow{\scriptstyle\partial} \\ H_{q-1}(A) & \xrightarrow{(f|A)_*} & H_{q-1}(B) \end{array}$$

the composition ∂f_* is obtained by moving over and then down, the composition $(f|A)_*\partial$ by moving down and over. The axiom requires that the two homomorphisms have the same value on each element of $H_q(X,A)$.

Axiom 4 (Exactness axiom). *If (X,A) is admissible and $i\colon A \to X$, $j\colon X \to (X,A)$ are inclusion maps, then the lower sequence of groups and homomorphisms*

$$\cdots \xleftarrow{i_*} H_{q-1}(A) \xleftarrow{\partial} H_q(X,A) \xleftarrow{j_*} H_q(X) \xleftarrow{i_*} H_q(A) \xleftarrow{\partial}$$

is exact. This lower sequence is called the homology sequence of (X,A) (abbreviated H S of (X,A)).

To make the above statement precise, the groups and homomorphisms of the lower sequence must be indexed by integers. We choose $H_0(X,A)$ as the 0th group, i.e. $G_{3q} = H_q(X,A)$, $\phi_{3q} = \partial$, $G_{3q+1} = H_q(X)$, etc.

Axiom 5 (Homotopy axiom). *If the admissible maps $f_0, f_1\colon (X,A) \to (Y,B)$ are homotopic in $\mathfrak{a}$, then, for each q, the homomorphisms f_{0*}, f_{1*} of $H_q(X,A)$ into $H_q(Y,B)$ coincide.*

Axiom 6 (Excision axiom). *If U is an open subset of X whose closure $\bar{U}$ is contained in the interior of A (i.e. $\bar{U} \subset V \subset A$ for some open set V of X), and if the inclusion map $(X - U, A - U) \to (X,A)$ is admissible, then it induces isomorphisms $H_q(X - U, A - U) \approx H_q(X,A)$ for each q.*

An inclusion map i: $(X - U, A - U) \subset (X,A)$ where U is open in X and $\overline{U}$ is in the interior of A, will be called an *excision map* or just an *excision.*

Axiom 7 (DIMENSION AXIOM). *If P is an admissible space consisting of a single point, then $H_q(P) = 0$ for all $q \neq 0$.*

This concludes the list of axioms for a homology theory. In proofs of theorems, use will be made of Axioms 1, 2, and 3 without comment or explicit reference. The remaining axioms will be referred to by name

The reasons for the names of Axioms 4, 5, and 6 are apparent The reason for calling Axiom 7 the Dimension axiom is not so apparent Suppose $H_q(X,A),\partial,f_*$ is a homology theory satisfying Axioms 1 through 7 Define $H'_q(X,A) = H_{q-1}(X,A)$ Define ∂' and $f_{*'}$ in the natural way. Then the new homology theory satisfies Axioms 1 through 6 This is also true of the homology theory $H''_q(X,A) = H_q(X,A) + H_{q-1}(X,A)$. Thus, Axiom 7 tends to insure that the dimensional index q shall have a geometric meaning.

The consistency of the axioms is easily verified by choosing each $H_q(X,A) = 0$. The interest, naturally, lies in the existence of non-trivial homology theories The existence of such will be proved in Chapters VII and IX.

The Homotopy axiom can be put in the following form, which is sometimes more convenient:

Axiom 5'. *If (X,A) is admissible and g_0,g_1: $(X,A) \to (X,A) \times I$ are defined by $g_0(x) = (x,0)$, $g_1(x) = (x,1)$, then $g_{0*} = g_{1*}$*

Indeed Axiom 5 implies Axiom 5' since g_0 and g_1 are homotopic with the identity map of $(X,A) \times I$ as homotopy. On the other hand, if h is a homotopy of maps f_0,f_1: $(X,A) \to (Y,B)$, then $f_0 = hg_0$ and $f_1 = hg_1$ and by Axioms 2 and 5'

$$f_{0*} = h_*g_{0*} = h_*g_{1*} = f_{1*}.$$

The Excision axiom may be reformulated as follows:

Axiom 6'. *Let X_1 and X_2 be subsets of a space X such that X_1 is closed and $X =$ Int $X_1 \cup$ Int X_2. If i: $(X_1,X_1 \cap X_2) \subset (X_1 \cup X_2,X_2)$ is admissible, then it induces isomorphisms i_*: $H_q(X_1,X_1 \cap X_2) \approx H_q(X_1 \cup X_2,X_2)$ for each q.*

The equivalence of this axiom with the original formulation is easily seen by setting $A = X_2$ and $U = X - X_1$. The new formulation of the Excision axiom closely resembles the following theorem of group theory: If G_1 and G_2 are subgroups of a group G, then the inclusion map $(G_1, G_1 \cap G_2) \to (G_1 \cup G_2, G_2)$ induces an isomorphism of the factor groups $G_1/G_1 \cap G_2 \approx G_1 \cup G_2/G_2$ Here $G_1 \cup G_2$ denotes the least subgroup of G containing both G_1 and G_2.

In all the homology theories that will be constructed in the sequel it will be true that $H_q(X,A) = 0$ for $q < 0$. Because of the various uniqueness theorems that will be established later, this is also true for any homology theory provided that the pair (X,A) is triangulable (see II,5). Because of these facts one could, without limiting the theory in any essential fashion, incorporate the requirement that $H_q(X,A) = 0$ for $q < 0$ into the axioms. One could go a step farther and assume that the groups $H_q(X,A)$ are defined only for $q \geqq 0$ and that ∂: $H_q(X,A) \to H_{q-1}(A)$ is defined only for $q \geqq 1$ The Exactness axiom would be the only one that would require modification Indeed the homology sequence is then no longer infinite in both directions, but terminates on the left with

$$H_0(X,A) \overset{j_*}{\leftarrow} H_0(X) \overset{i_*}{\leftarrow} \cdots$$

The modified Exactness axiom would assert, in addition to the kernel = image property, that j_* maps $H_0(X)$ onto $H_0(X,A)$ This is the precise condition that the sequence remain exact when it is completed to a nonterminating sequence by adjoining trivial groups and homomorphisms This modification of the basic definition is mathematically trivial since by defining $H_q(X,A) = 0$ for $q < 0$ one obtains a homology theory in the unmodified sense

The remainder of Chapter I is devoted to general theorems concerning a homology theory on an admissible category This hypothesis will be omitted from the statement of each theorem. In addition, the assumption that the pairs and maps occurring in theorems are admissible will be omitted unless there is a special reason for calling attention to this fact

3c. AXIOMS FOR COHOMOLOGY

A cohomology theory on an admissible category $\mathfrak{a}$ is a collection of three functions as follows The first is a function $H^q(X,A)$ defined for each (X,A) in $\mathfrak{a}$ and each integer q In each case the value of the function is an abelian group. It is called the *q-dimensional relative cohomology group of* X mod A

The second function, f^*, is defined for each admissible map

$$f: \quad (X,A) \to (Y,B)$$

and each integer q, and its value is a homomorphism

$$f^*: \quad H^q(Y,B) \to H^q(X,A)$$

called the homomorphism *induced* by f.

The third function, $\delta(q,X,A)$, is defined for $(X,A) \in \mathfrak{A}$ and an integer q Its value is a homomorphism

$$\delta: \quad H^{q-1}(A) \to H^q(X,A)$$

called the *coboundary operator*.

NOTE: f_* and f^* have reverse directions, likewise ∂ and δ. The index distinction between homology and cohomology groups is easily remembered by the following rule: H_q means homology since the index below suggests that the corresponding operator (boundary) lowers the dimension, H^q means cohomology since the index above suggests that the corresponding operator (coboundary) increases the dimension.

These three functions have the following properties (the statements are abbreviated since they parallel the corresponding homology axioms, in particular, conditions that the appropriate pairs and maps be admissible are omitted):

Axiom 1c. *If f = identity, then f^* = identity.*

Axiom 2c. $(gf)^* = f^*g^*$.

Axiom 3c. $\delta(f|A)^* = f^*\delta$.

Axiom 4c. *If i $A \to X$, j. $X \to (X,A)$ are inclusion maps, then the upper sequence of groups and homomorphisms*

$$\cdots \xrightarrow{i^*} H^{q-1}(A) \xrightarrow{\delta} H^q(X,A) \xrightarrow{j^*} H^q(X) \xrightarrow{i^*} H^q(A) \xrightarrow{\delta} \cdots$$

is exact. This upper sequence is called the cohomology sequence of (X,A) (abbreviated. C S. of (X,A))

Axiom 5c. *If the maps f,g: $(X,A) \to (Y,B)$ are homotopic, then $f^* = g^*$.*

Axiom 6c. *If U is open in X, and $\overline{U}$ is contained in the interior of A, then the inclusion map. $(X - U, A - U) \to (X,A)$, if admissible, induces isomorphisms: $H^q(X,A) \approx H^q(X - U, A - U)$.*

Axiom 7c. *If P is a point, then $H^q(P) = 0$ for $q \neq 0$.*

There is a duality relating homology and cohomology It is based on the Pontrjagin theory of character groups. Precisely, let $H^q(X,A),f^*,\delta$ be a cohomology theory satisfying Axioms 1c through 7c; and suppose

$H^q(X,A)$ is always a discrete abelian group (R = the ring of integers) or always a compact abelian group. Let $H_q(X,A)$ be the character group of $H^q(X,A)$, and let f_*,∂ be the homomorphisms dual to f^*,δ. Then it is readily shown, by the use of standard properties of character groups, that $H_q(X,A),f_*,\partial$ satisfy Axioms 1 through 7. It follows that the dual of each theorem about $\{H_q,f_*,\partial\}$ is a true theorem about $\{H^q,f^*,\delta\}$. When passing from a theorem to its dual, arrows are reversed, subgroups are replaced by factor groups and vice versa.

In case the values of $H^q(X,A)$ are R-modules, with R unrestricted, the duality has only a semiformal status. A partial duality is obtained by treating $H^q(X,A)$ as a discrete abelian group, and ignoring the operations of R. Then the strict duality above shows that the dual of each theorem of homology is a true theorem of cohomology at least in so far as the *additive* structures of the groups $H^q(X,A)$ are involved.

Because of this duality, theorems on cohomology will not be proved, and will be left to the reader as exercises. At the end of each section on homology in Chapters I and III, the dual definitions and theorems for cohomology will be stated.

4. HOMOMORPHISMS OF HOMOLOGY SEQUENCES

A map f: $(X,A) \to (Y,B)$ defines maps

$$f_1\colon\ X \to Y, \qquad f_2\colon\ A \to B.$$

The map f_2 is simply the map $f|A$ considered before.

THEOREM 4.1. *The collection of homomorphisms f_*, f_{1*}, and f_{2*} forms a homomorphism of the homology sequence of (X,A) into that of (Y,B). It will be denoted by f_{**}.*

PROOF. Consider the diagram

$$\begin{array}{ccccccccc}
\longleftarrow & H_q(X,A) & \xleftarrow{j_*} & H_q(X) & \xleftarrow{i_*} & H_q(A) & \xleftarrow{\partial} & H_{q+1}(X,A) & \longleftarrow \\
 & \downarrow f_* & & \downarrow f_{1*} & & \downarrow f_{2*} & & \downarrow f_* & \\
\longleftarrow & H_q(Y,B) & \xleftarrow{j'_*} & H_q(Y) & \xleftarrow{i'_*} & H_q(B) & \xleftarrow{\partial} & H_{q+1}(Y,B) & \longleftarrow
\end{array}$$

where i, j, i', j' are appropriate inclusions. We must verify the commutativity relations

$$f_*j_* = j'_*f_{1*}, \qquad f_{1*}i_* = i'_*f_{2*}, \qquad f_{2*}\partial = \partial f_*.$$

The first two relations follow from Axiom 2 since $fj = j'f_1$ and $f_1 i = i'f_2$. The third relation is a restatement of Axiom 3.

THEOREM 4.2 *If* $f\colon (X,A) \to (Y,B)$ *and* $f_{1*}\colon H_q(X) \approx H_q(Y)$, $f_{2*}\colon H_q(A) \approx H_q(B)$ *for all dimensions* q, *then* $f_*\colon H_q(X,A) \approx H_q(Y,B)$ *for all dimensions* q, *and* f_{**} *is an isomorphism.*

This theorem is a consequence of the following group theoretic lemma:

LEMMA 4.3 (THE "FIVE" LEMMA). *Let* $\{\psi_q\}$ *be a homomorphism of an exact lower sequence* $\{G_q,\phi_q\}$ *into an exact lower sequence* $\{G'_q,\phi'_q\}$. *If for some index* q, ψ_{q+2}, ψ_{q+1}, ψ_{q-1}, *and* ψ_{q-2} *are all isomorphisms, then* ψ_q *is also an isomorphism.*

This lemma is a consequence of the following two lemmas:

LEMMA 4.4. *If the kernel of* ψ_{q-1} *is zero, and* ψ_{q+2} *is onto, then*

$$\text{kernel } \psi_q = \phi_{q+1}(\text{kernel } \psi_{q+1}).$$

LEMMA 4.5. *If* ψ_{q+1} *is onto, and the kernel of* ψ_{q-2} *is zero, then*

$$\text{image } \psi_q = \phi_q'^{-1}(\text{image } \psi_{q-1}).$$

PROOFS. Both lemmas are proved by chasing around the following diagram:

$$\begin{array}{ccccccccc}
G_{q-2} & \xleftarrow{\phi_{q-1}} & G_{q-1} & \xleftarrow{\phi_q} & G_q & \xleftarrow{\phi_{q+1}} & G_{q+1} & \xleftarrow{\phi_{q+2}} & G_{q+2} \\
\downarrow{\scriptstyle \psi_{q-2}} & & \downarrow{\scriptstyle \psi_{q-1}} & & \downarrow{\scriptstyle \psi_q} & & \downarrow{\scriptstyle \psi_{q+1}} & & \downarrow{\scriptstyle \psi_{q+2}} \\
G'_{q-2} & \xleftarrow{\phi'_{q-1}} & G'_{q-1} & \xleftarrow{\phi'_q} & G'_q & \xleftarrow{\phi'_{q+1}} & G'_{q+1} & \xleftarrow{\phi'_{q+2}} & G'_{q+2}
\end{array}$$

Since $\psi_q\phi_{q+1} = \phi'_{q+1}\psi_{q+1}$, it follows that

$$\phi_{q+1}(\text{kernel } \psi_{q+1}) \subset \text{kernel } \psi_q.$$

To prove the inverse inclusion, assume that the hypotheses of 4.4 are satisfied and consider an element $g \in G_q$ with $\psi_q g = 0$. Then $\phi'_q\psi_q g = 0$, so $\psi_{q-1}\phi_q g = 0$. Since the kernel ψ_{q-1} is zero, it follows that $\phi_q(g) = 0$. By exactness, there is a $g_1 \in G_{q+1}$ such that $\phi_{q+1}g_1 = g$ Then $\phi'_{q+1}\psi_{q+1}g_1 = \psi_q\phi_{q+1}g_1 = \psi_q g = 0$, and, by exactness, there is a $g'_2 \in G'_{q+2}$ such that $\phi'_{q+2}g'_2 = \psi_{q+1}g_1$ Since ψ_{q+2} is onto, there is a $g_2 \in G_{q+2}$ such that $\psi_{q+2}g_2 = g'_2$. Then, on the one hand, $\psi_{q+1}(g_1 - \phi_{q+2}g_2) = \psi_{q+1}g_1 - \phi'_{q+2}\psi_{q+2}g_2 = \psi_{q+1}g_1 - \phi'_{q+2}g'_2 = 0$, while, on the other hand, $\phi_{q+1}(g_1 - \phi_{q+2}g_2) = \phi_{q+1}g_1 = g$. Thus $g \in \phi_{q+1}(\text{kernel } \psi_{q+1})$.

The proof of 4.5 is similar.

THEOREM 4.1c. *The collection of homomorphisms* f^*, f^*_1, f^*_2 *forms a homomorphism of the C.S. of* (Y,B) *into that of* (X,A). *It will be denoted by* f^{**}.

THEOREM 4.2c. *If* $f\colon (X,A) \to (Y,B)$ *and* $f^*_1\colon H^q(Y) \approx H^q(X)$,

f_2^*: $H^q(B) \approx H^q(A)$ for all dimensions q, then f^*: $H^q(Y,B) \approx H^q(X,A)$ for all q, and f^{**} is an isomorphism

5. INVARIANCE OF THE HOMOLOGY GROUPS

Two admissible pairs $(X,A),(Y,B)$ are *homeomorphic* if there exist admissible maps $f: (X,A) \to (Y,B)$ and $g: (Y,B) \to (X,A)$ such that both fg and gf are identity maps. The map f is then called a *homeomorphism*, and $g = f^{-1}$ is the inverse of f.

THEOREM 5.1 *A homeomorphism f of (X,A) onto (Y,B) induces isomorphisms f_*: $H_q(X,A) \approx H_q(Y,B)$*

PROOF. Since $f^{-1}f$ = identity, $(f^{-1}f)_* = (f^{-1})_*f_*$ = identity. Similarly $f_*(f^{-1})_*$ = identity. Therefore f_* has an inverse $(f_*)^{-1} = (f^{-1})_*$.

THEOREM 5.2. *A homeomorphism f of (X,A) onto (Y,B) induces an isomorphism of the H S of (X,A) onto that of (Y,B).*

PROOF. Since f defines homeomorphisms f_1: $X \to Y$ and f_2: $A \to B$, the theorem follows from 4.1 and 5.1.

THEOREM 5.1c. *A homeomorphism f of (X,A) onto (Y,B) induces isomorphisms f^* $H^q(Y,B) \approx H^q(X,A)$*

THEOREM 5.2c. *A homeomorphism f of (X,A) onto (Y,B) induces an isomorphism of the C.S. of (Y,B) onto that of (X,A).*

6. THE BASE POINT P_0 AND THE COEFFICIENT GROUP G

DEFINITION 6.1 Let P_0 denote a fixed reference point and also the (admissible) space consisting of this single point. The group $H_0(P_0)$ is called the *coefficient group* of the homology theory and is denoted by G. Its elements are denoted by g,g', etc

One can construct the coefficient group without choosing a base point by the following process. Let M be a set, and, for each $\alpha \varepsilon M$, let G_α be a group, and, for every ordered pair $\alpha,\beta \varepsilon M$, let π^α_β be an isomorphism $G_\alpha \approx G_\beta$ such that (1) π^α_α is the identity map of G_α for each α, and (2) if $\alpha,\beta,\gamma \varepsilon M$, then $\pi^\beta_\gamma\pi^\alpha_\beta = \pi^\alpha_\gamma$. Such a collection will be called a *transitive* system of groups To such a system we assign a single group G as follows An element $g \varepsilon G$ is a function assigning to each $\alpha \varepsilon M$ an element $g_\alpha \varepsilon G_\alpha$ such that $\alpha,\beta \varepsilon M$ implies $\pi^\alpha_\beta g_\alpha = g_\beta$ Addition is defined by adding functional values $(g + g')_\alpha = g_\alpha + g'_\alpha$. Then G is a group uniquely isomorphic to each G_α under the projection $g \to g_\alpha$.

If M is the collection of all spaces in $\mathfrak{A}$ each consisting of a single point, the G_α are the 0th homology groups of these spaces, and the π_α^β are the isomorphisms induced by the unique maps of one such space into another, then the above conditions are satisfied and the associated group G could be taken as the coefficient group

DEFINITION 6.1c. The group $H^0(P_0)$ is called the *coefficient group* of the cohomology theory and is denoted by G, its elements by g, g', etc.

7. THE REDUCED 0-DIMENSIONAL HOMOLOGY GROUP

DEFINITION 7.1. If $x \in X$ and $g \in G$, let $(gx)_X$ denote the image of g in $H_0(X)$ under the homomorphism induced by the map f: $P_0 \to X$ defined by $f(P_0) = x$. The image of G in $H_0(X)$ under f_* is denoted by $(Gx)_X$

THEOREM 7 2 *If f. $X \to Y$, $x \in X$, $y = f(x)$, and $g \in G$, then $f_*(gx)_X = (gy)_Y$ Thus f_* maps $(Gx)_X$ onto $(Gy)_Y$.*

The proof is trivial.

DEFINITION 7.3. If the unique map f: $X \to P_0$ is admissible, the space X is said to be *collapsible* In such a case the kernel of the homomorphism f_*: $H_0(X) \to G$ is defined It is called the *reduced 0-dimensional homology group of X*, and is denoted by $\tilde{H}_0(X)$.

LEMMA 7 4. *If f: $X \to Y$ is admissible and Y is collapsible, then X is collapsible If (X,A) is admissible, and X is collapsible, then A is collapsible and the map $(X,A) \to (P_0,P_0)$ is admissible.*

The first statement is a trivial consequence of property (3) of an admissible category. If (X,A) is admissible, the inclusion map $A \to X$ is admissible If, in addition, X is collapsible, the composite map $A \to X \to P_0$ is admissible, so A is collapsible. Since $X \to P_0$ is admissible, so also is $(X,X) \to (P_0,P_0)$. Therefore $(X,A) \to (X,X) \to (P_0,P_0)$ is admissible.

REMARK. For each of the first two examples of admissible categories given in §1, it is true that each admissible space is collapsible. However in the third example $\mathcal{A}_{LC}$ of locally compact spaces, the only collapsible spaces are the compact spaces.

THEOREM 7.5. *If P is a space consisting of a single point, then $\tilde{H}_0(P) = 0$ and $H_0(P) = (GP)_P$.*

This follows from 5 1 since $P \to P_0$ is a homeomorphism.

THEOREM 7 6. *If X is collapsible and $x \in X$, then $H_0(X)$ decomposes into the direct sum*

$$H_0(X) = \tilde{H}_0(X) + (Gx)_X$$

and the correspondence $g \to (gx)_X$ maps G isomorphically onto $(Gx)_X$.

PROOF. Let f_1 $P_0 \to X$ be defined by $f_1(P_0) = x$, and let f_2: $X \to P_0$. Since $f_2 f_1$ is the identity map of P_0, the composition of the homomorphisms f_{1*}. $G \to H_0(X)$ and f_{2*}. $H_0(X) \to G$ is the identity. Therefore f_{1*} maps G isomorphically onto $(Gx)_X$ and f_{2*} maps $(Gx)_X$ isomorphically onto G. Since $\tilde{H}_0(X) =$ kernel f_{2*}, it follows that $\tilde{H}_0(X)$ and $(Gx)_X$ have only zero in common. If $h \in H_0(X)$, let $h' = f_{1*}f_{2*}(h)$ and $h'' = h - h'$. It follows immediately that $h' \in (Gx)_X$, $h'' \in \tilde{H}_0(X)$, and $h = h' + h''$.

THEOREM 7.7. *If f. $X \to Y$, Y is collapsible, $x \in X$, and $y = f(x)$, then X is collapsible and f_* maps $\tilde{H}_0(X)$ into $\tilde{H}_0(Y)$ and maps $(Gx)_X$ isomorphically onto $(Gy)_Y$.*

PROOF Let f_1: $P_0 \to X$ be defined by $f_1(P_0) = x$ and let f_2: $Y \to P_0$. Then $f_2 f f_1$ is the identity map of P_0. Therefore the composition $G \to (Gx)_X \to (Gy)_Y \to G$ is the identity. By definition and 7 2, each homomorphism is onto; therefore each is an isomorphism. Since $f_2 f$ collapses X into P_0, $\tilde{H}_0(X) =$ kernel $(f_2 f)_* = f_*^{-1}$(kernel f_{2*}) $= f_*^{-1}(\tilde{H}_0(Y))$.

DEFINITION 7.8. If f: $X \to Y$ and Y is collapsible, the map of $\tilde{H}_0(X)$ into $\tilde{H}_0(Y)$, defined by f_* (see 7 7), is denoted by $\tilde{f}_*$.

COROLLARY 7 9 Kernel $\tilde{f}_* =$ kernel f_*.

DEFINITION 7.1c. If $x \in X$, $h \in H^0(X)$, and f: $P_0 \to X$ is defined by $f(P_0) = x$, let $f^*(h) \in G$ be denoted by $h(x)$, and let the kernel of f^*: $H^0(X) \to G$ be denoted by $\tilde{H}^0_x(X)$.

THEOREM 7.2c *If f· $X \to Y$, $x \in X$, $y = f(x)$, and $h \in H^0(Y)$, then $(f^*h)(x) = h(y)$. Thus f^* maps $\tilde{H}^0_y(Y)$ into $\tilde{H}^0_x(X)$, and $\tilde{H}^0_y(Y)$ contains the kernel of f^*.*

DEFINITION 7.3c. If f: $X \to P_0$ is admissible (i.e. X is collapsible), the image of G in $H^0(X)$ under f^* is denoted by G_X. The factor group $\tilde{H}^0(X) = H^0(X)/G_X$ is called the *reduced 0-dimensional cohomology group of X*

THEOREM 7 5c. *If P is a space consisting of a single point, then $H^0(P) = G_P$ and $\tilde{H}^0(P) = 0$.*

THEOREM 7.6c. *If X is collapsible and $x \in X$, then $H^0(X)$ decomposes into the direct sum $\tilde{H}^0_x(X) + G_X$, and the map $X \to P_0$ induces an isomorphism $G \approx G_X$.*

THEOREM 7.7c. *If f: $X \to Y$, Y is collapsible, $x \in X$, and $y = f(x)$, then X is collapsible and f^* maps $\tilde{H}^0_y(Y)$ into $\tilde{H}^0_x(X)$ and maps G_Y isomorphically onto G_X.*

DEFINITION 7.8c If f: $X \to Y$ and Y is collapsible, the map of $\tilde{H}^0(Y)$ into $\tilde{H}^0(X)$ induced by f^* is denoted by $\tilde{f}^*$.

COROLLARY 7.9c. *Under the natural map $H^0(Y) \to \tilde{H}^0(Y)$ the kernel of f^* is mapped isomorphically onto the kernel of $\tilde{f}^*$.*

8. THE REDUCED HOMOLOGY SEQUENCE

LEMMA 8.1. *For each space X and each integer q, $H_q(X,X) = 0$.*

PROOF. Let i: $X \to X$ and j: $X \to (X,X)$ be inclusions. Consider the section of the H.S. of (X,X) around $H_q(X,X)$:

$$H_{q-1}(X) \xleftarrow{i_*} H_{q-1}(X) \xleftarrow{\partial} H_q(X,X) \xleftarrow{j_*} H_q(X) \xleftarrow{i_*} H_q(X).$$

Since each i_* is an isomorphism onto, it follows by exactness that $0 = \text{kernel } i_{*q-1} = \text{image } \partial$. Therefore $H_q(X,X) = \text{kernel } \partial$. Similarly $H_q(X) = \text{image } i_{*q} = \text{kernel } j_{*q}$. Therefore, image $j_* = 0$. Since, by exactness, image $j_* = \text{kernel } \partial$, it follows that $H_q(X,X) = 0$.

We shall give a second proof of 8.1 which uses the Excision axiom, and uses only the part of the Exactness axiom which asserts that $j_*i_* = 0$. This proof will be of importance later when the full Exactness axiom is not available.

The inclusion map $(0,0) \subset (X,X)$ is an excision and therefore $H_q(X,X) \approx H_q(0)$. In the homology sequence of $(0,0)$ we have the homomorphisms

$$H_q(0) \xleftarrow{j_*} H_q(0) \xleftarrow{i_*} H_q(0).$$

Since i and j are identity maps, we have $i_* = j_* = \text{identity}$. Since $j_*i_* = 0$, it follows that $H_q(0) = 0$.

THEOREM 8.2. *If X is collapsible, then ∂ maps $H_1(X,A)$ into $\tilde{H}_0(A)$.*

PROOF. By 7.4, the map f: $(X,A) \to (P_0,P_0)$ is admissible. If $h \in H_1(X,A)$, then $f_*(h) \in H_1(P_0,P_0) = 0$ by 8.1. Therefore $(f|A)_*\partial(h) = \partial f_*(h) = 0$. Hence, by definition, $\partial(h) \in \tilde{H}_0(A)$.

DEFINITION 8.3. If f: $(X,A) \to (P_0,P_0)$ is admissible (i.e. X is collapsible), the kernel of f_{**} is called the *reduced homology sequence* of (X,A).

THEOREM 8.4. *The reduced homology sequence of (X,A) differs from the homology sequence of (X,A) only in that the section*

$$H_0(X,A) \xleftarrow{j_*} H_0(X) \xleftarrow{i_*} H_0(A) \xleftarrow{\partial} H_1(X,A)$$

of the latter has been replaced by

$$H_0(X,A) \xleftarrow{\tilde{j}_*} \tilde{H}_0(X) \xleftarrow{\tilde{i}_*} \tilde{H}_0(A) \xleftarrow{\tilde{\partial}} H_1(X,A)$$

where $\tilde{j}_*,\tilde{i}_*,\tilde{\partial}$ are the maps defined by j_*,i_*,∂ respectively.

COROLLARY 8.5. *If* f: $(X,A) \to (Y,B)$, *then* f_{**} *maps the reduced homology sequence of* (X,A) *into that of* (Y,B)

PROOF By 8.1 and the Dimension axiom, the H S. of (P_0,P_0) has just two nonzero terms. $H_0(P_0) = H_0(P_0)$. Therefore the kernel of f_{**} coincides with the H.S of (X,A) except for the two corresponding terms which, by 7 3, are $\tilde{H}_0(X)$ and $\tilde{H}_0(A)$ The corollary follows from 7 7

THEOREM 8 6. *If the unique map* f. $(X,A) \to (P_0,P_0)$ *is admissible (i.e.* X *is collapsible, see 7 4),* A *is nonvacuous and* $x \in A$, *then the H.S. of* (X,A) *decomposes into the direct sum of two exact subsequences.* (1) *the reduced* H.S. *of* (X,A) (*i e. the kernel of* f_{**}), *and* (2) *the isomorphic image of the* H S *of* (P_0,P_0) *under* g_{**} *where* g: $(P_0,P_0) \to (X,A)$ *is defined by* $g(P_0) = x$.

Observe that g is admissible since $(P_0,P_0) \to (A,A) \to (X,A)$ are admissible

Since fg is the identity map of (P_0,P_0), this theorem is a special case of the following one.

THEOREM 8.7. *Let* f: $(X,A) \to (Y,B)$ *and* g: $(Y,B) \to (X,A)$ be *admissible and such that* fg *is the identity map of* (Y,B). *Then the homology sequence of* (X,A) *decomposes into the direct sum of two exact subsequences.* (1) *the kernel of* f_{**}, *and* (2) *the isomorphic image of the homology sequence of* (Y,B) *under* g_{**}.

This theorem in turn is a consequence of a purely algebraic proposition·

LEMMA 8.8 *If* C,C' *are two lower sequences and* ψ $C \to C'$, ψ': $C' \to C$ *are homomorphisms such that* $\psi\psi'$: $C' \to C'$ *is the identity, then* C *decomposes into the direct sum of two subsequences:* (1) *the kernel of* ψ, *and* (2) *the isomorphic image of* C' *under* ψ' *If* C *is exact, then* C' *and the subsequences* (1) *and* (2) *are exact*

PROOF. Let K = kernel ψ, and L = image ψ'. Since $\psi\psi'$ is the identity, ψ' maps C' isomorphically onto L. For convenience, suppose C,C' are so indexed that ψ_q $C_q \to C'_q$ and therefore ψ'_q $C'_q \to C_q$. If $l \in L_q$, there exists a $c' \in C'_q$ such that $\psi'_q(c') = l$ If $\psi_q(l) = 0$, then $\psi_q\psi'_q(c') = c' = 0$, and therefore $l = 0$ Thus $K_q \cap L_q = 0$ Suppose $c \in C_q$ Let $l = \psi'_q\psi_q(c)$ Then $l \in L_q$, and $\psi_q(l) = \psi_q\psi'_q\psi_q(c) = \psi_q(c)$. Therefore $k = c - l$ is in K_q, and $c = k + l$. This proves that $C_q = K_q + L_q$ Suppose now that C is exact. This implies that $\phi_q\phi_{q+1} = 0$ (i e $\phi^2 = 0$); hence $(\phi|K)^2 = 0 = (\phi|L)^2$. Suppose that $k \in K_q$ and $\phi(k) = 0$. Since C is exact, there exists a $c \in C_{q+1}$ such that $\phi(c) = k$ Let $c = k_1 + l_1$ where $k_1 \in K_{q+1}$ and $l_1 \in L_{q+1}$. Then $k = \phi(k_1) + \phi(l_1)$. Thus $\phi(l_1) = k - \phi(k_1)$ lies in both K_q and L_q. Therefore $\phi(l_1) = 0$, and $k = \phi(k_1)$ A similar argument shows that, if $l \in L_q$

and $\phi(l) = 0$, then there is an $l_1 \varepsilon L_{q+1}$ such that $\phi(l_1) = l$. Thus K and L are exact Since $L \approx C'$, it follows that C' is exact.

LEMMA 8.1c. *For each space X and each integer q*, $H^q(X,X) = 0$.

THEOREM 8 2c *If X is collapsible, then* G_A *lies in the kernel of* δ: $H^0(A) \to H^1(X,A)$ *Therefore* δ *induces a homomorphism* $\tilde{\delta}$: $\tilde{H}^0(A) \to H^1(X,A)$.

DEFINITION 8.3c. If f: $(X,A) \to (P_0,P_0)$ is admissible (i e X is collapsible), the factor sequence (see definition §2) of the C.S of (X,A) by the image of f^{**} is called the *reduced C S of* (X,A).

THEOREM 8.4c. *The reduced C.S. of* (X,A) *differs from the C.S. of* (X,A) *only in that the section*

$$H^0(X,A) \xrightarrow{j^*} H^0(X) \xrightarrow{i^*} H^0(A) \xrightarrow{\delta} H^1(X,A)$$

of the latter has been replaced by

$$H^0(X,A) \xrightarrow{\tilde{j}^*} \tilde{H}^0(X) \xrightarrow{\tilde{i}^*} \tilde{H}^0(A) \xrightarrow{\tilde{\delta}} H^1(X,A)$$

where $\tilde{j}^*,\tilde{i}^*,\tilde{\delta}$ *are the coset mappings induced by* j^*,i^*,δ *respectively.*

COROLLARY 8.5c. *If* f: $(X,A) \to (Y,B)$, *then* f^{**} *induces a homomorphism of the reduced C.S. of* (Y,B) *into that of* (X,A).

THEOREM 8 6c *If* f: $(X,A) \to (P_0,P_0)$ *is admissible, and* $x \varepsilon A$, *then the C S. of* (X,A) *decomposes into the direct sum of two exact subsequences:* (1) *the isomorphic image of the C S of* (P_0,P_0) *under* f^{**}, *and* (2) *the kernel of* g^{**} *where* g. $(P_0,P_0) \to (X,A)$ *is defined by* $g(P_0) = x$. *Furthermore the second maps isomorphically onto the reduced C S of* (X,A) *under factorization of the C.S. of* (X,A) *by the first subsequence Thus the reduced C.S. of* (X,A) *is exact.*

THEOREM 8.7c *Let* f· $(X,A) \to (Y,B)$ *and* g: $(Y,B) \to (X,A)$ *be admissible and such that* fg *is the identity map of* (Y,B). *Then the C.S. of* (X,A) *decomposes into the direct sum of two exact subsequences:* (1) *the kernel of* g^{**}, *and* (2) *the isomorphic image of the C S of* (Y,B) *under* f^{**}.

LEMMA 8.8c. (*Replace "lower" by "upper" in* 8.8).

9. HOMOLOGICALLY TRIVIAL SPACES

All spaces in this section are assumed to be collapsible (see 7 3). However, the results stated hold for noncollapsible spaces except possibly in the dimensions 0 and 1.

DEFINITION 9.1. A space X is said to be *homologically trivial* if

$H_q(X) = 0$ for $q \neq 0$ and $\tilde{H}_0(X) = 0$. If A is not vacuous, (X,A) is said to be *homologically trivial* if $H_q(X,A) = 0$ for all q.

The reason for the exception $q = 0$ in the definition is found in 7.6. If the coefficient group is not zero, $H_0(X)$ cannot be zero for a nonempty space X.

THEOREM 9.2 *If (X,A) is homologically trivial and A is nonvacuous, then the inclusion map i: $(A,0) \to (X,0)$ induces isomorphisms*

$$i_*: \quad H_q(A) \approx H_q(X) \qquad \textit{for all } q,$$
$$\tilde{i}_*: \quad \tilde{H}_0(A) \approx \tilde{H}_0(X).$$

Conversely, if these relations hold, (X,A) is homologically trivial.

THEOREM 9.3 *If A is homologically trivial and is a nonvacuous subset of X, then the inclusion map j: $(X,0) \to (X,A)$ induces isomorphisms*

$$j_*: \quad H_q(X) \approx H_q(X,A) \qquad \textit{for } q \neq 0,$$
$$\tilde{j}_*: \quad \tilde{H}_0(X) \approx H_0(X,A).$$

Conversely, if these relations hold, A is homologically trivial.

THEOREM 9.4. *If X is homologically trivial and A is a nonvacuous subset, then the boundary operator of (X,A) is an isomorphism in each dimension*

$$\partial: \quad H_q(X,A) \approx H_{q-1}(A) \qquad \textit{for } q \neq 1,$$
$$\tilde{\partial}: \quad H_1(X,A) \approx \tilde{H}_0(A).$$

Conversely, if these relations hold, X is homologically trivial.

COROLLARY 9.5. *If both X and A are homologically trivial, so also is (X,A)*

These results depend on two algebraic propositions.

LEMMA 9.6 *If $C = \{C_q,\phi_q\}$ is an exact lower sequence, and, for some q, $C_q = 0$ and $C_{q+3} = 0$, then ϕ_{q+2}: $C_{q+2} \approx C_{q+1}$.*

PROOF. Since $C_{q+3} = 0$, image $\phi_{q+3} =$ kernel $\phi_{q+2} = 0$. Since $C_q = 0$, image $\phi_{q+1} = 0$. Therefore $C_{q+1} =$ kernel $\phi_{q+1} =$ image ϕ_{q+2}.

LEMMA 9.7. *If $C = \{C_q,\phi_q\}$ is an exact lower sequence, and, for some q, ϕ_{q+2}: $C_{q+2} \approx C_{q+1}$ and ϕ_{q-1}: $C_{q-1} \approx C_{q-2}$, then $C_q = 0$*

PROOF. Since kernel $\phi_{q-1} = 0$, it follows that image $\phi_q = 0$. Therefore $C_q =$ kernel $\phi_q =$ image ϕ_{q+1}. But image $\phi_{q+2} = C_{q+1}$ implies kernel $\phi_{q+1} = C_{q+1}$. Therefore $C_q =$ image $\phi_{q+1} = 0$.

PROOFS OF 9.2, 3, AND 4. The hypothesis of 9.2 is that every third term of the H.S. of (X,A) is zero (i.e. $H_q(X,A) = 0$). By exactness and 9.6, this implies that the remaining adjacent pairs are isomorphic under the mappings of the sequence. The converse of 9.2 is obtained from 9.7. Theorems 9.3 and 4 are proved in a similar way using the reduced H.S. of (X,A) instead of the H.S. of (X,A). The exactness of the reduced H.S. was proved in 8.6.

DEFINITION 9.1c. A space X is said to be *cohomologically trivial* if $H^q(X) = 0$ for $q \neq 0$ and $\tilde{H}^0(X) = 0$. If A is nonvacuous, (X,A) is said to be *cohomologically trivial* if $H^q(X,A) = 0$ for all q.

THEOREM 9.2c. *If (X,A) is cohomologically trivial and A is nonvacuous, then i^*: $H^q(X) \approx H^q(A)$ for all q. Conversely, if these relations hold, (X,A) is cohomologically trivial.*

THEOREM 9.3c. *If A is cohomologically trivial and is a nonvacuous subset of X, then*

$$j^*: \; H^q(X,A) \approx H^q(X) \qquad \textit{for } q \neq 0,$$
$$\tilde{j}^*: \; H^0(X,A) \approx \tilde{H}^0(X)$$

Conversely, if these relations hold, A is cohomologically trivial.

THEOREM 9.4c. *If X is cohomologically trivial and A is a nonvacuous subset, then*

$$\delta: \; H^{q-1}(A) \approx H^q(X,A) \qquad \textit{for } q \neq 1,$$
$$\tilde{\delta}: \; \tilde{H}^0(A) \approx H^1(X,A).$$

Conversely, if these relations hold, X is cohomologically trivial.

COROLLARY 9.5c. *If both X and A are cohomologically trivial, so also is (X,A).*

10. THE HOMOLOGY SEQUENCE OF A TRIPLE (X,A,B)

Suppose that $X \supset A \supset B$ and that the inclusions

$$\bar{i}: \; (A,B) \subset (X,B), \qquad \bar{j}: \; (X,B) \subset (X,A)$$

are admissible. Then (X,A,B) is called an admissible *triple*. The inclusion maps and boundary operators associated with the pairs $(X,A),(X,B),(A,B)$ are denoted by

$$\begin{array}{lll} i: A \to X, & j: X \to (X,A), & \partial: H_q(X,A) \to H_{q-1}(A), \\ i': B \to X, & j': X \to (X,B), & \partial': H_q(X,B) \to H_{q-1}(B), \\ i'': B \to A, & j'': A \to (A,B), & \partial'': H_q(A,B) \to H_{q-1}(B). \end{array}$$

A comparison of these maps with $\bar{i}$ and $\bar{j}$ suggests

DEFINITION 10.1. $\bar{\partial} = j''_*\partial$ is called the *boundary* operator of the triple (X,A,B).

$$\bar{\partial}: \; H_q(X,A) \to H_{q-1}(A,B).$$

The lower sequence of groups

$$\cdots \leftarrow H_{q-1}(A,B) \xleftarrow{\bar\partial} H_q(X,A) \xleftarrow{\bar j_*} H_q(X,B) \xleftarrow{\bar i_*} H_q(A,B) \leftarrow \cdots$$

is called the *homology sequence* of the triple (X,A,B). It is indexed so that $H_0(X,A)$ is the 0th group

THEOREM 10 2 *The homology sequence of a triple is exact.*

The proof of this theorem is a lengthy and tricky exercise in the use of axioms 1, 2, 3, and 4 The reader may save himself the effort of following the proof if he is willing to replace the Exactness axiom by the stronger proposition 10.2 It reduces to the former when $B = 0$.

PROOF Reference to the diagram below will assist the reader in following the proof

$$\begin{array}{ccccccc}
H_q(A) & \xrightarrow{i_*} & H_q(X) & & & & \\
\downarrow{\scriptstyle j''_*} & & \downarrow{\scriptstyle j'_*} & & & & \\
H_q(A,B) & \xrightarrow{\bar i_*} & H_q(X,B) & \xrightarrow{\bar j_*} & H_q(X,A) & & \\
 & & \downarrow{\scriptstyle \partial'} & & \downarrow{\scriptstyle \partial} & & \\
 & & H_{q-1}(B) & \xrightarrow{i''_*} & H_{q-1}(A) & \xrightarrow{i_*} & H_{q-1}(X) \\
 & & & & \downarrow{\scriptstyle j''_*} & & \downarrow{\scriptstyle j'_*} \\
 & & & & H_{q-1}(A,B) & \xrightarrow{\bar i_*} & H_{q-1}(X,B)
\end{array}$$

Commutativity relations hold in each square of the diagram. In addition to the homomorphisms displayed in the diagram, we have:

$$j_* = \bar j_* j'_*, \qquad i'_* = i_* i''_*, \qquad \partial'' = \partial' \bar i_*, \qquad \bar\partial = j''_* \partial.$$

The proof breaks up into the proofs of six propositions:

(1) $\bar j_* \bar i_* = 0$, (2) $\bar\partial \bar j_* = 0$, (3) $\bar i_* \bar\partial = 0$.

(4) If $x \in H_q(X,B)$ and $\bar j_* x = 0$, there exists an $x' \in H_q(A,B)$ such that $\bar i_* x' = x$

(5) If $y \in H_q(X,A)$ and $\bar\partial y = 0$, there exists a $y' \in H_q(X,B)$ such that $\bar j_* y' = y$

(6) If $z = H_{q-1}(A,B)$ and $\bar i_* z = 0$, there exists a $z' \in H_q(X,A)$ such that $\bar\partial z' = z$

PROOF OF (1). The map $\bar j \bar i$. $(A,B) \to (X,A)$ can be expressed as the composition of the inclusion maps k. $(A,B) \subset (A,A)$ and l:

$(A,A) \subset (X,A)$. By §1, $H_q(A,A) = 0$. Therefore $k_* = 0$. It follows that $\bar{j}_*\bar{i}_* = (\bar{j}\bar{i})_* = (lk)_* = l_*k_* = 0$

PROOF OF (2). $\bar{\partial}\bar{j}_* = j''_*\partial\bar{j}_* = j''_*i''_*\partial'$. Since the H.S. of (A,B) is exact, $j''_*i''_* = 0$

PROOF OF (3) $\bar{i}_*\bar{\partial} = \bar{i}_*j''_*\partial = j'_*i_*\partial$. Since the H S. of (X,A) is exact, $i_*\partial = 0$.

PROOF OF (4). Since $\bar{j}_*x = 0$, it follows that

$$i''_*\partial'x = \partial\bar{j}_*x = 0.$$

By the exactness of the H S. of (A,B), the element $\partial'x$ of the kernel of i''_* is the image of an element

$$x_1 \varepsilon H_q(A,B), \qquad \partial''x_1 = \partial'x.$$

Then

$$\partial'(x - \bar{i}_*x_1) = \partial'x - \partial'\bar{i}_*x_1 = \partial'x - \partial''x_1 = 0.$$

By the exactness of the H.S. of (X,B), the element $x - \bar{i}_*x_1$ of the the kernel of ∂' is the image of an element

$$u \varepsilon H_q(X), \qquad j'_*u = x - \bar{i}_*x_1.$$

Then

$$j_*u = \bar{j}_*j'_*u = \bar{j}_*(x - \bar{i}_*x_1) = \bar{j}_*x - \bar{j}_*\bar{i}_*x_1 = 0$$

by (1) and hypothesis By exactness of the H S of (X,A), the element u of the kernel of j_* is the image of an element

$$v \varepsilon H_q(A), \qquad i_*v = u.$$

Define

$$x' = x_1 + j''_*v.$$

Then

$$\bar{i}_*x' = \bar{i}_*x_1 + \bar{i}_*j''_*v = \bar{i}_*x_1 + j'_*i_*v$$
$$= \bar{i}_*x_1 + j'_*u = \bar{i}_*x_1 + x - \bar{i}_*x_1 = x.$$

PROOF OF (5). Since

$$j''_*\partial y = \bar{\partial}y = 0,$$

by the exactness of the H S of (A,B), the element ∂y of the kernel of j''_* is the image of an element

$$u \varepsilon H_{q-1}(B), \qquad i''_*u = \partial y.$$

Then, by exactness of the H.S of (X,A), $i_*\partial = 0$ So

$$i'_*u = i_*i''_*u = i_*\partial y = 0.$$

By exactness of the H S. of (X,B), the element u of the kernel of i'_* is the image of an element

$$y_1 \varepsilon H_q(X,B), \qquad \partial' y_1 = u.$$

Then

$$\partial(y - \bar{j}_*y_1) = \partial y - \partial\bar{j}_*y_1 = \partial y - i''_*\partial' y_1 = \partial y - i''_*u = 0.$$

By exactness of the H S. of (X,A), the element $y - \bar{j}_*y_1$ of the kernel of ∂ is the image of an element

$$v \varepsilon H_q(X), \qquad j_*v = y - \bar{j}_*y_1.$$

Define

$$y' = y_1 + j'_*v.$$

Then

$$\bar{j}_*y' = \bar{j}_*y_1 + \bar{j}_*j'_*v = \bar{j}_*y_1 + j_*v = \bar{j}_*y_1 + y - \bar{j}_*y_1 = y.$$

Proof of (6). Since $\bar{i}_*z = 0$, it follows that

$$\partial''z = \partial'\bar{i}_*z = 0.$$

By exactness of the H S. of (A,B), the element z of the kernel of ∂'' is the image of an element

$$u \varepsilon H_{q-1}(A), \qquad j''_*u = z.$$

Then

$$j'_*i_*u = \bar{i}_*j''_*u = \bar{i}_*z = 0.$$

By exactness of the H S. of (X,B), the element i_*u of the kernel of j'_* is the image of an element

$$v \varepsilon H_{q-1}(B), \qquad i'_*v = i_*u$$

Since $i_*i''_* = i'_*$, it follows that

$$i_*(u - i''_*v) = i_*u - i_*i''_*v = i_*u - i'_*v = 0.$$

By exactness of the H.S. of (X,A), the element $u - i''_*v$ of the kernel of i_* is the image of an element

$$z' \varepsilon H_q(X,A), \qquad \partial z' = u - i''_*v.$$

Then

$$\bar{\partial} z' = j''_* \partial z' = j''_*(u - i''_* v) = j''_* u - j''_* i''_* v = j''_* u = z.$$

This completes the proof.

A *map* $f\colon (X_1,A_1,B_1) \to (X,A,B)$ of one triple into another is a map of X_1 into X which carries A_1 into A, and B_1 into B. The map f defines maps

$$f_1\colon (X_1,A_1) \to (X,A), \qquad f_2\colon (A_1,B_1) \to (A,B),$$
$$f_3\colon (X_1,B_1) \to (X,B).$$

The map f is *admissible* if f_1, f_2, and f_3 are admissible In this case, the induced homomorphisms f_{1*}, f_{2*}, and f_{3*} map the groups of the H.S. of (X_1,A_1,B_1) into those of the H.S. of (X,A,B).

THEOREM 10.3. *A map $f\colon (X_1,A_1,B_1) \to (X,A,B)$ induces a homomorphism of the homology sequence of (X_1,A_1,B_1) into the homology sequence of (X,A,B).*

The proof requires only the verification of three commutativity relations and is left as an exercise for the reader.

THEOREM 10.4. *If B is nonvacuous, and any one of the three pairs $(X,A),(X,B),(A,B)$ is homologically trivial, then the homology groups of the remaining two pairs are isomorphic under the maps of the H S. of (X,A,B). Explicitly:*

(1) *If $H_q(X,A) = 0$ for each q, then $\bar{i}_*\colon H_q(A,B) \approx H_q(X,B)$ for each q.*

(2) *If $H_q(X,B) = 0$ for each q, then $\bar{\partial}\colon H_q(X,A) \approx H_{q-1}(A,B)$ for each q.*

(3) *If $H_q(A,B) = 0$ for each q, then $\bar{j}_*\colon H_q(X,B) \approx H_q(X,A)$ for each q.*

Conversely, any one of the three conclusions implies the corresponding hypothesis.

The proof is the same as that of 9.2 except for notation.

THEOREM 10 5. *Let (X,A,B) be an admissible triple. If $B \subset A$ induces isomorphisms $H_q(B) \approx H_q(A)$ for all values of q, then $(X,B) \subset (X,A)$ also induces isomorphisms $H_q(X,B) \approx H_q(X,A)$ for all q. Similarly, if $A \subset X$ induces $H_q(A) \approx H_q(X)$ for all q, then $(A,B) \subset (X,B)$ induces $H_q(A,B) \approx H_q(X,B)$ for all q.*

PROOF. For the first assertion, the hypothesis and the exactness of the H.S. of (A,B) imply that $H_q(A,B) = 0$ for all q The result follows now from 10.4.

Another proof is obtained by applying 4.2 to the inclusion map $(X,B) \subset (X,A)$.

The proof of the second part of the theorem is similar to that of the first part.

DEFINITION 10.1c. $\bar{\delta} = \delta j''^*$ is called the *coboundary operator of the triple* (X,A,B):

$$\bar{\delta}: H^{q-1}(A,B) \to H^q(X,A).$$

The upper sequence of groups

$$\cdots \to H^{q-1}(A,B) \xrightarrow{\bar{\delta}} H^q(X,A) \xrightarrow{\bar{j}^*} H^q(X,B) \xrightarrow{\bar{i}^*} H^q(A,B) \to \cdots$$

is called the *cohomology sequence of the triple* (X,A,B).

THEOREM 10 2c. *The cohomology sequence of a triple is exact.*

THEOREM 10.3c *A mapping* f: $(X_1,A_1,B_1) \to (X,A,B)$ *induces a homomorphism of the C S of* (X,A,B) *into that of* (X_1,A_1,B_1).

THEOREM 10.4c *If B is nonvacuous and any one of the pairs* $(X,A),(X,B),(A,B)$ *is cohomologically trivial, then the cohomology groups of the remaining two pairs are isomorphic under the maps of the C.S of* (X,A,B).

THEOREM 10 5c *Let* (X,A,B) *be an admissible triple. If* $B \subset A$ *induces isomorphisms* $H^q(A) \approx H^q(B)$ *for all q, then* $(X,B) \subset (X,A)$ *induces isomorphisms* $H^q(X,A) \approx H^q(X,B)$ *for all q. Similarly, if* $A \subset X$ *induces isomorphisms* $H^q(X) \approx H^q(A)$ *for all q, then* $(A,B) \subset (X,B)$ *induces isomorphisms* $H^q(X,B) \approx H^q(A,B)$.

11. HOMOTOPY EQUIVALENCE AND CONTRACTIBILITY

In this article use will be made for the first time of the Homotopy axiom. It is assumed here that all pairs, maps, and homotopies are in $\mathfrak{A}$.

DEFINITION 11.1 Two pairs (X,A) and (Y,B) are said to be *homotopically equivalent* (in $\mathfrak{A}$) if there exist maps

$$f\ (X,A) \to (Y,B), \qquad g.\ (Y,B) \to (X,A)$$

such that gf is homotopic to the identity map of (X,A), and fg is homotopic to the identity map of (Y,B). The pair of maps f,g is called a *homotopy equivalence* Frequently each of the maps f,g separately will be referred to as a homotopy equivalence.

THEOREM 11.2 *A homotopy equivalence induces isomorphisms* $f_*\cdot\ H_q(X,A) \approx H_q(Y,B)$, *and* $(f_*)^{-1} = g_*$, *for all q.*

PROOF. Since gf is homotopic to the identity, it follows from the Homotopy axiom that $(gf)_* = g_*f_* =$ identity. Similarly $f_*g_* =$ identity. Therefore f_* has the inverse g_*.

THEOREM 11 3. *A homotopy equivalence induces isomorphisms of the ordinary and reduced homology sequences of* (X,A) *with the corresponding sequences of* (Y,B)

PROOF. A homotopy equivalence f,g of (X,A) and (Y,B) defines homotopy equivalences of X and Y, and of A and B. Hence 11.2 implies that f induces an isomorphism of the H S of (X,A), and that g induces the inverse isomorphism. The proposition for the reduced H.S. now follows from 7 7.

LEMMA 11.4. *If a space X is contractible on itself to a point, then X is homotopically equivalent to a point.*

PROOF. Let $h(x,t)$ be a homotopy $h\colon X \times I \to X$ such that $h(x,0) = x$, and $h(x,1) = x_0$ for each $x \in X$. Let $f\colon X \to (x_0)$, and define $g\colon (x_0) \to X$ by $g(x_0) = x_0$. Then $fg =$ identity map of (x_0) and h is a homotopy connecting gf and the identity map of X. Therefore f,g form a homotopy equivalence.

THEOREM 11.5. *Every space contractible to a point over itself is homologically trivial.*

PROOF. 11 4 and 11 3 show that it suffices to prove that a space consisting of a single point P is homologically trivial By 7.5, $\tilde{H}_0(P) = 0$ and by the Dimension axiom $H_q(P) = 0$ for $q \neq 0$; hence P is homologically trivial.

DEFINITION 11.6. A pair (X',A'), contained in a pair (X,A), is called a *retract* of (X,A) if there exists a map $f\colon (X,A) \to (X',A')$ such that $f(x) = x$ for each $x \in X'$ It is called a *deformation retract* if there is a retraction f and the composition of f and the inclusion map $(X', A') \subset (X,A)$ is homotopic to the identity map of (X,A) It is called a *strong deformation retract* if the latter homotopy can be chosen to leave fixed each point of X', i e $h(x,t) = x$ for $x \in X'$ and all t. (These definitions depend on the category $\mathfrak{a}$ since all maps must belong to $\mathfrak{a}$).

LEMMA 11.7. *If (X',A') is a deformation retract of (X,A), the inclusion map $g\colon (X',A') \subset (X,A)$ and the retraction f form a homotopy equivalence of (X,A) and (X',A').*

By the definition of a deformation retract, gf is homotopic to the identity, and fg is the identity.

THEOREM 11.8. *If (X',A') is a deformation retract of (X,A), then the inclusion map $(X',A') \subset (X,A)$ induces an isomorphism of the H.S. of (X',A') onto that of (X,A)*

This follows immediately from 11 7 and 11.3.

THEOREM 11.2c. *A homotopy equivalence induces isomorphisms $f^*\colon H^q(Y,B) \approx H^q(X,A)$, and $(f^*)^{-1} = g^*$, for all q.*

THEOREM 11.3c. *A homotopy equivalence induces isomorphisms of the ordinary and reduced cohomology sequences of (Y,B) with the corresponding sequences of (X,A).*

THEOREM 11.5c. *Every space contractible to a point over itself is cohomologically trivial.*

THEOREM 11.8c *If (X',A') is a deformation retract of (X,A), then the inclusion map $(X',A') \subset (X,A)$ induces an isomorphism of the C.S. of (X,A) onto the C.S. of (X',A').*

12. EXCISION

The Excision axiom asserts that an excision map induces isomorphisms of the homology groups. If the condition $\bar{U} \subset \text{Int } A$ on an excision map is relaxed to $U \subset A$, then the conclusion is not generally valid. However, it does hold in certain useful cases. We shall give two such here. A full discussion of the various ways of strengthening the Excision axiom will be given in x,5.

THEOREM 12.1. *Let (X,A) be a pair, and let U,V be open subsets of X such that $\bar{V} \subset U \subset A$, the inclusion maps*

$$(X - U, A - U) \xrightarrow{f} (X - V, A - V) \xrightarrow{g} (X,A)$$

are admissible, and $(X - U, A - U)$ is a deformation retract of $(X - V, A - V)$. Then the inclusion map gf induces isomorphisms of the homology groups in all dimensions.

PROOF. Since g is an excision map, g_* is isomorphic. By 11.8, f_* is likewise isomorphic. Hence $(gf)_* = g_*f_*$ is isomorphic.

THEOREM 12.2. *Let (X,A) be a pair and let U be an open subset of X with $U \subset A$. If there exists a subset B of X containing A such that (i) the inclusion maps*

$$\begin{array}{ccc} (X - U, A - U) & \xrightarrow{f} & (X,A) \\ \downarrow{\scriptstyle l} & & \downarrow{\scriptstyle k} \\ (X - U, B - U) & \xrightarrow{m} & (X,B) \end{array}$$

are admissible, (ii) $\bar{U} \subset \text{Int } B$, (iii) A is a deformation retract of B, and (iv) $A - U$ is a deformation retract of $B - U$, then f induces isomorphisms in all dimensions.

PROOF. The Excision axiom and (ii) imply that m_* is an isomorphism. From (iii), (iv), and 11.8 it follows that the inclusion maps induce isomorphisms $H_q(A) \approx H_q(B)$ and $H_q(A - U) \approx H_q(B - U)$. From 10.5, it now follows that l_* and k_* are isomorphisms. This implies that $f_* = k_*^{-1}m_*l_*$ is an isomorphism.

It is to be noted that if $A - U$ is a *strong* deformation retract of

$B - U$, the homotopy h: $(B - U) \times I \rightarrow A - U$ can be extended to a homotopy $B \times I \rightarrow A$ by defining $h(x,t) = x$ for $x \in U$ If the extended homotopy is admissible, then A is a strong deformation retract of B. This is the case for the main applications of 12.2.

The theorems of this section hold as stated with cohomology in place of homology.

13. THE DIRECT SUM THEOREM

LEMMA 13.1. *In the diagram*

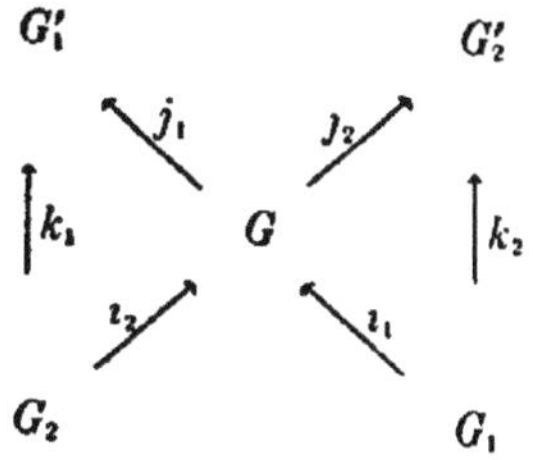

of groups and homomorphisms, assume that commutativity holds in each triangle, image i_α = kernel j_α $(\alpha = 1,2)$, *and* k_1,k_2 *are isomorphisms. Then* i_1,i_2 *are the components of an injective direct sum representation* i: $G_1 + G_2 = G$, *and* j_1,j_2 *are the components of a projective direct sum representation* j: $G = G_1' + G_2'$.

PROOF. Since $j_1i_2 = k_1$ is an isomorphism, it follows that j_1 maps the image of i_2 isomorphically. Thus (kernel j_1) $\cap$ (image i_2) $= 0$ Since image i_2 = kernel j_2, it follows that

(i) $$(\text{kernel } j_1) \cap (\text{kernel } j_2) = 0$$

If $g \in G$, let $\bar{g} = i_1k_2^{-1}j_2g + i_2k_1^{-1}j_1g$ Then $j_1\bar{g} = j_1i_2k_1^{-1}j_1g = k_1k_1^{-1}j_1g = j_1g$ and similarly $j_2\bar{g} = j_2g$ Thus $\bar{g} = g$ by (i). This proves that g has a representation $g = i_1g_1 + i_2g_2$. For any such representation, we have $j_2g = j_2i_1g_1 + j_2i_2g_2 = k_2g_1$, thus $g_1 = k_2^{-1}j_2g$. Similarly $g_2 = k_1^{-1}j_1g$, hence the representation is unique.

To prove the second part, consider $g_1' \in G_1'$, $g_2' \in G_2'$, and $g = i_2k_1^{-1}g_1' + i_1k_2^{-1}g_2'$. Then $j_1g = j_1i_2k_1^{-1}g_1' = k_1k_1^{-1}g_1' = g_1'$ and similarly $j_2g = g_2'$. This exhibits an element $g \in G$ with $j_\alpha g = g_\alpha'$ $(\alpha = 1,2)$. The uniqueness of g follows from (i).

REMARK. The proof as given remains valid if the hypothesis image i_1 = kernel j_1 is weakened to image $i_1 \subset$ kernel j_1. The relation image i_1 = kernel j_1 is a consequence.

THEOREM 13.2. *Let $X = X_1 \cup \cdots \cup X_n$ be the union of disjoint sets each of which is closed (and therefore open) in X. Let $A_i \subset X_i$ and $A = A_1 \cup \cdots \cup A_n$. Assume that all pairs formed of the sets X_i, A_i and their unions are admissible and all inclusion maps of such pairs are admissible. Let i_α: $(X_\alpha, A_\alpha) \subset (X,A)$ $(\alpha = 1, \cdots, n)$. Then the homomorphisms $i_{\alpha*}$: $H_q(X_\alpha, A_\alpha) \to H_q(X,A)$ yield an injective representation of $H_q(X,A)$ as a direct sum, i.e. each u in $H_q(X,A)$ can be expressed uniquely as* $u = \sum_\alpha i_{\alpha*} u_\alpha$, where $u_\alpha \in H_q(X_\alpha, A_\alpha)$.

PROOF. The theorem is trivial for $n = 1$. Assume the validity of the theorem for $n - 1$. Let $X' = X_1 \cup \cdots \cup X_{n-1}$, let $A' = A_1 \cup \cdots \cup A_{n-1}$, and consider the diagram of inclusion maps

$$\begin{array}{ccccc}
(X,X_n \cup A) & & & & (X,X' \cup A_n)\\
& \nwarrow{\scriptstyle j_n} & & \nearrow{\scriptstyle j'} & \\
\uparrow{\scriptstyle k'} & & (X,A) & & \uparrow{\scriptstyle k_n}\\
& \nearrow{\scriptstyle l'} & & \nwarrow{\scriptstyle l_n} & \\
(X' \cup A_n, A) & & & & (X_n \cup A, A)\\
\uparrow{\scriptstyle h'} & & & & \uparrow{\scriptstyle h_n}\\
(X',A') & & & & (X_n,A_n)
\end{array}$$

The maps h_n, $k_n h_n$, h', $k'h'$ are excisions, and therefore h_{n*}, k_{n*}, h'_*, k'_* are isomorphisms. Further, the exactness of the homology sequences of the triples $(X, X' \cup A_n, A)$ and $(X, X_n \cup A, A)$ implies

$$\text{kernel } j'_* = \text{image } l'_*, \qquad \text{kernel } j_{n*} = \text{image } l_{n*}.$$

Thus the conditions of 13.1 are fulfilled and l'_* and l_{n*} yield an injective direct sum representation of $H_q(X,A)$. It follows that i'_* and i_{n*}, where i': $(X',A') \subset (X,A)$, also yield a representation of $H_q(X,A)$ as a direct sum. By the inductive hypothesis, the maps $i'_{\alpha*}$: $H_q(X_\alpha, A_\alpha) \to H_q(X',A')$ where i'_α: $(X_\alpha, A_\alpha) \subset (X',A')$, $\alpha = 1, \cdots, n-1$, yield an injective representation of $H_q(X',A')$ as a direct sum. Since $i_{\alpha*} = i'_* i'_{\alpha*}$ for $i = 1, \cdots, n - 1$, it follows that $i_{\alpha*}$ $(i = 1, \cdots, n)$ form an injective direct sum representation.

THEOREM 13.2c. *Under the conditions of 13.2 the homomorphisms i^*_α: $H^q(X,A) \to H^q(X_\alpha, A_\alpha)$ yield a projective representation of $H^q(X,A)$ as a direct sum, i.e. for each sequence $u_\alpha \in H^q(X_\alpha, A_\alpha)$, $\alpha = 1, \cdots, n$, there is a unique element $u \in H^q(X,A)$ with $i^*_\alpha u = u_\alpha$, $\alpha = 1, \cdots, n$.*

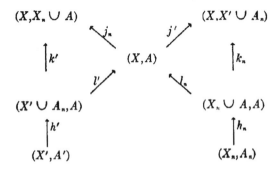

14. TRIADS

DEFINITION 14.1. A *triad* (X,X_1,X_2) consists of a space X and two subsets X_1,X_2 of X such that X, X_1, X_2, $X_1 \cup X_2$, $X_1 \cap X_2$ and all pairs formed from these are admissible, and all their inclusion maps are admissible The triad (X,X_1,X_2) is called *proper* if the inclusion maps k_2. $(X_1,X_1 \cap X_2) \subset (X_1 \cup X_2,X_2)$, k_1 $(X_2,X_1 \cap X_2) \subset (X_1 \cup X_2,X_1)$ induce isomorphisms of the homology groups in all dimensions. Note that (X,X_2,X_1) is a triad distinct from $(X;X_1,X_2)$ unless $X_1 = X_2$.

A triad may be proper with respect to one homology theory and not another, but certain triads are proper for all theories For example, if

$$X_1 = \overline{X}_1, \quad X_2 = \overline{X}_2, \quad \overline{X_1 - (X_1 \cap X_2)} \cap \overline{X_2 - (X_1 \cap X_2)} = 0,$$

where closure is taken in $X_1 \cup X_2$, then k_1 and k_2 are excision maps, and, by the Excision axiom, the triad is proper relative to any homology theory. Still weaker conditions for independence of the homology theory can be based on the results of §12

Observe that, if $(X;X_1,X_2)$ is a proper triad, then $(X_1 \cup X_2;X_1,X_2)$ is also proper.

THEOREM 14 2 *A triad (X,X_1,X_2) is proper if and only if the inclusion maps i_α $(X_\alpha,X_1 \cap X_2) \rightarrow (X_1 \cup X_2, X_1 \cap X_2)$ yield, for each q, an injective representation of $H_q(X_1 \cup X_2, X_1 \cap X_2)$ as a direct sum, i e if and only if every $u \in H_q(X_1 \cup X_2, X_1 \cap X_2)$ can be expressed uniquely as $u = i_{1*}u_1 + i_{2*}u_2$ for $u_\alpha \in H_q(X_\alpha,X_1 \cap X_2)$, $\alpha = 1,2$.*

PROOF. Consider the diagram

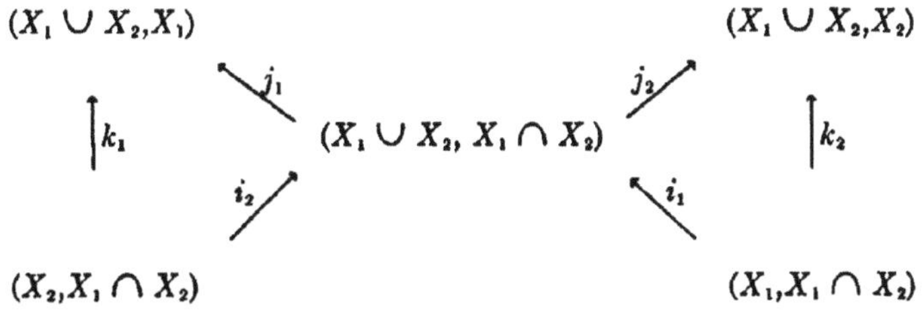

The relation kernel $j_{\alpha*}$ = image $i_{\alpha*}$ follows from the exactness of the H.S. of the triple $(X_1 \cup X_2, X_\alpha, X_1 \cap X_2)$. If the triad is proper, then k_{1*} and k_{2*} are isomorphisms, and 13 1 implies the direct sum conclusion

Conversely assume that the direct sum decomposition holds. Then i_{1*} has kernel zero Thus, in the H.S. of the triple $(X_1 \cup X_2, X_1, X_1 \cap X_2)$, we have $\partial = 0$, and therefore, by exactness, j_{1*} is onto.

Now let $u \in H_q(X_1 \cup X_2,X_1)$. Then $u = j_{1*}v$ for some $v \in$

$H_q(X_1 \cup X_2, X_1 \cap X_2)$ By the direct sum assumption, we have $v = i_{1*}u_1 + i_{2*}u_2$ so that

$$u = j_{1*}i_{1*}u_1 + j_{1*}i_{2*}u_2 = k_{1*}u_2,$$

and k_{1*} is onto.

Assume $u \in H_q(X_2,X_1 \cap X_2)$ and $k_{1*}u = 0$ Then $j_{1*}i_{2*}u = 0$. Thus by exactness, there is a $v \in H_q(X_1,X_1 \cap X_2)$ with $i_{1*}v = i_{2*}u$. Then $i_{1*}(-v) + i_{2*}u = 0$ By the direct sum assumption, this implies $u = 0$. Therefore k_{1*} is an isomorphism By symmetry, the same holds for k_{2*} Thus the triad (X,X_1,X_2) is proper.

DEFINITION 14 3. Let (X,X_1,X_2) be a proper triad The composition of the homomorphisms

$$H_q(X,X_1 \cup X_2) \xrightarrow{\partial} H_{q-1}(X_1 \cup X_2) \xrightarrow{l_{2*}} H_{q-1}(X_1 \cup X_2,X_2) \xrightarrow{k_{2*}^{-1}} H_{q-1}(X_1,X_1 \cap X_2)$$

where k_2,l_2 are inclusion maps, is called the *boundary* operator of the proper triad and is denoted, ambiguously, by ∂ (Note that this new boundary operator is just the boundary operator of the triple $(X, X_1 \cup X_2, X_2)$ followed by an excision) The lower sequence

$$\leftarrow H_{q-1}(X_1,X_1 \cap X_2) \xleftarrow{\partial} H_q(X,X_1 \cup X_2) \xleftarrow{j_*} H_q(X,X_2) \xleftarrow{i_*} H_q(X_1,X_1 \cap X_2) \leftarrow$$

where i,j are inclusions, is called the *homology sequence* of the proper triad

THEOREM 14 4 *The homology sequence of a proper triad is exact.*

PROOF. As shown by the following diagram

$$\begin{array}{ccccccc} H_q(X,X_2) & \xleftarrow{i'_*} & H_q(X_1 \cup X_2,X_2) & \xleftarrow{\partial'} & H_{q+1}(X,X_1 \cup X_2) & \leftarrow & H_{q+1}(X,X_2) \\ & \nwarrow^{i_*} & \uparrow k_{2*} & \swarrow_{\partial} & & & \\ & & H_q(X_1,X_1 \cap X_2) & & & & \end{array}$$

the H.S. of $(X;X_1,X_2)$ is obtained from that of $(X, X_1 \cup X_2, X_2)$ by replacing the group $H_q(X_1 \cup X_2,X_2)$ by the isomorphic group $H_q(X_1,X_1 \cap X_2)$ under k_{2*}, and defining ∂ so that $k_{2*}\partial = \partial'$ Thus the H S of (X,X_1,X_2) is isomorphic to the H S of $(X, X_1 \cup X_2, X_2)$. Since the latter is exact, so is the former.

Note that, if $(X;X_1,X_2)$ is a triad with $X_1 \supset X_2$, then $(X;X_1,X_2)$ is a proper triad and its homology sequence reduces to the homology sequence of the triple (X,X_1,X_2).

THEOREM 14 5. *If (X,X_1,X_2) and $(Y;Y_1,Y_2)$ are proper triads, and*

f: $(X;X_1,X_2) \to (Y,Y_1,Y_2)$, then f induces a homomorphism f_{**} of the homology sequence of (X,X_1,X_2) into that of (Y,Y_1,Y_2). In particular, the boundary operator for proper triads commutes with induced homomorphisms

PROOF. Since i_*, j_*, i'_*, j'_* are induced by inclusion maps, commutativity obviously holds in the squares not involving ∂. The latter square extends into

$$\begin{array}{ccccc}
H_q(X_1,X_1 \cap X_2) & \xleftarrow{k_*^{-1}} & H_q(X_1 \cup X_2,X_2) & \xleftarrow{\partial'} & H_{q+1}(X,X_1 \cup X_2) \\
\downarrow f_{1*} & & \downarrow f_{2*} & & \downarrow f_{3*} \\
H_q(Y_1,Y_1 \cap Y_2) & \xleftarrow{k_*'^{-1}} & H_q(Y_1 \cup Y_2,Y_2) & \xleftarrow{\partial'} & H_{q+1}(Y,Y_1 \cup Y_2)
\end{array}$$

where f_1,f_2,f_3 are defined by f By 10.3, commutativity holds on the right Since k,k' are inclusions, we have $k'_*f_{1*} = f_{2*}k_*$ Since k'_*,k_* are isomorphisms, this yields $f_{1*}k_*^{-1} = k_*'^{-1}f_{2*}$

THEOREM 14 6 *Let $(X;X_1,X_2)$ be a proper triad with $X = X_1 \cup X_2$, $A = X_1 \cap X_2$ Let f_1,f_2,f be maps $(X,A) \to (Y,B)$ such that*

$$f_1|X_1 = f|X_1, \qquad f_1(X_2) \subset B,$$
$$f_2|X_2 = f|X_2, \qquad f_2(X_1) \subset B.$$

Then $f_* = f_{1*} + f_{2*}$.

PROOF. Consider the diagram

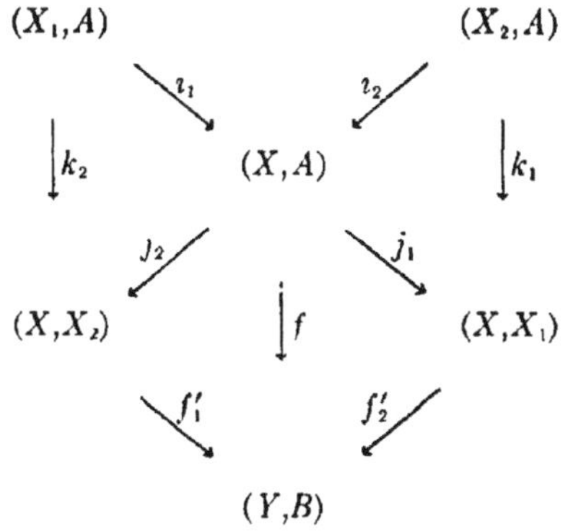

where f'_i is defined by f_i. Observe that commutativity does not hold in the lower two triangles; however we have

$$f_1 = f'_1j_2, \qquad f_2 = f'_2j_1, \qquad fi_1 = f'_1k_2, \qquad fi_2 = f'_2k_1.$$

If we pass to the induced homomorphisms on the homology groups, we observe that the part of the diagram not involving the group $H_q(Y,B)$ is a diagram of the type considered in 13.1. Thus, for each $u \in H_q(X,A)$, we have

$$u = i_{1*}k_{2*}^{-1}j_{2*}u + i_{2*}k_{1*}^{-1}j_{1*}u.$$

Then

$$\begin{aligned} f_*u &= f_*i_{1*}k_{2*}^{-1}j_{2*}u + f_*i_{2*}k_{1*}^{-1}j_{1*}u = f'_{1*}k_{2*}k_{2*}^{-1}j_{2*}u + f'_{2*}k_{1*}k_{1*}^{-1}j_{1*}u \\ &= f'_{1*}j_{2*}u + f'_{2*}j_{1*}u = f_{1*}u + f_{2*}u, \end{aligned}$$

and the proof is complete

The definitions of a triad and a proper triad relative to a cohomology theory is an obvious analog of 14.1

THEOREM 14 2c *A triad (X,X_1,X_2) is proper if and only if, for each q, the homomorphisms i_α^*: $H^q(X_1 \cup X_2, X_1 \cap X_2) \to H^q(X_\alpha,X_1 \cap X_2)$ yield a projective representation of $H^q(X_1 \cup X_2, X_1 \cap X_2)$ as a direct sum.*

DEFINITION 14.3c. The *coboundary operator* of a proper triad $(X;X_1,X_2)$ is the composition

$$H^{q-1}(X_1,X_1 \cap X_2) \xrightarrow{k_2^{*-1}} H^{q-1}(X_1 \cup X_2,X_2) \xrightarrow{l_2^*} H^{q-1}(X_1 \cup X_2) \xrightarrow{\delta} H^q(X,X_1 \cup X_2),$$

and is also denoted by δ The *cohomology sequence* of the proper triad is the upper sequence

$$\to H^{q-1}(X_1,X_1 \cap X_2) \xrightarrow{\delta} H^q(X,X_1 \cup X_2) \xrightarrow{j^*} H^q(X,X_2) \xrightarrow{i^*} H^q(X_1,X_1 \cap X_2) \to$$

THEOREM 14 4c. *The cohomology sequence of a proper triad is exact.*

THEOREM 14 5c. *If f. $(X;X_1,X_2) \to (Y;Y_1,Y_2)$ is a map of one proper triad into another, then f induces a homomorphism f^{**} of the cohomology sequence of the second into that of the first*

THEOREM 14.6c. *Under the conditions of 14.6, $f^* = f_1^* + f_2^*$.*

15. THE MAYER-VIETORIS SEQUENCE OF A TRIAD

Let $(X;X_1,X_2)$ denote a proper triad such that

(1) $$X = X_1 \cup X_2, \qquad A = X_1 \cap X_2.$$

The appellation "Mayer-Vietoris" is usually applied to a formula relating the Betti numbers (ranks of the homology groups) of X, X_1, X_2, A. It provides a method of calculating the Betti numbers of a space obtained by assembling two spaces A purely group theoretic generalization of this formula has been given by several authors. It is shown below that this group theoretic formulation can be simplified to the statement that a certain lower sequence is exact. The purely algebraic part of the construction is based on the following lemma, which we refer to as the "hexagonal lemma."

LEMMA 15.1. *If, in the following diagram of groups and homomorphisms,*

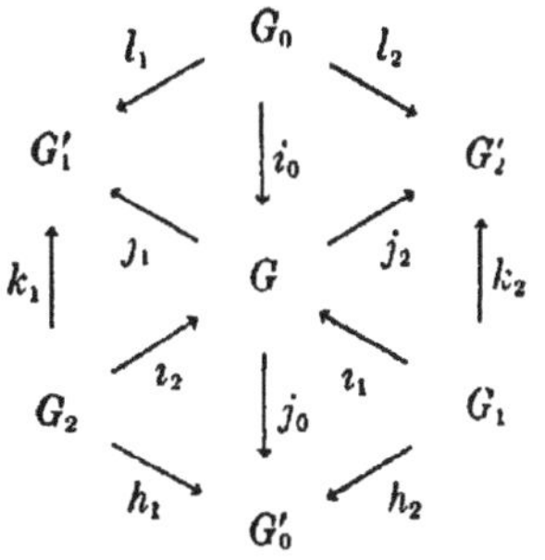

commutativity holds in each triangle, k_1 and k_2 are isomorphisms onto, image i_α = kernel j_α *for* $\alpha = 1,2$, *and* $j_0 i_0 = 0$, *then the two homomorphisms of G_0 into G'_0 obtained by skirting the sides of the hexagon differ in sign only Explicitly*

$$h_1 k_1^{-1} l_1 g = -h_2 k_2^{-1} l_2 g \qquad \textit{for each } g \in G_0.$$

PROOF. By 13.1, G decomposes into a direct sum of the isomorphic images of G_1 and G_2; and for any $g \in G_0$,

$$\begin{aligned} i_0 g &= i_2 k_1^{-1} j_1 i_0 g + i_1 k_2^{-1} j_2 i_0 g \\ &= i_2 k_1^{-1} l_1 g + i_1 k_2^{-1} l_2 g. \end{aligned}$$

Applying j_0 to both sides, and using $j_0 i_0 = 0$ and commutativity relations, we obtain the desired result.

The construction and proof of exactness of the Mayer-Vietoris sequence is based on the diagram below, in which all homomorphisms other than ∂ are induced by inclusion maps We assume that relations (1) hold Observe that the lower hexagon satisfies the hypotheses of 15.1.

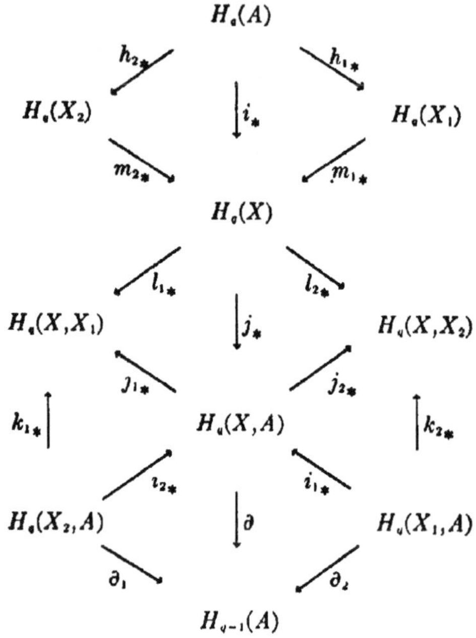

Definition 15.2. The *Mayer-Vietoris sequence of a proper triad* $(X;X_1,X_2)$ *with* $X = X_1 \cup X_2$ and $A = X_1 \cap X_2$ is the lower sequence

$$\cdots \leftarrow H_{q-1}(A) \overset{\Delta}{\leftarrow} H_q(X) \overset{\phi}{\leftarrow} H_q(X_1) + H_q(X_2) \overset{\psi}{\leftarrow} H_q(A) \leftarrow \cdots$$

where Δ,ϕ,ψ are defined as follows

$$\begin{aligned} \psi u &= (h_{1*}u, -h_{2*}u), && u \in H_q(A), \\ \phi(v_1,v_2) &= m_{1*}v_1 + m_{2*}v_2, && v_1 \in H_q(X_1),\ v_2 \in H_q(X_2), \\ \Delta w &= -\partial_1 k_{1*}^{-1} l_{1*} w, && w \in H_q(X), \\ &= \partial_2 k_{2*}^{-1} l_{2*} w, && \text{by 15 1.} \end{aligned}$$

Theorem 15.3 *The Mayer-Vietoris sequence of a proper triad* $(X;X_1,X_2)$ *with* $X = X_1 \cup X_2$ *is exact*

Proof. We must prove the usual six propositions.

(1) If $u \in H_q(A)$, then

$$\phi\psi u = \phi(h_{1*}u, -h_{2*}u) = m_{1*}h_{1*}u - m_{2*}h_{2*}u = i_*u - i_*u = 0.$$

(2) If $v = (v_1,v_2) \in H_q(X_1) + H_q(X_2)$, then, by exactness, $l_{1*}m_{1*}v_1 = 0$. This implies

$$\Delta m_{1*}v_1 = -\partial_1 k_{1*}^{-1} l_{1*} m_{1*} v_1 = 0.$$

By 15.1, $\Delta = \partial_2 k_{2*}^{-1} l_{2*}$. Therefore $\Delta m_{2*}v_2 = 0$. Thus

$$\Delta\phi v = \Delta(m_{1*}v_1 + m_{2*}v_2) = \Delta m_{1*}v_1 + \Delta m_{2*}v_2 = 0.$$

(3) If $w \in H_q(X)$, then

$$h_{2*}\Delta w = -h_{2*}\partial_1 k_{1*}^{-1} l_{1*} w = 0$$

since $h_{2*}\partial_1 = 0$ by exactness. Using 15.1, we obtain similarly that $h_{1*}\Delta = 0$. Hence $\psi\Delta = 0$.

(4) If $w \in H_q(X)$ and $\Delta w = 0$, then

$$\partial_\alpha k_{\alpha*}^{-1} l_{\alpha*} w = 0, \qquad \alpha = 1,2.$$

By exactness, there exists $v'_\alpha \in H_q(X_\alpha)$ such that

$$n_{1*}v'_1 = k_{2*}^{-1} l_{2*} w, \qquad n_{2*}v'_2 = k_{1*}^{-1} l_{1*} w$$

where n_α: $X_\alpha \subset (X_\alpha,A)$. By 13.1, j_*w splits into a sum which now becomes

$$\begin{aligned} j_*w &= i_{1*}n_{1*}v'_1 + i_{2*}n_{2*}v'_2 \\ &= j_*m_{1*}v'_1 + j_*m_{2*}v'_2 \end{aligned}$$

By exactness of the H.S. of (X,A), there is a $u \in H_q(A)$ such that

$$i_*u = w - m_{1*}v'_1 - m_{2*}v'_2.$$

If we let $v_1 = v'_1 + h_{1*}u$ and $v_2 = v'_2$, it follows that $\phi(v_1,v_2) = w$.

(5) If $v = (v_1,v_2) \in H_q(X_1) + H_q(X_2)$, and $\phi v = 0$, then

$$j_*\phi v = j_*m_{1*}v_1 + j_*m_{2*}v_2 = i_{1*}n_{1*}v_1 + i_{2*}n_{2*}v_2 = 0$$

By 13.1, the images of i_{1*} and i_{2*} have only the zero in common. Therefore $i_{\alpha*}n_{\alpha*}v_\alpha = 0$ $(\alpha = 1,2)$. Again by 13.1, $i_{\alpha*}$ has kernel zero. Therefore $n_{\alpha*}v_\alpha = 0$ $(\alpha = 1,2)$. By exactness, there exists a $u_\alpha \in H_q(A)$ such that $h_{\alpha*}u_\alpha = v_\alpha$ $(\alpha = 1,2)$. Therefore

$$\phi v = m_{1*}h_{1*}u_1 + m_{2*}h_{2*}u_2 = i_*(u_1 + u_2)$$

Since $\phi v = 0$, by exactness there is an $x \in H_{q+1}(X,A)$ such that $\partial x = u_1 + u_2$. By 13.1, there exists $x_\alpha \in H_{q+1}(X_\alpha,A)$ $(\alpha = 1,2)$ such that $x = i_{1*}x_1 + i_{2*}x_2$. Then

$$\partial x = \partial_2 x_1 + \partial_1 x_2 = u_1 + u_2.$$

Let $u = u_1 - \partial_2 x_1$. Then $u = -(u_2 - \partial_1 x_2)$. It follows now that

$h_{1*}u = h_{1*}u_1 - h_{1*}\partial_2 x_1 = h_{1*}u_1 = v_1$. Similarly $h_{2*}u = -v_2$. Therefore $\psi u = v$

(6) If $u \in H_q(A)$ and $\psi(u) = 0$, then $h_{1*}u = 0$ and $h_{2*}u = 0$. By exactness, there exists $x_\alpha \in H_{q+1}(X_\alpha, A)$ $(\alpha = 1,2)$ such that

$$\partial_1 x_2 = u, \qquad \partial_2 x_1 = -u.$$

This implies

$$\partial i_{1*}x_1 + \partial i_{2*}x_2 = \partial(i_{1*}x_1 + i_{2*}x_2) = 0.$$

By exactness, there is a $w \in H_{q+1}(X)$ such that

$$j_*w = -i_{1*}x_1 - i_{2*}x_2.$$

Then

$$\begin{aligned}\Delta w &= -\partial_1 k_{1*}^{-1} l_{1*} w = -\partial_1 k_{1*}^{-1} j_{1*} j_* w \\ &= \partial_1 k_{1*}^{-1} j_{1*}(i_{1*}x_1 + i_{2*}x_2) \\ &= 0 + \partial_1 k_{1*}^{-1} j_{1*} i_{2*} x_2 = \partial_1 x_2 = u\end{aligned}$$

since $k_{1*} = j_{1*}i_{2*}$ This completes the proof of 15 3.

THEOREM 15.4 *If $(X;X_1,X_2),(Y;Y_1,Y_2)$ are proper triads with $X = X_1 \cup X_2$, $Y = Y_1 \cup Y_2$, and $f.\ (X;X_1,X_2) \to (Y;Y_1,Y_2)$, then f induces a homomorphism of the Mayer-Vietoris sequence of $(X;X_1,X_2)$ into that of $(Y;Y_1,Y_2)$*

If one observes that f induces a map of the diagram preceding 15.2 into the analogous diagram for $(Y;Y_1,Y_2)$, it is seen that the desired commutativity relations follow from standard relations. The proofs are left to the reader.

LEMMA 15 5 *If $(X;X_1,X_2)$ is a proper triad, then commutativity holds in the diagram*

$$\begin{array}{ccc} H_{q+1}(X,X_1 \cup X_2) & \xrightarrow{\partial} & H_q(X_1,X_1 \cap X_2) \\ \downarrow{\scriptstyle \partial_1} & & \downarrow{\scriptstyle \partial_2} \\ H_q(X_1 \cup X_2) & \xrightarrow{\Delta} & H_{q-1}(X_1 \cap X_2) \end{array}$$

where ∂ is the boundary operator of the H.S. of the triad $(X;X_1,X_2)$ and Δ is the boundary operator of the Mayer-Vietoris sequence of the triad $(X_1 \cup X_2;X_1,X_2)$

PROOF. Consider the diagram

$$H_{q+1}(X,X_1 \cup X_2) \xrightarrow{\partial_1} H_q(X_1 \cup X_2) \xrightarrow{l_{2*}} H_q(X_1 \cup X_2,X_2)$$

$$\xleftarrow{k_{2*}} H_q(X_1,X_1 \cap X_2) \xrightarrow{\partial_2} H_{q-1}(X_1 \cap X_2)$$

where l_2 and k_2 are inclusion maps. Then, by 14.3 and 15 2,

$$\partial_2\partial = \partial_2 k_{2*}^{-1} l_{2*}\partial_1 = \Delta\partial_1,$$

which is the desired result.

We now proceed to define the *relative* Mayer-Vietoris sequence of a proper triad (X,X_1,X_2), (it is not assumed that $X = X_1 \cup X_2$). The definition and proof of exactness are based on the diagram below, analogous to the one preceding 15.2. We note that ∂, j_* are two consecutive homomorphisms in the H S of the triple $(X,X_1 \cup X_2,X_1 \cap X_2)$, and therefore the lower hexagon satisfies the hypotheses of 15.1.

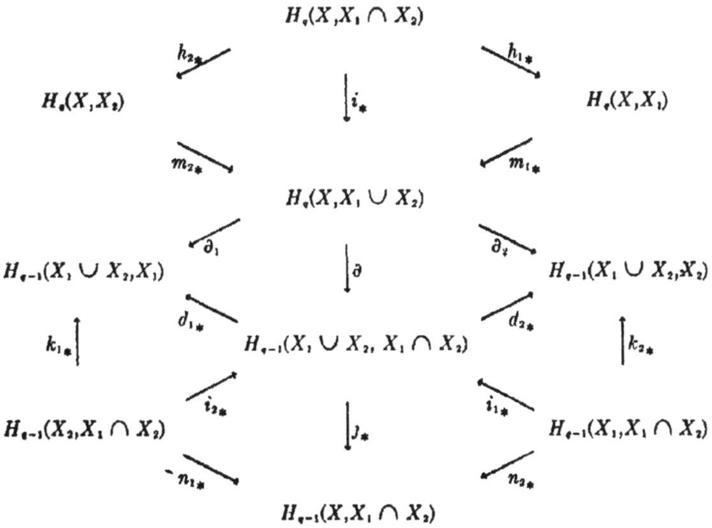

DEFINITION 15.6. The *relative Mayer-Vietoris sequence of the proper triad* (X,X_1,X_2) is the lower sequence

$$\overset{\phi}{\leftarrow} H_q(X,X_1) + H_q(X,X_2) \overset{\psi}{\leftarrow} H_q(X,X_1 \cap X_2) \leftarrow \cdots$$

$$\cdots \leftarrow H_{q-1}(X,X_1 \cap X_2) \overset{\Delta}{\leftarrow} H_q(X,X_1 \cup X_2)$$

where Δ,ϕ,ψ are defined as follows.

$$\begin{aligned} \psi u &= (h_{1*}u, -h_{2*}u), & u &\in H_q(X, X_1 \cap X_2), \\ \phi(v_1,v_2) &= m_{1*}v_1 + m_{2*}v_2, & v_1 &\in H_q(X,X_1),\ v_2 \in H_q(X,X_2), \\ \Delta w &= n_{2*}k_{2*}^{-1}\partial_2 w, & w &\in H_q(X, X_1 \cup X_2), \\ &= -n_{1*}k_{1*}^{-1}\partial_1 w, & &\text{by 15 1.} \end{aligned}$$

THEOREM 15.7 *The relative Mayer-Vietoris sequence of a proper triad is exact*

The proof is analogous to that of 15 3 and is left to the reader. The same applies to

THEOREM 15 8 *If f $(X;X_1,X_2) \to (Y,Y_1,Y_2)$ is a mapping of proper triads, then f induces a homomorphism of the relative Mayer-Vietoris sequence of (X,X_1,X_2) into that of (Y,Y_1,Y_2)*

DEFINITION 15.2c The *Mayer-Vietoris cohomology sequence of a proper triad* (X,X_1,X_2) with $X = X_1 \cup X_2$, $A = X_1 \cap X_2$ is the upper sequence

$$\cdots \to H^{q-1}(A) \xrightarrow{\Delta} H^q(X) \xrightarrow{\phi} H^q(X_1) + H^q(X_2) \xrightarrow{\psi} H^q(A) \to \cdots$$

where

$$\begin{aligned} \Delta u &= -l_1^* k_1^{*-1}\delta_1 u, & u &\in H^{q-1}(A), \\ &= l_2^* k_2^{*-1}\delta_2 u, & &\text{by 15.1,} \\ \phi w &= (m_1^* w, m_2^* w), & w &\in H^q(X), \\ \psi(v_1,v_2) &= h_1^* v_1 - h_2^* v_2, & v_1 &\in H^q(X_1),\ v_2 \in H^q(X_2). \end{aligned}$$

THEOREM 15 3c *The Mayer-Vietoris cohomology sequence of a proper triad $(X;X_1,X_2)$ with $X = X_1 \cup X_2$ is exact*

THEOREM 15.4c. *If f. $(X,X_1,X_2) \to (Y;Y_1,Y_2)$ is a map of one proper triad into another, and $X = X_1 \cup X_2$, $Y = Y_1 \cup Y_2$, then f induces a homomorphism of the Mayer-Vietoris cohomology sequence of the second triad into that of the first*

LEMMA 15 5c *If $(X;X_1,X_2)$ is a proper triad, then commutativity holds in the diagram*

$$\begin{array}{ccc} H^{q+1}(X, X_1 \cup X_2) & \xleftarrow{\ \delta\ } & H^q(X_1, X_1 \cap X_2) \\ \uparrow{\scriptstyle \delta_1} & & \uparrow{\scriptstyle \delta_2} \\ H^q(X_1 \cup X_2) & \xleftarrow{\ \Delta\ } & H^{q-1}(X_1 \cap X_2) \end{array}$$

where δ is from the C.S. of the triad $(X;X_1,X_2)$ and Δ is from the Mayer-Vietoris sequence of the triad $(X_1 \cup X_2, X_1,X_2)$

DEFINITION 15.6c. The *relative Mayer-Vietoris cohomology sequence of the proper triad* $(X;X_1,X_2)$ is the upper sequence

$$\cdots \to H^{q-1}(X,X_1 \cap X_2) \xrightarrow{\Delta} H^q(X,X_1 \cup X_2)$$
$$\xrightarrow{\phi} H^q(X,X_1) + H^q(X,X_2) \xrightarrow{\psi} H^q(X,X_1 \cap X_2) \to \cdots$$

where (using the diagram preceding 15 6)

$$\begin{aligned} \Delta u &= \delta_2 k_2^{*-1} n_2^* u, & u &\in H^{q-1}(X,X_1 \cap X_2), \\ &= -\delta_1 k_1^{*-1} n_1^* u, & & \\ \phi w &= (m_1^* w_1, m_2^* w), & w &\in H^q(X,X_1 \cup X_2), \\ \psi(v_1,v_2) &= h_1^* v_1 - h_2^* v_2, & v_1 &\in H^q(X,X_1),\ v_2 \in H^q(X,X_2). \end{aligned}$$

THEOREM 15.7c. *The relative Mayer-Vietoris cohomology sequence of a proper triad is exact*

THEOREM 15.8c. *If* f: $(X;X_1,X_2) \to (Y;Y_1,Y_2)$ *is a mapping of proper triads, then* f *induces a homomorphism of the relative Mayer-Vietoris cohomology sequence of* $(Y;Y_1,Y_2)$ *into that of* (X,X_1,X_2).

16. CELLS AND SPHERES

Let R^n be the euclidean n-space. Coordinates $(x_1, \cdots, x_n)$ will sometimes be abbreviated by the vector symbol x, and the norm of x is defined to be $||x|| = (\sum_1^n x_i^2)^{\frac{1}{2}}$. Each of the following six symbols will denote the subset of R^n defined by the algebraic condition to its right:

n-cell	E^n :	$\lVert x\rVert \leqq 1,$	
$(n-1)$-sphere	S^{n-1}:	$\lVert x\rVert = 1,$	
upper cap	E_+^{n-1}:	$\lVert x\rVert = 1,$	$x_n \geqq 0,$
lower cap	E_-^{n-1}	$\lVert x\rVert = 1,$	$x_n \leqq 0,$
$(n-2)$-sphere	S^{n-2}:	$\lVert x\rVert = 1,$	$x_n = 0,$
$(n-1)$-cell	E^{n-1}.	$\lVert x\rVert \leqq 1,$	$x_n = 0.$

Clearly, $S^{n-1} = E_+^{n-1} \cup E_-^{n-1}$ and $S^{n-2} = E_+^{n-1} \cap E_-^{n-1}$.

S^0 consists of two points, one of which is E_+^0 and the other E_-^0. We also define S^{-1} to be the vacuous set. R^0 and E^0 are single points.

Let R^{n-1} be the euclidean subspace of R^n defined by $x_n = 0$, and let f be the projection of R^n into R^{n-1}:

$$f(x_1, \cdots, x_n) = (x_1, \cdots, x_{n-1}, 0).$$

The function f defines maps

$$f_+: E_+^{n-1} \to E^{n-1}, \qquad f_-: E_-^{n-1} \to E^{n-1}.$$

It is easy to verify that f_+ and f_- are homeomorphisms

We shall assume throughout this section that all the pairs and maps employed are admissible. In particular all spaces will be assumed to be collapsible. This is the case for all the admissible categories that we shall have occasion to consider in the sequel.

LEMMA 16.1. *E^n, E_-^{n-1}, and (E^n, E_-^{n-1}) are homologically trivial.*

PROOF. Let $x \in E^n$, $0 \leq t \leq 1$. Using vector notation, define $h(x,t) = (1 - t)x$. Then $h(x,0) = x$, $h(x,1) = 0$. Therefore E^n is contractible on itself to a point. Thus, by 11.5, E^n is homologically trivial. Since E_-^{n-1} is homeomorphic to E^{n-1}, it is also homologically trivial. Finally (E^n, E_-^{n-1}) is homologically trivial by virtue of 9.5.

LEMMA 16.2. *The triads $(E^n,E_+^{n-1}, E_-^{n-1})$ and $(S^{n-1}, E_+^{n-1},E_-^{n-1})$ are proper*

PROOF. We must show that the inclusion maps

$$k_2\colon (E_+^{n-1},S^{n-2}) \subset (S^{n-1},E_-^{n-1}), \qquad k_1 \quad (E_-^{n-1},S^{n-2}) \subset (S^{n-1},E_+^{n-1})$$

induce isomorphisms of the homology groups in all dimensions. Because of symmetry, it suffices to consider k_2 only.

Let V be the subset of R^n determined by the conditions

$$||x|| = 1, \qquad x_n < -\tfrac{1}{2}.$$

Clearly V is an open subset of S^{n-1} and $\overline{V}$ lies in the interior of E_-^{n-1} relative to S^{n-1}. The homotopy

$$F(x,t) = x \qquad x \in E_+^{n-1}, 0 \leq t \leq 1$$

$$F(x,t) = \frac{(1 - t)x + tf_-(x)}{||(1 - t)x + tf_-(x)||} \qquad x \in E_-^{n-1} - V, 0 \leq t \leq 1$$

shows that (E_+^{n-1},S^{n-2}) is a strong deformation retract of $(S^{n-1} - V,E_-^{n-1} - V)$. Now apply 12 1 with $X = S^{n-1}$, $A = E_-^{n-1}$, $U = E_-^{n-1} - S^{n-2}$ to obtain that k_2 induces isomorphisms in all dimensions.

THEOREM 16.3 *The homology sequence of the triad $(E^n;E_+^{n-1},E_-^{n-1})$ reduces to the isomorphism*

$$\partial\colon \ H_n(E^n,S^{n-1}) \approx H_{n-1}(E_+^{n-1},S^{n-2})$$

All other groups in the sequence are trivial. The above isomorphism is called the incidence isomorphism and is denoted by $[E^n\!:\!E_+^{n-1}]$ The isomorphism $[E^n\!:\ E_-^{n-1}]$ is defined similarly using the triad $(E^n,E_-^{n-1},E_+^{n-1})$.

THEOREM 16 4 *The homology groups of (E^n,S^{n-1}) are as follows:*

$$H_n(E^n,S^{n-1}) \approx G,$$
$$H_q(E^n,S^{n-1}) = 0 \qquad \textit{for } q \neq n.$$

PROOF OF 16 3 AND 16.4. By 16.1, (E^n,E_-^{n-1}) is homologically trivial, thus $H_q(E^n,E_-^{n-1}) = 0$; and, by the exactness of the H.S. of the triad

$(E^n;E^{n-1}_+,E^{n-1}_-)$, we have $\partial\colon H_q(E^n,S^{n-1}) \approx H_{q-1}(E^{n-1}_+,S^{n-2})$. Since the pairs (E^{n-1}_+,S^{n-2}) and (E^{n-1},S^{n-2}) are homeomorphic, we have $H_q(E^n,S^{n-1}) \approx H_{q-1}(E^{n-1},S^{n-2})$. Iterating this result yields $H_q(E^n,S^{n-1}) \approx H_{q-n}(E^0,S^{-1}) = H_{q-n}(E^0)$. Since E^0 is a single point, 16.4 follows from the Dimension axiom and the definition of the coefficient group G. This also establishes 16.3.

THEOREM 16.5. *Commutativity holds in the diagram*

$$\begin{array}{ccc} H_n(E^n,S^{n-1}) & \xrightarrow{[E^n,E^{n-1}_+]} & H_{n-1}(E^{n-1}_+,S^{n-2}) \\ \downarrow{\scriptstyle \partial_1} & \Delta & \downarrow{\scriptstyle \partial_2} \\ H_{n-1}(S^{n-1}) & \longrightarrow & H_{n-2}(S^{n-2}) \end{array}$$

where Δ is the boundary operator in the Mayer-Vietoris sequence of the triad $(S^{n-1};E^{n-1}_+,E^{n-1}_-)$. For $n > 2$ all four homomorphisms are isomorphisms For $n = 2$ the same holds provided $H_{n-2}(S^{n-2}) = H_0(S^0)$ is replaced by $\tilde{H}_0(S^0)$. In particular

$$\Delta\colon\ H_n(S^n) \approx H_{n-1}(S^{n-1}), \qquad n > 1,$$
$$\Delta\colon\ H_1(S^1) \approx \tilde{H}_0(S^0).$$

THEOREM 16.6. *The homology groups of S^n are as follows:*

$$\begin{array}{lll} H_0(S^n) \approx G, & \tilde{H}_0(S^n) = 0, & n > 0, \\ H_0(S^0) \approx G + G, & \tilde{H}_0(S^0) \approx G, & \\ H_n(S^n) \approx G, & & n > 0, \\ H_p(S^n) = 0, & & p \neq n,0. \end{array}$$

PROOF. The commutativity relation of 16.5 is a consequence of 15.5 Since E^n and E^{n-1}_+ are homologically trivial, it follows from 9.4 that

$$\begin{array}{lll} \partial_1\colon & H_q(E^n,S^{n-1}) \approx H_{q-1}(S^{n-1}), & q > 1, \\ \partial_1\colon & H_1(E^n,S^{n-1}) \approx \tilde{H}_0(S^{n-1}), & \\ \partial_2\colon & H_q(E^{n-1}_+,S^{n-2}) \approx H_{q-1}(S^{n-2}), & q > 1, \\ \partial_2\colon & H_1(E^{n-1}_+,S^{n-2}) \approx \tilde{H}_0(S^{n-2}). & \end{array}$$

These give all the conclusions of 16.5 Combining these isomorphisms with 16.4 yields all the results of 16.6 except those concerning $H_0(S^n)$. Since S^n is nonempty for $n \geqq 0$, it follows from 7.6 that $H_0(S^n) \approx \tilde{H}_0(S^n) + G$ This concludes the proof

The formulations and proofs of analogous results for cohomology are left to the reader.

NOTES

The origins of the basic concepts The basic machinery for constructing homology groups (namely: complexes and incidence numbers) is due to Poincaré [*Analysis Situs*, Jour. de l'Ec. Polyt. (2) 1 (1895), 1-123]. Using these he defined directly the Betti numbers and torsion numbers. They are the numerical invariants which characterize the homology groups based on the coefficient group of integers. For many years, they were the primary source of interest It was during the period 1925-1935 that attention shifted from the numerical invariants to the groups themselves. This shift was due in part to the influence of E. Noether. It was also enforced by two directions of generalization: (1) from complexes to more general spaces where the homology groups are not characterized by numerical invariants, and (2) from integer coefficients to arbitrary coefficients where, again, numerical invariants are inadequate. Thus, although Poincaré did not speak of the homology groups themselves, he is to be credited with the origin of the concept.

The concept of relative homology (modulo a subcomplex) is due to Lefschetz [Proc. Nat Acad. 13 (1927), 614-622] The operator ∂ was used by Lefschetz. It is not clear who first gave it formal recognition. The origin of the induced homomorphism f_* is likewise obscure. It has been used, in a sense, since the time of Poincaré, but for at least thirty-five years it had no name nor any formal status. This lack of formal recognition of ∂ and f_* is a natural consequence of the failure to accord the homology groups a formal status.

Each of our axioms is a theorem of classical homology theory. In most cases it is not clear who first stated and proved them. The Axioms 1, 2, 3, and 7 are, perhaps, too basic and too well understood to warrant such explicit treatment. One must be interested in an axiomatic development before one thinks of writing them down.

The first formal recognition of the homology sequence and its exactness is due to Hurewicz [Bull. Amer. Math. Soc 47 (1941), 562]. It was subsequently exploited by Kelley and Pitcher [Annals of Math. 48 (1947), 682-709] It should be noted though that the six parts of exactness were well known and had occurred frequently in the proofs of other less obvious propositions.

The excision property was implicit in Lefschetz's construction of the relative groups. He often used the expression "the homology groups of $X - A$" instead of "the homology groups of X mod A."

The homotopy axiom has had no more formal recognition than f_*. However the proposition "homotopic cycles are homologous" has been known and used for many years.

The origin of the cohomology groups is thoroughly confused The "co" terminology is due to Whitney [Duke Math. Jour. 3(1937), 35-45]. The existence of cohomology groups was implicit in the duality theorem of Alexander [Trans. Amer. Math. Soc. 23 (1922), 333-349]. The complete group invariant form of the duality theorem was proved by Pontrjagin [Annals of Math. 35 (1934), 904-914]. Cocycles made their first formal appearance under the name *pseudocycles* in the book of Lefschetz [Colloq. Publ. Amer. Math. Soc., vol. 12, 1930]. The first intrinsic definition of the cohomology groups was given by Alexander at the Moscow conference in 1936.

Homotopy groups. The homotopy groups of Hurewicz are quite similar to homology groups in that there are concepts analogous to those of homology theory and they obey similar axioms. The qth homotopy group depends not only on a pair (X,A) but also on a reference point $x_0 \in A$ and is denoted by $\pi_q(X,A,x_0)$. It is defined when $q \geqq 2$. If A is a single point, it is also defined for $q = 1$ and is called the fundamental group. The groups are abelian for $q \geqq 3$, and $\pi_2(X,x_0)$ is also abelian A boundary operator $\partial.\ \pi_q(X,A,x_0) \to \pi_{q-1}(A,x_0)$ exists for $q \geqq 2$ If f: $(X,A,x_0) \to (Y,B,y_0)$, there are induced homomorphisms f_*: $\pi_q(X,A,x_0) \to \pi_q(Y,B,y_0)$.

When suitably modified, all of our Axioms 1 through 7 hold for homotopy groups with the sole exception of the Excision axiom This is the fundamental property distinguishing homology from homotopy. It accounts for the computability of homology groups as contrasted with our meager knowledge of homotopy groups.

As a simple example, let the n-sphere S^n be divided into upper and lower hemispheres, E^n_+, E^n_-, by an equator S^{n-1}. Using a fixed reference point $x_0 \in S^{n-1}$ for all groups, we obtain the diagram

$$\begin{array}{ccc} \pi_q(E^n_+, S^{n-1}) & \xrightarrow{i_*} & \pi_q(S^n, E^n_-) \\ \downarrow \partial & & \uparrow j_* \\ \pi_{q-1}(S^{n-1}) & \xrightarrow{E} & \pi_q(S^n) \end{array}$$

where i and j are inclusion maps. All homotopy groups of a cell are zero. This and exactness imply that ∂ and j_* are isomorphisms onto. Then we define E by $E = j_*^{-1} i_* \partial^{-1}$.

Now i is an excision map, and we wish to state properties of i_*. We have introduced the equivalent homomorphism E since it can be proved to coincide with the suspension (*Einhangung*) homomorphism introduced by Freudenthal [Composito Math. 5 (1937), 299-314]. Since $\pi_2(S^1) = 0$ and $\pi_3(S^2)$ is infinite cyclic, it follows that E, and therefore i_*, is not always an isomorphism onto. Hence the Excision axiom fails for homotopy groups in a very simple case. Freudenthal has shown [ibid] that E is an isomorphism for each $q < 2n - 1$. We may interpret his results as asserting that the excision property does hold in certain restricted cases. Most of the little we know about $\pi_q(S^n)$ is based on the results of Freudenthal concerning E.

The problem of axiomatizing the homotopy groups has not been solved.* One would naturally seek a substitute for the Excision axiom. There is a reasonable candidate. Let B be a fibre bundle over the base space X with projection $p: B \to X$, let $x_0 \in X$, $Y_0 = p^{-1}(x_0)$, and $y_0 \in Y_0$. Then, for $q \geqq 2$, p_* maps $\pi_q(B,Y_0,y_0)$ isomorphically onto $\pi_q(X,x_0)$. This is proved using the covering homotopy theorem for bundles. If π_q is replaced by H_q, then p_* is usually not an isomorphism onto. We have therefore a simple and useful property of homotopy groups which may serve as a substitute for the excision property.

In axiomatizing the homotopy groups one would need an additional basic concept, namely, the isomorphisms $\pi_q(X,A,x_0) \approx \pi_q(X,A,x_1)$ assigned to a homotopy class of paths in A from x_0 to x_1. One would deal not with single groups but with systems of groups connected by isomorphisms assigned to the fundamental groupoid of A. The latter would include the operations of $\pi_1(A,x_0)$ on $\pi_q(X,A,x_0)$.

Cohomotopy groups. The Borsuk-Spanier cohomotopy groups [Annals of Math. 50 (1949), 203-245] are similar to cohomology groups. If (X,A) is a compact pair of finite dimension n, then the qth cohomotopy group $\pi^q(X,A)$ is defined for each integer $q > (n+1)/2$ and is an abelian group. The elements of $\pi^q(X,A)$ are homotopy classes of maps of (X,A) into (S^q,y_0) where S^q is a q-sphere and y_0 is a point. The set $\pi^q(X,A)$ is defined for $q \geqq 0$; but the addition is only defined for $q > (n+1)/2$. A mapping $\delta: \pi^{q-1}(A) \to \pi^q(X,A)$ is defined for $q > 0$ and is homomorphic when both sides are groups. If $f: (X,A) \to (Y,B)$, then $f^*: \pi^q(Y,B) \to \pi^q(X,A)$ is defined for all q and is homomorphic when both sides are groups. Spanier has shown [ibid] that the cohomotopy groups satisfy the analogs of *all* the cohomology axioms—insofar as they are meaningful.

The axiomatization of the cohomotopy groups has not been attempted. The chief distinction between cohomology and cohomotopy is the absence of a group structure in $\pi^q(X,A)$ for $q \leqq (n+1)/2$. This makes it impossible to compute cohomotopy groups by an induction starting with $q = 0$. Our knowledge of cohomotopy groups is as meager as that of homotopy groups.

*An axiomatization has subsequently been given by J. Milnor [Annals of Math. 63 (1956), 272-284].

EXERCISES

A. Elementary properties of exact sequences.

Establish the following propositions for an exact lower sequence $\{G_q,\phi_q\}$.

1. $\phi_{q+1} = 0$ if and only if kernel $\phi_q = 0$.
2. $\phi_{q-1} = 0$ if and only if $\phi_q(G_q) = G_{q-1}$.
3. $\phi_{q+1} = 0$ and $\phi_{q-1} = 0$ if and only if ϕ_q: $G_q \approx G_{q-1}$
4. $G_q = 0$ if and only if $\phi_{q+1} = 0$ and $\phi_q = 0$.
5. $G_q = 0$ if and only if $\phi_{q+2}(G_{q+2}) = G_{q+1}$ and kernel $\phi_{q-1} = 0$

B. The Axioms.

1. Show that Axiom 1 is a consequence of Axioms 2, 3, and 4.
2. Show that Axiom 1 is a consequence of Axioms 2 and 6.
3. Assume that, for every admissible pair (X,A), the map p: $(X,A) \times I \to (X,A)$ defined by $p(x,t) = x$ is admissible. Show, using Axioms 1 and 2, that the Homotopy axiom is equivalent to either of the following if suitable maps are admissible

Axiom 5″. p_* is an isomorphism

Axiom 5‴. The kernel of p_* is zero.

In the following problems, assume that the homology theory is defined on the category of all pairs and maps.

C. Retracts.

1. If ϕ: $G \to H$, ψ: $H \to G$, and $\psi\phi$ = identity, then ϕ has kernel zero and H decomposes into a direct sum

$$H = \text{image}\,\phi + \text{kernel}\,\psi$$

2. If r: $X \to A$ is a retraction, and i: $A \to X$ is the inclusion, then $H_q(X)$ decomposes into the direct sum

$$H_q(X) = \text{image}\,i_{*q} + \text{kernel}\,r_{*q},$$

and

$$H_q(X) \approx H_q(A) + H_q(X,A).$$

D. The 0-dimensional groups.

1. If $x \in X$, then $H_0(X,x) \approx \tilde{H}_0(X)$.
2. $\tilde{H}_0(X) = 0$ if and only if $H_0(X,A) = 0$ for each nonvacuous subset A of X. Assume that $H_{-1}(A) = 0$.
3. If X is an arcwise connected space, $x,x' \in X$, and $g \in G$, then $(gx)_X = (gx')_X$ and $(Gx)_X = (Gx')_X$.
4. If X is a Hausdorff space consisting of just two points, then $\tilde{H}_0(X) \approx G$.
5. If $G \neq 0$ and $\tilde{H}_0(X) = 0$, then X is connected.

E. TRIVIAL MAPS.

1. If $f.\ (X,A) \to (Y,B)$ is such that $f(X) \subset B$, then $f_{*q} = 0$ for each q.

2. If $f\colon\ X \to (Y,B)$ is such that $f(X)$ is a point of Y, then $f_{*q} = 0$ for $q \neq 0$ and $\tilde{f}_{*0} = 0$.

F. DIRECT SUM.

1. Prove 13 2 with the condition that the sets X_α are disjoint replaced by the following conditions (1) X is normal, (2) $X_\alpha \cap A = A_\alpha$, and (3) the sets $X_\alpha - A_\alpha$ have disjoint closures

G. MAYER-VIETORIS SEQUENCES.

1 Let $(X;X_1,X_2)$ be a proper triad Show that in the diagram

$$\begin{array}{ccc} H_{q+1}(X,X_1 \cup X_2) & \xrightarrow{\partial_1} & H_q(X_1,X_1 \cap X_2) \\ \downarrow{\scriptstyle \partial_2} & & \downarrow{\scriptstyle \partial} \\ H_q(X_2,X_1 \cap X_2) & \xrightarrow{\partial} & H_{q-1}(X_1 \cap X_2) \end{array}$$

where ∂_1 and ∂_2 are the boundary operators of the triads $(X;X_1,X_2)$ and $(X;X_2,X_1)$ the anticommutativity relation

$$\partial\partial_1 + \partial\partial_2 = 0$$

holds.

2 Let $(X;X_1,X_2)$ be a proper triad; prove that ∂ maps the relative Mayer-Vietoris sequence into the Mayer-Vietoris sequence of the triad $(X_1 \cup X_2,X_1,X_2)$.

3. For any space X, show that the Mayer-Vietoris relative boundary operator Δ (defined in 15 6) for the triad $(X \times I;\ X \times 0,\ X \times 1)$ is an isomorphism, thereby establishing an isomorphism $H_q(X \times I, X \times 0 \cup X \times 1) \approx H_{q-1}(X)$ Prove this last result without explicit use of the Mayer-Vietoris sequence

4. Let $(X;X_1,X_2)$ be a proper triad such that $X = X_1 \cup X_2$, let $A = X_1 \cap X_2$, $I = [0,1]$, and let

$$Y = (X_1 \times 0) \cup (A \times I) \cup (X_2 \times 1)$$
$$B = (X_1 \times 0) \cup (X_2 \times 1)$$

be the indicated subsets of $X \times I$ Show, using the isomorphism of problem 3, that the Mayer-Vietoris sequence of $(X;X_1,X_2)$ is isomorphic to the H S. of (Y,B).

5 Show that the relative Mayer-Vietoris sequence of $(X;X_1,X_2)$ is

isomorphic to the homology sequence of a triple consisting of $X \times I$ and suitable subspaces.

6. Examine the Mayer-Vietoris sequence of the triad $(X;X,X)$.

7. Consider a proper triad $(X;X_1,X_2)$ with $X = X_1 \cup X_2$, $A = X_1 \cap X_2 \neq 0$. Define the reduced Mayer-Vietoris sequence and prove its exactness.

H. Ranks of Groups.

It will be assumed that all groups considered are D-modules over a domain of integrity D (i.e a commutative ring D with a unit element, and such that $d_1 \neq 0$, $d_2 \neq 0$ imply $d_1d_2 \neq 0$).

Definition. The elements $g_1, \cdots, g_r$ of a D-module G are called *linearly independent*, provided any relation $d_1g_1 + \cdots + d_rg_r = 0$ where $d_1, \cdots, d_r \in D$ implies $d_1 = \cdots = d_r = 0$. The maximum number of linearly independent elements in G is called the *rank* of G and is denoted by $r(G)$ (or $r_D(G)$ if we wish to indicate D). If this maximum does not exist, we set $r(G) = \infty$. If D is a field, then $r(G)$ is the dimension of G over D.

1. If H is a submodule of G, show that $r(G) = r(H) + r(G/H)$.

2 Let $\{G_q,\phi_q\}$ be an exact lower sequence of D-modules such that each $r(G_q)$ is finite. Let $\pi_q = r(\text{kernel } \phi_q)$ Then for $m \leqq n$

$$\sum_{q=m}^{n} (-1)^q r(G_q) = (-1)^n \pi_n - (-1)^{m-1}\pi_{m-1}.$$

I. Betti numbers.

Definition Assume in this section that a homology theory is given such that $H_q(X,A)$ are all D-modules and f_*,∂ are linear over D. The rank $r[H_q(X,A)]$ is then denoted by $R_q(X,A)$ and is called the rth Betti number of (X,A).

1. Let (X,A) be a pair. Consider the three sequences $\{R_q(A)\}$, $\{R_q(X)\}$, $\{R_q(X,A)\}$. Show that, if two of these sequences contain only finite numbers, then so does the third one Show that, if in two of these sequences only a finite number of terms are different from zero, then the same holds in the third sequence

2. Show that (assuming all numbers involved are finite)

$$\sum_{k=n}^{m} (-1)^q R_q(X) - \sum_{k=n}^{m} (-1)^q R_q(A) - \sum_{k=n}^{m} (-1)^q R_q(X,A) = (-1)^{n-1} w_{n-1} - (-1)^m w_m$$

where w_q is the rank of the kernel of the homomorphism $H_q(A) \to H_q(X)$ induced by the inclusion $A \subset X$.

Definition. If the integers $\{R_q(X,A)\}$ are all finite and only a finite number of them are different from zero then the integer

$$\chi(X,A) = \sum_{-\infty}^{\infty} (-1)^q R_q(X,A)$$

is called the Euler characteristic of (X,A).

3. If (X,A) is a pair and two of the Euler characteristics $\chi(A)$, $\chi(X)$, $\chi(X,A)$ are defined, then so is the third, and

$$\chi(X) = \chi(A) + \chi(X,A).$$

4. Let (X,X_1,X_2) be a proper triad with $X = X_1 \cup X_2$, $A = X_1 \cap X_2$. Let N_q be the intersection of the kernels of the homomorphisms

$$H_q(A) \to H_q(X_1), \qquad H_q(A) \to H_q(X_2)$$

induced by inclusion maps $A \subset X_1$, $A \subset X_2$. Prove the Mayer-Vietoris formula

$$R_q(X_1 \cup X_2) + R_q(X_1 \cap X_2) = R_q(X_1) + R_q(X_2) + r(N_q) + r(N_{q-1})$$

and derive from it the formula

$$\chi(X_1 \cup X_2) + \chi(X_1 \cap X_2) = \chi(X_1) + \chi(X_2)$$

provided all four Euler characteristics involved are defined.

5. Formulate the Mayer-Vietoris formula for cohomology. Formulate *relative* Mayer-Vietoris formulas for homology and cohomology.

CHAPTER II

Simplicial complexes

1. INTRODUCTION

This chapter develops the analytical geometric tools needed in subsequent chapters. These are: simplex, complex, subcomplex, simplicial map, triangulation, and simplicial approximation. Homology theory is not mentioned, its study is resumed in Chapter III. The reason for this hiatus is the necessity of singling out a class of spaces (triangulable spaces) sufficiently simple that an algorithm can be given for computing their homology groups. The nature of this class is nearly predictable on the basis of the results of Chapter I. Knowing the groups of a point, the groups of a contractible space are determined. We choose a class of contractible spaces (i.e. simplexes) and form more complicated spaces (i.e. complexes) by assembling these in a smooth fashion. Then the groups of the latter spaces can be computed by the use of Mayer-Vietoris sequences or similar devices.

It is not enough to be able to compute groups; it is also necessary to compute homomorphisms. This requires singling out a simple class of maps—simplicial maps. Although these are quite restricted, it is shown that any map of one triangulable space in another is homotopic to one such; indeed, simplicial maps are dense in the function space of all maps. This is achieved by barycentric subdivision and the simplicial approximation theorem.

Although triangulable spaces appear to form a rather narrow class, a major portion of the spaces occurring in applications of topology to geometry and analysis are of this type. Furthermore, it is shown in Chapter X that any compact space can be expressed as a limit of triangulable spaces in a reasonable sense. In this sense, triangulable spaces are dense in the family of compact spaces.

2. SIMPLEXES

DEFINITION 2.1. An *n-simplex* s is a set of $n + 1$ objects called vertices, usually denoted by $\{A\}$, together with the set of all real-valued functions α defined on $\{A\}$ satisfying

(1) $$\sum_A \alpha(A) = 1, \qquad \alpha(A) \geqq 0.$$

A single function α is called a point of s. The values of α on the vertices of s are called the *barycentric coordinates* of the point α. The distance $\rho(\alpha,\beta)$ of two points α,β of s is defined by

$$\rho(\alpha,\beta) = [\sum_A (\alpha(A) - \beta(A))^2]^{\frac{1}{2}}.$$

The topological space thus defined is denoted by $|s|$ and is called the space of s. Clearly the barycentric coordinates are continuous functions on $|s|$.

The use of the term "barycenter" stems from the fact that, if the vertices $\{A\}$ of s are points of a euclidean space, then the point α corresponds to the center of gravity of the system of masses obtained by assigning to each vertex A the mass $\alpha(A)$.

DEFINITION 2.2. A simplex s together with a simple ordering $A^0 < \cdots < A^n$ of its vertices is called an *ordered simplex*. Let R^{n+1} be the cartesian space of all coordinates $(x_0, \cdots, x_n)$. The correspondence $\alpha \to (\alpha(A^0), \cdots, \alpha(A^n))$ is then an isometric map $s \to R^{n+1}$, called the *canonical imbedding* of s in R^{n+1}. The image of s in R^{n+1} is denoted by Δ^n and is called the unit simplex of R^{n+1}. Clearly Δ^n is the intersection of the plane $\sum_1^n x_i = 1$ with the sector $x_i \geqq 0$ for $i = 1, \cdots, n$.

Since Δ^n is closed and bounded in R^{n+1}, it is compact. Hence $|s|$ is compact for any simplex s. Clearly, the dimension of $|s|$, in the euclidean sense, is n. Observe that an unordered n-simplex has $(n + 1)!$ canonical imbeddings in R^{n+1} corresponding to the various orderings of its vertices.

DEFINITION 2.3. A *q-face* s' of an n-simplex s is a q-simplex whose vertices form a subset of the vertices of s.

A point α' of s' is a function defined over a subset of the vertices of s. It can be extended to a function α on all vertices by setting $\alpha(A) = \alpha'(A)$ if A is in s', and $\alpha(A) = 0$ otherwise. Then α is clearly a point of s. In view of the conditions (1) on α,α', the extension α of α' is unique. The map $\alpha' \to \alpha$ imbeds $|s'|$ isometrically in $|s|$. Following the custom, we identify α' with α so that $|s'|$ is a subset of $|s|$. It is surely a closed subset of $|s|$. It is defined by the equations $\alpha(A) = 0$ for A not in s'.

A 0-simplex has just one vertex A, and just one point $\alpha(A) = 1$. It is customary to identify the vertex with the point and to denote either by A. With this convention the vertices A of s are the 0-faces of s, and are points of s. As a result, $A(B)$ is defined for any two vertices, and $A(B) = 0$ if $A \neq B$, and $A(A) = 1$ for each A. In the unit simplex Δ^n in R^{n+1} the vertices appear as the unit points on the coordinate axes. In addition Δ^n is the smallest convex set in R^{n+1}

containing these unit points. For this reason the simplex is said to *span* its vertices.

DEFINITION 2.4. The *point set boundary of s*, denoted by $|\dot{s}|$, is the set of points of s having at least one coordinate equal to zero. Clearly $|\dot{s}|$ is a closed subset of $|s|$. The open set $|s| - |\dot{s}|$ is called the *interior* of s and is also referred to as the *open simplex*.

If $\alpha^0, \cdots, \alpha^q$ are points of an n-simplex s, and $w_0, \cdots, w_q$ are non-negative real numbers such that $w_0 + \cdots + w_q = 1$, then the function α defined by

$$\alpha = w_0\alpha^0 + \cdots + w_q\alpha^q$$

is again a point of s. For clearly $\alpha(A) \geqq 0$ for each vertex A of s, and

$$\sum_A \alpha(A) = \sum_A \sum_i w_i\alpha^i(A) = \sum_i w_i \sum_A \alpha^i(A) = \sum_i w_i = 1.$$

LEMMA 2.5. *If $\alpha^0, \cdots, \alpha^q$ are points of a simplex s, and if $w_0, \cdots, w_q$ are positive numbers of sum 1 such that the point $\sum w_i\alpha^i$ is a vertex A of s, then each $\alpha^i = A$.*

PROOF. For any vertex $B \neq A$ of s, we have $\sum w_i\alpha^i(B) = A(B) = 0$. Since each $w_i > 0$, each $\alpha^i(B) = 0$. As this holds for each B different from A, it follows that $\alpha^i = A$.

3. SIMPLICIAL COMPLEXES

DEFINITION 3.1. A *simplicial complex K* is a collection of faces of a simplex s satisfying the condition that every face of a simplex in the collection is likewise in the collection. The *space* $|K|$ of K is the subset of $|s|$ consisting of those points which belong to simplexes of K.

The same simplicial complex K may lie in two different simplexes s_1 and s_2. In such a case K also lies in the simplex s spanning the vertices common to s_1 and s_2. Since the topology of s is the subspace topology of both $|s_1|$ and $|s_2|$, it follows that the topology of K is independent of the particular simplex s in terms of which it is defined.

The collection of all faces of s including s itself is a simplicial complex. This complex is also denoted by s. The collection of all faces of s, excluding s itself, is a simplicial complex and is denoted by $\dot{s}$.

DEFINITION 3.2. A simplicial complex K is said to be *n-dimensional* (briefly: an n-complex) provided K contains an n-simplex but no $(n + 1)$-simplex (and, therefore, no simplex of dimension $>n$).

DEFINITION 3.3. If K is a simplicial complex, a *subcomplex L* of K is a subcollection of the simplexes of K such that each face of a simplex in L is also in L. Clearly, L is a simplicial complex.

LEMMA 3.4. *If $\alpha^0, \cdots, \alpha^n$ are points of K, where K is a simplicial* *complex, and* $w_0, \cdots, w_n$ *are positive numbers such that* $w_0 + \cdots + w_n = 1$, *then the point* $\alpha = w_0\alpha^0 + \cdots + w_n\alpha^n$ *is in K if and only if the points* $\alpha^0, \cdots, \alpha^n$ *are in a simplex of K.*

PROOF Let s be the simplex containing K and let s' be the lowest dimensional face of s containing the points $\alpha^0, \cdots, \alpha^n$. Then $\alpha \in s'$.

If $\alpha^0, \cdots, \alpha^n$ are in a simplex of K, then s' is in K and $\alpha \in |K|$. Thus the condition is sufficient

Suppose now that $\alpha \in |K|$ If $A^0, \cdots, A^q$ are the vertices of s', then

$$\alpha = \alpha(A^0)A^0 + \cdots + \alpha(A^q)A^q,$$

where

$$\alpha(A^i) = \sum_j w_j\alpha^j(A^i).$$

Since $w_i > 0$ and s' is the least simplex containing $\alpha^0, \cdots, \alpha^n$, it follows that $\alpha(A^i) > 0$ and α is in the interior of s'. Hence $\alpha \in |K|$ implies that s' is in K and thus $\alpha^0, \cdots, \alpha^n$ are in a simplex of K

COROLLARY 3 5 *If* $\alpha^0, \cdots, \alpha^n \in |K|$, *then all the points*

$$\alpha = w_0\alpha^0 + \cdots + w_n\alpha^n$$

where $w_0 + \cdots + w_n = 1$, $w_i > 0$ *are in K if and only if* $\alpha^0, \cdots, \alpha^n$ *are in a simplex of K*

DEFINITION 3.6 For each vertex A of K, the *open star* of A is the *subset* st(A) of K defined by the condition $\alpha(A) > 0$

An equivalent definition is that st(A) is the union of all open simplexes of K having A as a vertex

LEMMA 3 7 *If* $A^0, \cdots, A^n$ *are distinct vertices of the simplicial complex K, then their open stars have a nonempty intersection if and only if* $A^0, \cdots, A^n$ *are the vertices of a simplex of K.*

PROOF If $A^0, \cdots, A^n$ are the vertices of a simplex s, then the interior of s is in st(A^i) for each i, so that $\bigcap$st$(A^i) \neq 0$. Conversely, suppose that $\alpha \in \bigcap$ st(A^i) Then $\alpha(A^i) > 0$ for $i = 0, \cdots, n$, and therefore $A^0, \cdots, A^n$ are the vertices of a simplex of K by virtue of 3 4.

4. LINEAR AND SIMPLICIAL MAPS

DEFINITION 4.1. If K, K' are simplicial complexes and $f. |K| \to |K'|$ is a map, then we say that f is *linear* (Notation f: $K \to K'$) if f is linear in terms of the barycentric coordinates Precisely, if $\alpha, \alpha^0, \cdots, \alpha^n$ are points of $|K|$ and

$$\alpha = w_0\alpha^0 + \cdots + w_n\alpha^n, \qquad \sum_0^n w_i = 1, \qquad w_i \geq 0,$$

then

$$f(\alpha) = w_0 f(\alpha^0) + \cdots + w_n f(\alpha^n).$$

A linear map which carries vertices into vertices is called *simplicial*

Let L,L' be subcomplexes of K,K' respectively. By a *linear* [*simplicial*] map f $(K,L) \to (K',L')$ is meant a map of $(|K|,|L|)$ into $(|K'|,|L'|)$ which defines a linear [simplicial] map of K into K'.

Two trivial consequences are

THEOREM 4 2. *The identity map* $(K,L) \to (K,L)$ *is simplicial*

THEOREM 4.3 *If* f $(K,L) \to (K',L')$ *and* g $(K',L') \to (K'',L'')$ *are both linear* [*simplicial*], *then* gf: $(K,L) \to (K'',L'')$ *is linear* [*simplicial*]

The following theorem expresses a fundamental property of simplicial complexes, namely. linear maps are described by their behavior on vertices.

THEOREM 4 4 *A linear map* f: $(K,L) \to (K',L')$ *is uniquely determined by its values on the vertices A map* ϕ *of the vertices of* K *into points of* K' *can be extended to a linear map* f $(K,L) \to (K',L')$ *if and only if the* ϕ*-image of the set of vertices of any simplex of* K *or* L *is contained in a simplex of* K' *or* L' *respectively If* ϕ *maps vertices into vertices, then* f *is simplicial*

PROOF. If f: $(K,L) \to (K',L')$ is linear, and α is in a simplex s of K, then

$$\alpha = \alpha(A^0)A^0 + \cdots + \alpha(A^n)A^n,$$

where $A^0, \cdots, A^n$ are the vertices of s Then

$$f(\alpha) = \alpha(A^0)f(A^0) + \cdots + \alpha(A^n)f(A^n),$$

so that $f(\alpha)$ is determined by its values on vertices Further, if α is an interior point of s, then $\alpha(A^i) > 0$ for $i = 0, \cdots, n$; and by 3 4, $f(A^0), \cdots, f(A^n)$ are in a simplex of K' Similarly, if s is a simplex of L, then $f(A^0), \cdots, f(A^n)$ are in a simplex of L' This proves that the condition stated in 4.4 is necessary for the extension of a vertex map ϕ.

To prove that the condition on ϕ is also sufficient, consider the simplexes s and s' spanning the vertices of K and K' respectively. Let $A^0, \cdots, A^n$ be the vertices of s. Given a vertex map ϕ satisfying the conditions of 4 4, for each α in s, define $f(\alpha)$ by

$$f(\alpha) = \alpha(A^0)\phi(A^0) + \cdots + \alpha(A^n)\phi(A^n).$$

It is readily seen that $f(\alpha)$ is in s', and f is a linear map $s \to s'$.

Now let $\alpha \in |K|$ (or $\alpha \in |L|$) and let $B^0, \cdots, B^q$ be the subsequence of $A^0, \cdots, A^n$ consisting of those vertices with $\alpha(A^i) > 0$. Then

$\alpha = \alpha(B^0)B^0 + \cdots + \alpha(B^q)B^q$ with $\alpha(B^i) > 0$, and by 3 4, the vertices $B^0, \cdots, B^q$ are those of a simplex of K (or of L) Hence the points $\phi(B^0), \cdots, \phi(B^q)$ are all in a simplex of K' (or of L'), so that

$$f(\alpha) = \alpha(B^0)\phi(B^0) + \cdots + \alpha(B^q)\phi(B^q)$$

is in K' (or in L'). Thus f defines a linear map $f\colon\ (K,L) \to (K',L')$, and the proof is complete

A trivial consequence of Definition 4 1 is

Theorem 4 5. *If $f\colon\ (K,L) \to (K',L')$ is linear [simplicial], and if $(\tilde{K},\tilde{L}),(\tilde{K}',\tilde{L}')$ are subcomplexes of $(K,L),(K',L')$, respectively, such that f maps $|\tilde{K}|$ into $|\tilde{K}'|$ and $|\tilde{L}|$ into $|\tilde{L}'|$, then the map of $(\tilde{K},\tilde{L})$ into $(\tilde{K}',\tilde{L}')$ defined by f is linear [simplicial]*

Lemma 4 6. *If $f\colon\ K \to K'$ is linear and maps $|K|$ onto $|K'|$, then each vertex of K' is the image of at least one vertex of K.*

Proof. Let B be a vertex of K'. Then $B = f(\alpha)$ where α is a point on K Since $\alpha = \sum \alpha(A^i)A^i$ where A^i are the vertices of K, it follows that $B = f(\alpha) = \sum \alpha(A^i)f(A^i)$ This implies that, if $\alpha(A^i) > 0$, then $B = f(A^i)$ by 2 5

Theorem 4 7 *If a simplicial map $f\colon\ K \to K'$ is a homeomorphism, then f^{-1} is simplicial*

Proof. It follows from the preceding lemma that f establishes a 1-1 correspondence between the vertices of K and K'. Let ϕ be the vertex map of the vertices of K' onto those of K given by f^{-1} Consider the vertices $B^0, \cdots, B^n$ of a simplex of K' and let β be a point in the interior of that simplex Then

$$\beta = v_0B^0 + \cdots + v_nB^n$$

where $\sum v_i = 1$ and $v_i > 0$ Also $f^{-1}(\beta)$ is a point of K, and

$$f^{-1}(\beta) = w_0A^0 + \cdots + w_mA^m$$

where $A^0, \cdots, A^m$ are distinct vertices of K, $\sum w_i = 1$, and $w_i > 0$. It follows that

$$\beta = w_0f(A^0) + \cdots + w_mf(A^m).$$

This implies that $m = n$ and $f(A^0), \cdots, f(A^n)$ is a permutation of $B^0, \cdots, B^n$. Since $A^0, \cdots, A^n$ are the vertices of a simplex of K, it follows that $\phi(B^0), \cdots, \phi(B^n)$ are the vertices of a simplex of K. Hence, by 4 4, ϕ extends to a simplicial map $g\colon\ K' \to K$. Clearly $g = f^{-1}$.

Theorem 4 8 *If $f\colon\ K \to K'$ is a homeomorphism, and both f and f^{-1} are linear, then f is simplicial*

Proof. Let A be a vertex of K. Then

$$f(A) = v_0B^0 + \cdots + v_nB^n,$$

where $B^0, \cdots, B^n$ are the vertices of a simplex of K', $\sum v_i = 1$, and $v_i > 0$. Applying f^{-1} we find

$$A = v_0 f^{-1}(B^0) + \cdots + v_n f^{-1}(B^n).$$

Hence, by 2.5, $A = f^{-1}(B^0) = \cdots = f^{-1}(B^n)$, and therefore $B^0 = \cdots = B^n$, and $f(A)$ is a vertex of K'.

5. TRIANGULATED SPACES

DEFINITION 5.1. Given a pair (X,A), a *triangulation* $T = \{t,(K,L)\}$ of (X,A) consists of a simplicial pair (K,L) and a homeomorphic map

$$t\colon \quad (|K|,|L|) \to (X,A).$$

The pair (X,A) together with a triangulation T is called a *triangulated* pair. If a triangulation of a pair (X,A) exists, the pair is called *triangulable*.

Let $(X,A),(X',A')$ have triangulations $T = \{t,(K,L)\}$, $T' = \{t',(K',L')\}$ respectively. A map

$$f\colon \quad (X',A') \to (X,A)$$

is called linear [or simplicial], with respect to T,T', if the map $t^{-1}ft'$ is a linear [or simplicial] map of (K',L') into (K,L).

The definition of triangulated triples and their maps is similar

As a direct consequence of the definition we have

LEMMA 5.2 *The identity map of a triangulated pair is simplicial. The composition of two linear or two simplicial maps is linear or simplicial.*

As consequences of 4.7 and 4.8 we have

LEMMA 5.3. *The inverse of a simplicial homeomorphism is simplicial.*

LEMMA 5.4 *A linear homeomorphism, whose inverse is linear, is simplicial.*

If $T = \{t,(K,L)\}$ is a triangulation of (X,A), then the various simplicial concepts in K can be carried over into X by means of the map t Thus a *simplex of the triangulation* T will mean the t-image of a simplex of K, etc.

DEFINITION 5.5. The *mesh* of a triangulation $T = \{t,K\}$ of a metric space X is the maximum of the diameters of the simplexes of T Precisely,

$$\text{mesh } T = \max\,[\text{diam } t(|s|)],$$

where s is any simplex of K.

DEFINITION 5.6. Let $T = \{t,K\}$ be a triangulation of a space X. A function $f\colon X \to R^n$, mapping X homeomorphically onto a subset of euclidean space R^n is called a *linear imbedding of X into R^n* (relative to T) provided the map ft: $K \to R^n$ is linear, i e. provided the cartesian coordinates of the point $ft(\alpha)$ are linear functions of the barycentric coordinates of the point α on K If f is a linear imbedding, then the distance function in X defined by

$$\rho(x,y) = |f(x) - f(y)|, \qquad x,y \in X$$

is called a linear metric in X (relative to T).

LEMMA 5.7 *Every space X with a triangulation $T = \{t,K\}$ has a linear imbedding*

PROOF The complex K is a subcomplex of a q-simplex s. Let l $|s| \to R^{q+1}$ be a canonical imbedding as defined in 2 2 Then $f = lt^{-1}$ is a linear imbedding of X in R^{q+1}

COROLLARY 5 8 *Every triangulated space has a linear metric.*

6. BARYCENTRIC SUBDIVISION

Let K be a simplicial complex, and let s be a simplex of K with vertices $A^0, \cdots, A^n$ The *barycenter* b_s of s is the point defined by

$$b_s = \frac{1}{n+1} A^0 + \cdots + \frac{1}{n+1} A^n.$$

In particular, if A is a vertex, then $b_A = A$

We assign to K a second complex, denoted by Sd K, defined as follows Its vertices are the barycenters of the simplexes of K For each sequence $s_0, s_1, \cdots, s_q$ of simplexes of K such that s_i is a face of s_{i+1} $(i = 0, \cdots, q - 1)$, the sequence of corresponding barycenters is the set of vertices of a simplex of Sd K Only simplexes obtained in this manner are in Sd K

With the notation as above, the vertices $b_{s_0}, \cdots, b_{s_q}$ of a simplex of Sd K lie in the simplex s_q of K Therefore the identity map of the vertices of Sd K into K has, by 4 4, a unique linear extension to a map

$$l_K\colon \ \text{Sd } K \to K$$

DEFINITION 6.1. The pair consisting of the complex Sd K and the linear map l_K is called the *barycentric subdivision* of K.

If L is a subcomplex of K, then Sd L is a subcomplex of Sd K and $l_L = l_K|\text{Sd } L$ The pair consisting of the pair (Sd K, Sd L) and the linear map

$$l_{(K,L)}\colon \ (\text{Sd } K, \text{Sd } L) \to (K,L)$$

defined by l_K, is called the *barycentric subdivision* of (K,L).

LEMMA 6.2 *The map l_K is 1-1 Therefore Sd K is (linearly) homeomorphic to K.*

PROOF. Let

$$(1) \qquad \alpha = w_0 b_{s_0} + \cdots + w_q b_{s_q}$$

be a point in the simplex of Sd K with vertices $b_{s_0}, \cdots, b_{s_q}$. Since $b_{s_0}, \cdots, b_{s_q}$ are in a simplex of K, the same symbol (1) represents a point of $|K|$. Thus, in this symbolism, l_K becomes an identity map. Without loss of generality we may assume that each s_i is an i-simplex and a face of s_{i+1} Let then $A^0, \cdots, A^q$ be the vertices of s_q ordered in such a fashion that $A^0, \cdots, A^i$ are the vertices of s_i, $i = 0, \cdots, q$ Then, by definition of the barycenter,

$$b_{s_i} = \frac{1}{i+1} A^0 + \cdots + \frac{1}{i+1} A^i.$$

Substituting this in (1) we find that

$$\alpha = \alpha(A^0)A^0 + \cdots + \alpha(A^q)A^q$$

where

$$(2) \qquad \alpha(A^i) = \sum_i^q \frac{1}{j+1} w_j$$

are the barycentric coordinates of α in K. From (2) we deduce

$$(3) \qquad \alpha(A^0) \geqq \cdots \geqq \alpha(A^q),$$

and that

$$(4) \qquad \begin{aligned} w_i &= (i+1)[\alpha(A^i) - \alpha(A^{i+1})] \qquad \text{for } i = 0, \cdots, q-1, \\ w_q &= (q+1)\alpha(A^q). \end{aligned}$$

These formulas show that l_K maps the simplex of Sd K with vertices $b_{s_0}, \cdots, b_{s_q}$ in a 1 1 fashion onto the portion of the simplex with vertices $A^0, \cdots, A^q$ determined by conditions (3) Any point α of $A^0 \cdots A^q$ lies in some set defined by (3) for some order of the A's If it lies in two such sets, then $\alpha(A^i) = \alpha(A^{i+1})$ for one such order. In this case $w_i = 0$, so the transformation defined by (4) is uniquely defined. It is obviously the inverse of l_K.

The following lemma is an immediate consequence of the definition of Sd K.

LEMMA 6 3 *If K is an n-complex, then* Sd K *is an n-complex.*

DEFINITION 6 4 Let $T = \{t,(K,L)\}$ be a triangulation of (X,A). Consider the barycentric subdivision (Sd K, Sd L) and its linear homeomorphism $l_{(K,L)}$ onto (K,L). The triangulation

$$\text{Sd } T = \{tl_{(K,L)}, (\text{Sd } K, \text{Sd } L)\}$$

is called the *barycentric subdivision of T*. The ith barycentric subdivision of T is defined inductively by

$$ {}^0T = T, \qquad {}^iT = \text{Sd}\,({}^{i-1}T), \qquad i = 1, 2, \cdots. $$

LEMMA 6.5 *For every triangulation $T = \{t,K\}$ of a metric space X, we have*

$$ \lim_{i\to\infty} \text{mesh}\ {}^iT = 0 $$

Since X is compact, the limit in question (if equal to zero) is independent of the choice of the metric. Hence, by 5.8, we may assume that the metric in X is linear. Suppose that K is an n-complex; then, by 6.3, the successive barycentric subdivisions of K also are n-complexes. Lemma 6.5 is thus a consequence of the following stronger lemma.

LEMMA 6.6 *Let X be a space with a metric linear relative to a triangulation $T = \{t,K\}$, where K is an n-complex. Then*

$$ \text{mesh Sd}\ T \leqq \frac{n}{n+1}\ \text{mesh}\ T. $$

This lemma follows readily from the following succession of elementary lemmas concerning points in a euclidean space.

Let $p_0, \cdots, p_n$ be a sequence of points of a euclidean space R^q. Consider the set $C(p_0, \cdots, p_n)$ of all points $x \in R^q$ of the form (in vector notation)

$$ x = w_0p_0 + \cdots + w_np_n, $$

where $w_i \geqq 0$ and $\sum w_i = 1$. Observe that, if $x,y \in C$, then

$$ (1) \qquad |y - x| \leqq |y - p_i| \qquad \text{for some } i = 0, \cdots, n, $$

because

$$ \begin{aligned} |y - x| &= |\sum (w_iy - w_ip_i)| \leqq \sum w_i\ |y - p_i| \\ &\leqq (\max |y - p_i|) \sum w_i = \max |y - p_i|. \end{aligned} $$

Applying (1) again, we find that $|y - x| \leqq |p_j - p_i|$ for some p_i and p_j. Hence

LEMMA 6.7. $\text{diam}\ C(p_0, \cdots, p_n) = \text{diam}\ (p_0, \cdots, p_n)$.

LEMMA 6.8 *If*

$$ b = \frac{1}{n+1}(p_0 + \cdots + p_n), \qquad b' = \frac{1}{i+1}(p_0 + \cdots + p_i), \quad i \leqq n, $$

then

$$ |b - b'| \leqq \frac{n}{n+1}\ \text{diam}\ (p_0, \cdots, p_n). $$

Indeed by (1) we have $|b - b'| \leq |b - p_j|$ for some $j = 0, \cdots, n$. Hence

$$
\begin{aligned}
|b - b'| &\leq |b - p_j| \\
&= \left|\frac{1}{n+1}(p_0 + \cdots + p_n) - p_j\right| = \frac{1}{n+1}\left|\sum_{i=0}^{n}(p_i - p_j)\right| \\
&\leq \frac{1}{n+1}\sum_{i=0}^{n}|p_i - p_j| \leq \frac{n}{n+1}\operatorname{diam}(p_0, \cdots, p_n).
\end{aligned}
$$

Lemmas 6.7 and 6.8 imply that, if s' is a simplex of the barycentric subdivision of an n-simplex s of T, then, in the linear metric, diam $s' \leq n/(n+1)$ diam s. This implies 6.6.

7. SIMPLICIAL APPROXIMATION

DEFINITION 7.1. Let X and X' be spaces with triangulations $T = \{t,K\}$ and $T' = \{t',K'\}$ respectively, and let a map $f\colon X \to X'$ be given. A simplicial map $g\colon X \to X'$ (with respect to T,T') is called a *simplicial approximation to* f if, for each $x \in X$, $g(x)$ lies on the closed simplex whose interior contains $f(x)$.

LEMMA 7.2. *A necessary and sufficient condition for a simplicial map* $g\colon X \to X'$ *to be a simplicial approximation to* f *is that, for every vertex* A *of* X,

$$f(\operatorname{st}(A)) \subset \operatorname{st}(g(A)).$$

PROOF. Suppose g is a simplicial approximation to f. Let $x \in \operatorname{st}(A)$, i.e. x has a positive barycentric coordinate relative to A. It follows that $g(x)$ has a positive barycentric coordinate relative to $g(A)$. Since $g(x)$ lies in the closure of the open simplex containing $f(x)$, it follows that $f(x)$ has a positive barycentric coordinate relative to $g(A)$, i.e. $f(x) \in \operatorname{st}(g(A))$.

Conversely, suppose that the condition of 7.2 is fulfilled. Let $x \in X$ and let s be a simplex of X (relative to T) containing x in its interior. Similarly let s' be a simplex of X' (relative to T') containing $f(x)$ in its interior For every vertex A of s we have $x \in \operatorname{st}(A)$, so that $f(x) \in \operatorname{st}(g(A))$. Thus $g(A)$ is a vertex of s' and g maps s onto a face of s'. Thus $g(x)$ lies in the closed simplex s'

The main result of this section is the following existence theorem:

THEOREM 7.3. *Let* X *be a triangulable metric space,* X' *a triangulated space with triangulation* $T' = \{t',K'\}$*, and let a map* $f\colon X \to X'$ *be given. There is a number* $\epsilon > 0$ *such that, for any triangulation* T *of* X *of mesh* $<\epsilon$*, there exists a simplicial map* $g\colon X \to X'$ *relative to* T,T' *which is a simplicial approximation to* f.

This theorem and 6.5 yield

COROLLARY 7.4. *Let X,X' be topological spaces with triangulations T,T' and let a map f: $X \to X'$ be given. There is an integer h_0 such that for every $h \geqq h_0$ there exists a simplicial map g: $X \to X'$ relative to ${}^hT,T'$ which is a simplicial approximation to f.*

The proof of 7.3 will be preceded by the following lemma:

LEMMA 7.5. *If X is a compact metric space and ϕ a collection of open sets covering X, then there exists a positive number ϵ such that each subset of X of diameter $<\epsilon$ is contained in at least one set of the family ϕ. The least upper bound of all such numbers ϵ is called the Lebesgue number of the covering ϕ.*

PROOF. Suppose this is false. Then, for each positive integer n, there exists a subset A_n of X of diameter $< 1/n$ not contained in any set of ϕ. Let $x_n \in A_n$. By the compactness of X, there is a point $x \in X$ such that each neighborhood of x contains infinitely many points of $\{x_n\}$. Let U be an open set of ϕ containing x, and let d be the distance from x to $X - U$. Select n so that $n > 2/d$ and $\rho(x,x_n) < d/2$. Then, for each $y \in A_n$,

$$\rho(y,x) \leqq \rho(y,x_n) + \rho(x,x_n) \leqq \frac{1}{n} + \frac{d}{2} < d.$$

Hence $y \in U$ and $A_n \subset U$, a contradiction.

PROOF OF 7.3. Let $B^1, \cdots, B^m$ be the vertices of the triangulation T', let $U_i = \text{st}(B^i)$ (in T'), and let $V_i = f^{-1}(U_i)$. Since $\{U_i\}$ is an open covering of X', $\{V_i\}$ is an open covering of X. Let η be the Lebesgue number of $\{V_i\}$, let $\epsilon = \eta/2$, and let $T = \{t,K\}$ be a triangulation of X with mesh $<\epsilon$. Then, for every vertex A of the triangulation T, the set $\text{st}(A)$ has diameter $<\eta$, so that $\text{st}(A) \subset V_i$ for some $i = 1, \cdots, m$. Choose such an i and define $g(A) = B^i$. Then

$$\text{st}(A) \subset f^{-1}(\text{st}(g(A))), \tag{1}$$

where $\text{st}(A)$ is in T and $\text{st}(g(A))$ is in T'. If $A^0, \cdots, A^n$ are the vertices of a simplex of T, then, by 3 7, $\bigcap \text{st}(A^i) \neq 0$ This and (1) imply $\bigcap \text{st}(g(A^i)) \neq 0$ Hence by 3.7 the vertices $g(A^0), \cdots, g(A^n)$ are in a simplex of T'. Therefore by 4 4 the vertex map g extends to a map g: $X \to X'$ which is simplicial relative to T,T'. To be precise, the vertex map g defines a vertex map $(t')^{-1}gt$ of the vertices of K into those of K'. This vertex map extends to a simplicial map $\bar{g}$: $K \to K'$, and g. $X \to X'$ is defined by $g = t'\bar{g}t^{-1}$. The inclusion (1) implies $f(\text{st}(A)) \subset \text{st}(g(A))$; so, by 7 2, g is a simplicial approximation to f.

THEOREM 7.6. *Let $(X,A),(X',A')$ be pairs with triangulations $T = \{t,(K,L)\}$, $T' = \{t',(K',L')\}$. Consider a map f: $(X,A) \to (X',A')$*

and the maps f_1: $X \to X'$, f_2: $A \to A'$ *defined by* f. *If* g_1 *is any simplicial approximation to* f_1, *(relative to* T,T'*), then* $g_1(A) \subset A'$ *and* g_1 *defines maps*

$$g: \ (X,A) \to (X',A'), \qquad g_2. \ A \to A'.$$

The map g *is simplicial on* T *to* T'. *The map* g_2 *is a simplicial approximation to* f_2 *relative to the induced triangulations of* A *and* A'. *Finally, and most important, the maps* f *and* g *are homotopic in such a way that, during the homotopy, the image of a point* $x \in X$ *stays in the simplex whose interior contains* $f(x)$

PROOF. If $x \in A$, then $f_1(x) \in A'$, and $f_1(x)$ is in a simplex of A' Hence $g_1(x) \in A'$. For each $x \in X$, the points $f(x)$ and $g(x)$ lie in the same simplex of X', hence the corresponding points $\bar{f}(x),\bar{g}(x)$ of K' lie in the same simplex of K'. Then the required homotopy is given by

$$h(x,r) = t'[r\bar{f}(x) + (1 - r)\bar{g}(x)], \qquad 0 \leqq r \leqq 1.$$

8. PRODUCTS OF SIMPLICIAL COMPLEXES

The cartesian product of two simplicial complexes is not directly a simplicial complex This fact introduces a difficulty in the study of the cartesian product of triangulable spaces. The difficulty is circumvented by the introduction of a simplicial product This type of product arises naturally in connection with cartesian products of spaces and their coverings by open sets

DEFINITION 8 1. Given two simplexes s_1 and s_2 of dimension p and q respectively, the *simplicial product* $s_1 \Delta s_2$ is the simplex of dimension $(p + 1)(q + 1) - 1$ whose vertices are the pairs (A,B) where A is a vertex of s_1 and B is a vertex of s_2. If K_1 and K_2 are subcomplexes of s_1 and s_2, then the *simplicial product* $K_1 \Delta K_2$ is the subcomplex of $s_1 \Delta s_2$ consisting of all simplexes $s_1' \Delta s_2'$ with $s_1' \subset K_1$, $s_2' \subset K_2$, and of all faces of such simplexes.

This last addition is due to the fact that a face of $s_1 \Delta s_2$ need not be the simplicial product of faces of s_1 and s_2, because the vertices of $s_1 \Delta s_2$ are the cartesian product of the vertices of s_1 and s_2, and a subset of the product need not be a product of subsets

The following five lemmas are obvious consequences of these definitions.

LEMMA 8 2. *A simplex of* $s_1 \Delta s_2$ *with vertices* $(A^0,B^0), \cdots, (A^n,B^n)$ *is in* $K_1 \Delta K_2$ *if and only if* $A^0, \cdots, A^n$ *are the vertices of a simplex of* K_1 *with repetitions allowed, and* $B^0, \cdots, B^n$ *are the vertices of a simplex of* K_2 *with repetitions allowed.*

LEMMA 8.3. *If L_1 and L_2 are subcomplexes of K_1 and K_2 respectively, then $L_1 \Delta L_2$ is a subcomplex of $K_1 \Delta K_2$. The simplicial product of the pairs (K_1,L_1) and (K_2,L_2) is defined then as*

$$(K_1,L_1) \Delta (K_2,L_2) = (K_1 \Delta K_2,\ L_1 \Delta K_2 \cup K_1 \Delta L_2).$$

In reading this formula it is understood that the operations Δ precede the operation $\cup$.

LEMMA 8.4. *If $f_1: (K_1,L_1) \to (K_1',L_1')$ and $f_2: (K_2,L_2) \to (K_2',L_2')$ are simplicial maps, then the vertex map $(A,B) \to (f_1(A),f_2(B))$ defines a simplicial map*

$$f_1 \Delta f_2: (K_1,L_1) \Delta (K_2,L_2) \to (K_1',L_1') \Delta (K_2',L_2').$$

LEMMA 8.5. *The vertex maps $(A,B) \to A$ and $(A,B) \to B$ define simplicial maps*

$$\pi_1: K_1 \Delta K_2 \to K_1, \qquad \pi_2: K_1 \Delta K_2 \to K_2,$$

which will be referred to as the projections of the simplicial product onto its factors.

LEMMA 8.6. *The vertex map $A \to (A,A)$ defines a simplicial map*

$$\Delta: (K,L) \to (K,L) \Delta (K,L),$$

called the diagonal map.

DEFINITION 8.7. A simplicial complex K is *ordered* if, for each simplex, a simple order of its vertices is given such that the order of each simplex agrees with the orders of its faces. The order in a subcomplex will always be assumed to be the order induced by the order in K.

Every complex K can be ordered by selecting a simple order of its vertices (i.e. ordering the least simplex containing K).

Clearly an order in a complex K is equivalent to a binary relation $A \leqq A'$ for the vertices of K subject to three conditions: (i) $A \leqq B$ and $B \leqq A$ imply $A = B$, (ii) A and B are vertices of a simplex of K if and only if $A \leqq B$ or $B \leqq A$, and (iii) if A,B,C are vertices of a simplex of K, and $A \leqq B$ and $B \leqq C$, then $A \leqq C$.

DEFINITION 8.8. Let K_1 and K_2 be ordered simplicial complexes. Define a partial order in the set of vertices of $K_1 \Delta K_2$ by $(A,B) \leqq (A',B')$ if $A \leqq A'$ and $B \leqq B'$. (In general this is not an order in $K_1 \Delta K_2$). The *cartesian product* $K_1 \times K_2$ is the complex consisting of those simplexes of $K_1 \Delta K_2$ whose vertices are simply ordered by this relation. If L_1 and L_2 are subcomplexes of K_1 and K_2, then $(K_1,L_1) \times (K_2,L_2)$ is defined as $(K_1 \times K_2,\ L_1 \times K_2 \cup K_1 \times L_2)$.

We proceed now to compare the cartesian product $K_1 \times K_2$ of ordered complexes K_1 and K_2 with the cartesian product $|K_1| \times |K_2|$ of the spaces $|K_1|$ and $|K_2|$. Let x be any point of $K_1 \Delta K_2$. Then the projections $\pi_1(x)$ and $\pi_2(x)$ are points of K_1 and K_2 respectively. Thus $\pi(x) = (\pi_1(x),\pi_2(x))$ yields a map

$$\pi\colon \quad |K_1 \Delta K_2| \to |K_1| \times |K_2|.$$

If K_1 and K_2 are ordered, the map π defines a map

$$\eta\colon \quad |K_1 \times K_2| \to |K_1| \times |K_2|.$$

LEMMA 8 9. *The map η is a homeomorphism, and $\{\eta,K_1 \times K_2\}$ is a triangulation of $|K_1| \times |K_2|$. If L_1 and L_2 are subcomplexes of K_1 and K_2, then η carries $|L_1 \times L_2|$ onto $|L_1| \times |L_2|$. Furthermore, this triangulation has the property that, for each vertex B of K_2, say, the correspondence $x \to (x,B)$ is a simplicial map of K_1 into $|K_1| \times |K_2|$.*

PROOF. We shall define a map $\bar{\eta}\colon \ |K_1| \times |K_2| \to |K_1 \times K_2|$ that will be inverse to η Let $\alpha \in |K_1|$, $\beta \in |K_2|$. We can express α and β uniquely in the form

$$\alpha = \alpha(A^0)A^0 + \cdots + \alpha(A^p)A^p$$
$$\beta = \beta(B^0)B^0 + \cdots + \beta(B^q)B^q$$

where $A^0 < \cdots < A^p$, $B^0 < \cdots < B^q$, $\alpha(A^i) > 0$, $\beta(B^j) > 0$, $\sum \alpha(A^i) = \sum \beta(B^j) = 1$. Let

$$a^m = \sum_{i=0}^{m} \alpha(A^i), \qquad b^n = \sum_{j=0}^{n} \beta(B^j)$$

where $0 \leqq m \leqq p$ and $0 \leqq n \leqq q$. Now let

$$c^0 \leqq \cdots \leqq c^{p+q} = c^{p+q+1} \ (=1)$$

be a single sequence obtained by rearranging the symbols $a^0, \cdots, a^p, b^0, \cdots, b^q$ in order of magnitude. For each $r = 0, \cdots, p + q$ let $C^r = (A^i,B^j)$ where i and j are the numbers of a's and b's, respectively, in $c^0, \cdots, c^{r-1}$. Then $i + j = r$ and C^{r+1} is either (A^{i+1},B^j) or (A^i,B^{j+1}), depending on whether c^r is a^i or b^j. It follows that $C^0 < \cdots < C^{p+q}$, and they are therefore the vertices of a simplex of $K_1 \times K_2$. Now set

$$\gamma = \sum_{i=0}^{p+q} (c^i - c^{i-1})C^i,$$

where $c^{-1} = 0$. Since $\sum_{i=0}^{p+q} (c^i - c^{i-1}) = c^{p+q} - c^{-1} = 1$, γ is a well-defined point of $K_1 \Delta K_2$

In the definition of C^r the following ambiguity may occur: If $a^i = b^j$, then C^{i+j+1} could be either (A^{i+1},B^j) or (A^i,B^{j+1}), depending on whether a^i or b^j was labeled as c^{i+j}. In either case $c^{i+j+1} - c^{i+j} = 0$, so that the choice of C^{i+j+1} has no effect on γ, which is thus defined without ambiguity. We define $\bar{\eta}$ by setting $\bar{\eta}(\alpha,\beta) = \gamma$.

To prove that $\bar{\eta}$ is the inverse of η, it suffices to establish the following two propositions:

(1) $\bar{\eta}$ maps $|K_1| \times |K_2|$ onto $|K_1 \times K_2|$,

(2) $\eta\bar{\eta}$ is the identity map of $|K_1| \times |K_2|$

To prove (1), consider any point γ in $K_1 \times K_2$ Then $\gamma = \sum_{r=0}^{s} d_r D^r$ where $D^r < D^{r+1}$. Let $A^0 < \cdots < A^p$ be the distinct first coordinates of $D^0, \cdots, D^s$ and similarly let $B^0 < \cdots < B^q$ be the second coordinates By adjoining extra D's with coefficient zero when necessary, it can be arranged that, if $D^r = (A^i,B^j)$, then $D^{r+1} = (A^i,B^{j+1})$ or $D^{r+1} = (A^{i+1},B^j)$. Define $c^r = \sum_0^r d_t$. Let $C^r = (A^i,B^j)$, if $C^{r+1} = (A^{i+1},B^j)$, then set $a^i = c^r$, if $C^{r+1} = (A^i,B^{j+1})$, then set $b^j = c^r$. This defines $a^0 \leqq \cdots \leqq a^p = 1$ and $b^0 \leqq \cdots \leqq b^q = 1$. Finally set $\alpha = \sum_0^p (a^i - a^{i-1})A^i$ and $\beta = \sum_0^q (b^j - b^{j-1})B^j$. It is then easy to verify that $\bar{\eta}(\alpha,\beta) = \gamma$

To prove (2) it suffices to show that if $\gamma = \bar{\eta}(\alpha,\beta)$, then $\pi_1(\gamma) = \alpha$ and $\pi_2(\gamma) = \beta$. Let $C^r, \cdots, C^{r+k}$ be all the vertices in $C^0, \cdots, C^{p+q}$ with A^i as first coordinate Then $C^{r+l} = (A^i,B^{j+l})$ where $r = i + j$ and $0 \leqq l \leqq k$, while $C^{r-1} = (A^{i-1},B^j)$ and $C^{r+k+1} = (A^{i+1},B^{j+k})$. Then from the definitions of π_1 and of γ,

$$\begin{aligned}\pi_1(\gamma)(A^i) &= \sum_{l=0}^{k} (c^{r+l} - c^{r+l-1}) = c^{r+k} - c^{r-1} \\ &= a^i - a^{i-1} = \alpha(A^i).\end{aligned}$$

Since this holds for every i, it follows that $\pi_1(\gamma) = \alpha$. Similarly $\pi_2(\gamma) = \beta$ This concludes the proof

COROLLARY 8 10 *If (X,A) and (Y,B) are triangulable pairs, then so are $(X \times Y, A \times B)$ and $(X,A) \times (Y,B) = (X \times Y, A \times Y \cup X \times B)$.*

The previous discussion shows that the simplicial product $K_1 \Delta K_2$ contains as subcomplexes the cartesian products $K_1 \times K_2$ for the various orderings of K_1 and K_2. The next lemma implies that $|K_1 \Delta K_2|$ and $|K_1 \times K_2|$ are homotopically equivalent.

LEMMA 8.11. *$|K_1 \times K_2|$ is a deformation retract of $|K_1 \Delta K_2|$.*

PROOF. Consider the maps π and η introduced above. Then $r = \eta^{-1}\pi$ is a retraction of $|K_1 \Delta K_2|$ into $|K_1 \times K_2|$. Further, for each x in $K_1 \Delta K_2$, the points x and $r(x)$ lie in a simplex of $K_1 \Delta K_2$. Therefore $(1 - t)x + tr(x)$ yields the desired homotopy.

Both the retraction r and the homotopy remain valid for subcomplexes L_1 and L_2 of K_1 and K_2. This implies that $(K_1 \times K_2,$

$L_1 \times L_2)$ is a deformation retract of $(K_1 \Delta K_2, L_1 \Delta L_2)$ and that $(K_1,L_1) \times (K_2,L_2)$ is a deformation retract of $(K_1,L_1) \Delta (K_2,L_2)$.

The closed unit interval $I = [0,1]$ may be regarded as an ordered 1-simplex with vertices $0 < 1$. Thus the above discussion yields the definition of $K \times I$ for any ordered simplicial complex K. If we write

$$A_0 = (A,0), \qquad A_1 = (A,1)$$

where A is any vertex of K, then each simplex of $K \times I$ has vertices

$$A_0^0, \cdots, A_0^i, A_1^i, \cdots, A_1^q$$

or

$$A_0^0, \cdots, A_0^i, A_1^{i+1}, \cdots, A_1^q, \qquad (0 \leqq i \leqq q)$$

where $A^0 < \cdots < A^q$ and are vertices of a simplex of K.

9. REGULAR NEIGHBORHOODS

We show in this section that, if L is a subcomplex of K, then $|L|$ is smoothly imbedded in $|K|$ in the sense that $|L|$ is a strong deformation retract (see I,11 6) of a closed triangulable neighborhood of $|L|$ in $|K|$.

DEFINITION 9.1. Let L be a subcomplex of K The open set

$$N(L) = \bigcup \mathrm{st}(A)$$

where the union is extended over all the vertices A of L, is called the *regular neighborhood* of L in K.

It should be observed that, in general, L is not a deformation retract of $N(L)$. For example, let $K = s$ be a 1-simplex and $L = \dot{s}$ its boundary, then $N(L) = |s|$; and the fact that $|s|$ is connected, and $|\dot{s}|$ is not, shows that $|\dot{s}|$ is not a deformation retract of $|s|$ Consideration of this example leads to the following definition:

DEFINITION 9.2. L is called a *full* subcomplex of K, if L contains each simplex of K whose vertices are all in L.

Clearly, in the above example L is not a full subcomplex

LEMMA 9 3. *Let L be a full subcomplex of K. For each $\alpha \in N(L)$, define*

$$v(\alpha) = \sum \alpha(A),$$
$$f(\alpha) = \sum \frac{1}{v(\alpha)} \alpha(A) A.$$

In each case, the sum is taken over the vertices of L. Then $f(\alpha) \in |L|$, and f is a retraction of $N(L)$ into $|L|$. Moreover, the homotopy h: $N(L) \times I \to N(L)$, defined by

$$h(\alpha,t) = tf(\alpha) + (1 - t)\alpha,$$

is a strong deformation retraction of $N(L)$ into $|L|$. Finally, the bary-

centric coordinate of $h(\alpha,t)$ with respect to a vertex A is a nondecreasing function of t if A is in L, and is a nonincreasing function if A is not in L.

PROOF. First note that $f(\alpha)$ is a point of the least simplex s containing α. Furthermore its barycentric coordinates are zero for vertices not in L. Thus $f(\alpha)$ belongs to a face s' of s, all of whose vertices are in L. Since L is full, $f(\alpha)$ is in L. If α is in L, then $v(\alpha) = 1$ and $f(\alpha) = \alpha$, so that f is a retraction. Since α and $f(\alpha)$ are in the same simplex s, the point $h(\alpha,t)$ also lies in s. The last statement of the lemma is a direct consequence of the formulas for f and h. It follows that $h(\alpha,t)$ is in $N(L)$ for all $\alpha \in N(L)$ and $t \in I$. This completes the proof.

The following lemma shows that the assumption of 9.3 that L is full is not a severe restriction from the topological point of view.

LEMMA 9.4. *If L is a subcomplex of K, then, in the barycentric subdivision, Sd L is a full subcomplex of Sd K.*

PROOF. A simplex $\bar{s}$ of Sd K has vertices $b_{s_0}, \cdots, b_{s_q}$ where s_i are simplexes of K and s_i is a face of s_{i+1}. If all the vertices of $\bar{s}$ are in Sd L, then the last vertex b_{s_q} of $\bar{s}$ is in Sd L. Then s_q is also in L. Hence $\bar{s}$ is in Sd s_q, and thus in Sd L.

Note that the deformation of 9.3 is not defined on the closure of $N(L)$. To obtain a closed neighborhood of which L is a deformation retract we resort to further subdivision.

DEFINITION 9.5. If L is a subcomplex of K, the kth regular neighborhood $N^k(L)$ is the image in K of the regular neighborhood $N(\text{Sd}^k L)$ under the natural map $\text{Sd}^k K \to K$. In particular $N(L) = N^0(L)$.

LEMMA 9.6. *If L is a full subcomplex of K, then a point α of K is in $N^1(L)$ if and only if there is a vertex A of L such that $\alpha(A) > \alpha(B)$ for each vertex B of K not in L.*

PROOF. Let $\alpha \in K$. Then α is in a simplex s whose vertices $A^0, \cdots, A^q$ may be labeled so that $\alpha(A^0) \geqq \cdots \geqq \alpha(A^q)$. Let s_i be the simplex spanning the vertices $A^0, \cdots, A^i$. The point $\alpha' = l_k^{-1}(\alpha)$ in Sd K, which corresponds to α (see proof of 6.2), is given by

$$\alpha' = w_0 b_{s_0} + \cdots + w_q b_{s_q}, \qquad w_i = (i + 1)[\alpha(A^i) - \alpha(A^{i+1})].$$

Let j be the least index such that $w_j > 0$.

If now $\alpha \in N^1(L)$, that is, if $\alpha' \in N(\text{Sd } L)$, then, for some index i, we must have $b_{s_i} \in$ Sd L and $w_i > 0$. Then s_i is in L and $A^0, \cdots, A^i$ are in L. Since $i \geqq j$, it follows that $A^0, \cdots, A^j$ are in L, and the condition of 9.6 is satisfied.

Conversely, if the condition of 9.6 is satisfied, then $A^0, \cdots, A^j$ are in L and, since L is full, the simplex s_j is in L. Thus b_{s_j} is in Sd L and, since $w_j > 0$, it follows that $\alpha' \in N(\text{Sd } L)$. Thus $\alpha \in N^1(L)$ and the lemma is proved.

LEMMA 9.7. *If L is a subcomplex of K, then the closure of $N^{k+1}(L)$ is in $N^k(L)$ for $k = 0,1, \cdots$.*

PROOF. It suffices to prove that the closure of $N^1(L)$ is in $N(L)$. If L is full, this follows from 9.6 since, in passing to closure, the inequality $\alpha(A) > \alpha(B)$ becomes $\alpha(A) \geqq \alpha(B)$ If L is not full, let L' be the subcomplex of all simplexes of K with vertices in L. Then L' is full, and $N(L) = N(L')$ and $N^1(L) \subset N^1(L')$.

As observed in 9.3, the homotopy h does not disturb the relation $\alpha(A) > \alpha(B)$ for A in L and B not in L. Consequently, if $\alpha \in N^1(L)$ and L is full, then $h(\alpha,t) \in N^1(L)$ for each t. Since the closure of $N^1(L)$ is in $N(L)$, we have

LEMMA 9.8. *If L is a full subcomplex of K, the homotopy h of 9.3 defines a strong deformation retraction of $\overline{N}^1(L)$ into $|L|$.*

Combining this with 9.4 we obtain the main objective of this section:

THEOREM 9.9. *If L is a subcomplex of K, then $|L|$ is a strong deformation retract of the closure of the second regular neighborhood $N^2(L)$.*

EXERCISES

A. SIMPLEXES.

1. If α,β are distinct points of a simplex s, show that $f\colon I \to s$ defined by $f(t) = (1 - t)\alpha + t\beta$ is a 1-1 linear map of I into s. It is called the *line segment* from α to β in s, and $f(t)$ is said to divide this segment in the ratio $t\colon (1 - t)$. Show that each point of s other than its barycenter lies on a unique line segment from the barycenter to a point of $|s|$.

2. If s is an n-simplex, and s' is an $(n - 1)$-face of s, show that the two triples $(s,\dot{s},s')$ and (E^n,S^{n-1}, E_+^{n-1}) are homeomorphic.

B. EUCLIDEAN COMPLEXES.

DEFINITION. The points $A^0, \cdots, A^q$ of euclidean n-space R^n are called *linearly independent* if they are not contained in a hyperplane of dimension $<q$.

1. Show that $A^0, \cdots, A^q$ are linearly independent if and only if the matrix

$$\begin{Vmatrix} a_1^0, & \cdots, & a_n^0, 1 \\ \cdots & \cdots & \cdots \\ a_1^q, & \cdots, & a_n^q, 1 \end{Vmatrix}$$

where $a_1^i, \cdots, a_n^i$ are the coordinates of A^i, has rank $q + 1$.

2. Show that, if $A^0, \cdots, A^q$ are linearly independent, then every subset of $A^0, \cdots, A^q$ is also linearly independent.

3. Let s^q be a q-simplex with vertices $A^0, \cdots, A^q$. Let f: $s^q \to R^n$ be a linear mapping. Show that f is a linear imbedding if and only if the points $f(A^0), \cdots, f(A^q)$ are linearly independent.

4. Show that R^n contains an infinite sequence of points each $n + 1$ of which are linearly independent.

5. Using 4, prove that each n-complex K admits a linear imbedding in R^{2n+1}.

DEFINITION. Let $A^0, \cdots, A^q$ be linearly independent points of R^n. The least convex set σ containing $A^0, \cdots, A^q$ is called the *euclidean q-simplex* with vertices $A^0, \cdots, A^q$. The *faces* of σ are the euclidean simplexes whose vertices form a subset of $A^0, \cdots, A^q$. A finite set $\Sigma = \{\sigma\}$ of euclidean simplexes in R^n is called a euclidean complex if the intersection of any two simplexes of Σ is either vacuous or is a common face of both of them.

6. Show that if f: $K \to R^n$ is a linear imbedding of a complex into R^n, then the images of the simplexes of K form a euclidean complex in R^n. Conversely, if Σ is a euclidean complex in R^n, then there is a simplicial complex K and a linear imbedding f: $K \to R^n$ such that the simplexes of Σ coincide with the images of the simplexes of K.

C. SPACES WITH OPERATORS.

DEFINITION. A group W (not necessarily abelian, and written multiplicatively) is said to *operate* on the space X if, for every $w \in W$ and $x \in X$, an element $wx \in X$ is defined such that $w_2(w_1x) = (w_2w_1)x$, $1x = x$, and that $x \to wx$ is a continuous map w: $X \to X$. If W operates on X and Y, then a map f: $X \to Y$ is called *equivariant* if $f(wx) = wf(x)$ for all $w \in W$, $x \in X$.

W is said to *operate* on the simplicial complex K if, for each $w \in W$, a simplicial map w: $K \to K$ is given such that the fixed points of w form a subcomplex of K and $w_2(w_1x) = (w_2w_1)x$, $1x = x$ for each $x \in |K|$.

If W operates on X, and $T = \{t,K\}$ is a triangulation of X, then T is said to be *invariant* if, for every $w \in W$, the map $t^{-1}wt$: $K \to K$ is simplicial and its fixed points form a subcomplex of K.

1. If W operates on X, and an invariant triangulation of X exists, then W contains an invariant subgroup W_0 such that W/W_0 is finite, and $w_0x = x$ for all $w_0 \in W_0$, $x \in X$.

2. If the triangulation T of X is invariant, then so is the barycentric subdivision of T.

3. In theorem 7.3 assume that W operates on both X and X', that the triangulations T and T' are invariant, and that f: $X \to X'$ is equivariant. Show that the simplicial approximation g can be chosen equivariant.

D. Joins of complexes

Definition. The join $K \circ M$ of two disjoint complexes is the complex whose vertices form the union of $A^1, \cdots A^r, B^1, \cdots, B^s$ of the vertices of K and M, respectively. A set of vertices are those of a simplex of the join if its subsets in K and M, if not vacuous, span simplexes there.

1. K and M are subcomplexes of $K \circ M$.

2. If s_1 is a p-simplex and s_2 a q-simplex, then $s_1 \circ s_2$ is a $(p + q + 1)$-simplex, and its boundary is $(s_1 \circ s_2) \cup (s_1 \circ s_2)$.

3. $|K|$ is a deformation retract of $|K \circ M| - |M|$.

4. Let L be the subset of points α of $K \circ M$ such that $\sum \alpha(A^i) = \sum \alpha(B^j) = 1/2$. Show that L is homeomorphic to $|K| \times |M|$. Show that $|K \circ M| - |K| - |M|$ is homeomorphic to the product of L with an open interval.

5. Let $0 \leqq p \leqq \dim K$. In Sd K, let L(resp. M) be the subcomplex of simplexes whose vertices are barycenters of simplexes of dimensions $\leqq p$(resp. $> p$). Show that Sd K is isomorphic to a subcomplex of $L \circ M$.

6. Define the join $X \circ Y$ of two spaces; and show that $|K| \circ |M|$ is homeomorphic to $|K \circ M|$.

7. If X is contractible on itself to a point, show that the same is true of $X \circ Y$.

8. Show that the join of a p-cell and a q-sphere is a $(p + q + 1)$-cell. Show that the join of a p-sphere and a q-sphere is a $(p + q + 1)$-sphere.

E. Elementary subdivision of a complex.

Definition. If s is a simplex of K, the *complement* $K \dot{-} s$ is the subcomplex consisting of all simplexes of K that do not have s as a face. The *boundary* of $K \dot{-} s$, written Bd(s), is the closed subcomplex consisting of simplexes in $K \dot{-} s$ which are faces of simplexes not in $K \dot{-} s$. The *elementary subdivision* of K with respect to s, written $\text{Sd}_s(K)$, is the complex $(K \dot{-} s) \cup (b_s \circ \text{Bd}(s))$, where b_s is the barycenter of s.

1. Show that the linear map of $\text{Sd}_s(K) \to K$ defined by the identity transformation of the vertices is a homeomorphism.

2. If the simplexes of K are ordered $s_1, s_2, \cdots, s_k$ so that dim $s_j \geqq$ dim s_{j+1} for each j, show that the barycentric subdivision of K is the result of the succession of elementary subdivisions

$$\text{Sd}_{s_k}(\text{Sd}_{s_{k-1}}(\cdots \text{Sd}_{s_1}(K)\cdots))$$

F. Infinite complexes.

Definition. Let W be an infinite set of objects called *vertices*. An *infinite complex* K with vertices in W is an infinite collection of (finite dimensional) simplexes whose vertices are in W, subject to the condition that a face of a simplex in the collection is also in the collection. Each point α of K is described by barycentric coordinates $\alpha(A) \geqq 0$ for $A \in W$, where $\alpha(A) > 0$ only for a finite number of elements of W, and $\sum_A \alpha(A) = 1$. The complex K is said to be *locally finite* if each vertex of K is a vertex of at most a finite number of simplexes of K.

If α and β are points of K, then $\rho(\alpha,\beta) = \{\sum_A [\alpha(A) - \beta(A)]^2\}^{\frac{1}{2}}$ is a distance function in K, and defines the *metric topology*.

The *weak topology* in K is obtained by considering as open all sets whose intersection with every (closed) simplex s of K is an open subset of $|s|$.

Thus to the infinite complex correspond two topological spaces $|K|_m$ and $|K|_w$.

1. The identity map $|K|_w \rightarrow |K|_m$ is continuous

2. The weak and the metric topologies coincide if and only if K is locally finite

3. Linear and simplicial maps (defined as in the finite case) are continuous in both topologies. In particular, for any vertex A, the barycentric coordinate $\alpha(A)$ is continuous in α for both topologies. A subcomplex is closed in either topology. For every vertex A of K, the set $\mathrm{st}(A)$ is open in either topology

DEFINITION A *subdivision* (K',l) of an infinite complex K is a pair consisting of an infinite complex K' and a linear map $l.\ K' \rightarrow K$ which is a homeomorphism in the weak topology

4. If (K',l) is a subdivision of K, and L is any finite subcomplex of K, then $l^{-1}(L)$ is a finite subcomplex of K'

5. Given a family $\{U\}$ of open sets covering the space $|K|_w$, there is a subdivision (K',l) of K such that, for every vertex A' of K', the set $l(\mathrm{st}(A'))$ is in one of the sets of the family $\{U\}$.

DEFINITION. An *infinite triangulation* of a space X is a pair $T = \{t,K\}$ where K is an infinite complex and $t\colon\ |K|_w \rightarrow X$ is a homeomorphic mapping of $|K|_w$ onto X

DEFINITION. If $T = \{t,K\}$ is a triangulation of X, and $\{U\}$ is a family of open sets covering X such that, for every vertex A of K, the set $t(\mathrm{st}(A))$ is in one of the sets of $\{U\}$, then the triangulation T is said to be finer than the covering $\{U\}$

6. (Simplicial approximation theorem). Let X be a space which is triangulable as an infinite complex, X' a space with (infinite) triangulation T', and let a map $f.\ X \rightarrow X'$ be given. There is a covering $\{U\}$ of X by open sets such that, for every triangulation T of X which is finer than $\{U\}$, there is a map $g.\ X \rightarrow X'$ which is a simplicial approximation (relative to T,T') to f.

7. If K and M are nonempty complexes one of which is infinite, then the join $K \circ M$ is not locally finite.

CHAPTER III

Homology theory of simplicial complexes

1. INTRODUCTION

In this chapter it is assumed that a homology theory is given such that the groups $H_q(X,A)$ are defined for all triangulable pairs and that f_* is defined for every map $f\colon (X,A) \to (Y,B)$ of one triangulable pair into another. In other words, let $\mathfrak{T}$ denote the category of triangulable pairs and maps of one such pair into another. It follows from II,8.10 that $\mathfrak{T}$ is an admissible category for homology theory as defined in I,1. It is assumed in this chapter that the homology (cohomology) theory considered is defined on an admissible category $\mathfrak{A}$ which contains $\mathfrak{T}$.

The first main objective of this chapter is the derivation from our axioms of the classical algorithm used to define and compute the homology groups of a simplicial complex. This motivates our use of this algorithm in the existence proofs in later chapters.

The second objective of this chapter is a uniqueness theorem which asserts that any two homology theories, with isomorphic coefficient groups, are isomorphic on the category $\mathfrak{T}$ of triangulable pairs. This asserts, to a certain extent, that the axioms are categorical, i.e. admitting an interpretation unique up to isomorphisms. This result will be extended later (Chapter XII) to a larger class of spaces.

As in Chapter I the statements concerning cohomology are listed at the end of each section, without proof.

2. EXCISION AND DIRECT SUM THEOREMS

It will be shown here that, for triangulable pairs, excision and direct sum theorems are valid in more general situations than those encountered in Chapter I

THEOREM 2.1. *If K_1 and K_2 are subcomplexes of a complex K, then the inclusion map*

$$i\colon (K_1,K_1 \cap K_2) \subset (K_1 \cup K_2,K_2)$$

induces isomorphisms in each dimension. Consequently $(|K|;|K_1|,|K_2|)$ is a proper triad (see I,14.1).

Proof. Without loss of generality we may assume that $K = K_1 \cup K_2$ Let $X = \overline{N}^2(K_1)$ be the closure of the second regular neighborhood of K_1, and let $A = X \cap |K_2|$. Then (X,A) is a triangulable pair, since both X and A correspond to subcomplexes of the second barycentric subdivision of K. The map i may be factored into inclusions

$$(|K_1|,|K_1 \cap K_2|) \xrightarrow{i_1} (X,A) \xrightarrow{i_2} (|K|,|K_2|).$$

Since $|K| = N^2(K_1) \cup (|K| - |K_1|) \subset \text{Int } X \cup \text{Int } |K_2|$, it follows that:

$$|K| = X \cup |K_2|, \qquad A = X \cap |K_2|, \qquad |K| = \text{Int } X \cup \text{Int } |K_2|.$$

Thus, by the Excision axiom, the map i_2 induces isomorphisms in all dimensions

By II,9.9, $|K_1|$ is a strong deformation retract of X. Throughout this deformation each point of $|K_2|$ remains in $|K_2|$. Thus the pair $(|K_1|,|K_1 \cap K_2|)$ is a deformation retract of (X,A). Hence, by I,11.8, the homomorphisms induced by i_1 are isomorphisms

Theorem 2.2. *Let (X,A) be a triangulable pair, and let U be an open subset of X such that $U \subset A$. If $(X - U, A - U)$ is a triangulable pair, then the inclusion map f: $(X - U, A - U) \subset (X,A)$ induces an isomorphism in each dimension.*

Proof. Let T be a triangulation of (X,A) (see II,5.1) and let V be the interior of A (relative to X). Then $X - V$ and $A - V$ are subcomplexes of X relative to the triangulation T, so that by 2.1 the inclusion map i: $(X - V, A - V) \subset (X,A)$ induces isomorphisms. Set $X' = X - U$, $A' = A - U$, and let V' be the interior of A' relative to X'. Then, by the same token, the map i': $(X' - V', A' - V') \subset (X',A')$ induces isomorphisms Since the pairs $(X' - V', A' - V')$ and $(X - V, A - V)$ coincide, and since $i' = fi$, it follows that $f_* = i'_* i_*^{-1}$ is an isomorphism.

Theorem 2.3. *Let K be a simplicial complex with subcomplexes $K_1, \cdots, K_r, L$ such that*

$$K = K_1 \cup \cdots \cup K_r \cup L, \qquad K_i \cap K_j \subset L, \quad \text{for } i \neq j.$$

Let $L_i = K_i \cap L$, and let

$$k_i\colon \ (K_i,L_i) \subset (K,L).$$

Then, the homomorphisms k_{i*}: $H_q(|K_i|,|L_i|) \to H_q(|K|,|L|)$ $(i = 1, \cdots, r)$ form an injective representation of $H_q(|K|,|L|)$ as a direct sum.

Proof. The case $r = 1$ is covered by 2.1 since $K = K_1 \cup L$ and $L_1 = K_1 \cap L$. Assume, inductively, that the theorem has been proved for $r = n \geqq 1$, and suppose that $r = n + 1$ Set $K' = K_1 \cup \cdots \cup K_n \cup L$ and consider the inclusion maps

$$(K_i,L_i) \xrightarrow{k'_i} (K',L) \xrightarrow{j} (K,L) \qquad (i = 1, \cdots, n),$$

$$(K_{n+1},L_{n+1}) \xrightarrow{k'} (K_{n+1} \cup L,L) \xrightarrow{j'} (K,L).$$

Clearly $k_{i*} = j_* k'_{i*}$ for $i = 1, \cdots, n$, and $k_{n+1*} = j'_* k'_*$. Further, by 2 1, k'_* is an isomorphism. Since $(|K|;|K'|,|K_{n+1} \cup L|)$ is a proper triad with

$$K' \cup (K_{n+1} \cup L) = K, \qquad K' \cap (K_{n+1} \cup L) = L,$$

it follows, from I.14 2, that j_* and j'_* yield an injective representation of $H_q(|K|,|L|)$ as a direct sum. By the inductive hypothesis, the maps k'_{i*} $(i = 1, \cdot \cdot, n)$ yield an injective representation of $H_q(|K'|,|L|)$ as a direct sum This implies the conclusion of 2 3.

THEOREM 2 3c *The maps* k^*_i *of* 2 3 *yield a projective representation of* $H^q(|K|,|L|)$ *as a direct sum.*

3. THE INCIDENCE ISOMORPHISM

Let s^q be a q-simplex, $\dot{s}^q$ its boundary, and s^{q-1} a $(q-1)$-face of s^q. Let A be the vertex of s^q opposite s^{q-1}, and c^{q-1} the closed star of A in $\dot{s}^q$, i e c^{q-1} is the subcomplex of s^q consisting of all faces of s^q excepting s^q and s^{q-1}. Thus

$$\dot{s}^q = s^{q-1} \cup c^{q-1}, \qquad \dot{s}^{q-1} = s^{q-1} \cap c^{q-1}.$$

For each $\alpha \in s^q$, the function

$$(1 - t)\alpha + tA, \qquad 0 \leqq t \leqq 1$$

is a homotopy contracting the pair $(|s^q|,|c^{q-1}|)$ into the point pair (A,A). Hence, by I,11.5 and I,9 5

LEMMA 3.1. $|s^q|,|c^{q-1}|$ *and* $(|s^q|,|c^{q-1}|)$ *are homologically trivial.*

Next we turn to the discussion of the triad $(|\dot{s}^q|;|s^{q-1}|,|c^{q-1}|)$. By 2.1, this is a proper triad.

THEOREM 3.2. *The homology sequence of the triad* $(|\dot{s}^q|;|s^{q-1}|,|c^{q-1}|)$ *reduces to the isomorphism*

$$\partial.\ H_q(|s^q|,|\dot{s}^q|) \approx H_{q-1}(|s^{q-1}|,|\dot{s}^{q-1}|).$$

All other groups in the sequence are zero.

Theorem 3 3 *The homology groups of* $(|s^q|,|\dot{s}^q|)$ *are as follows*·

$$H_q(|s^q|,|\dot{s}^q|) \approx G$$
$$H_p(|s^q|,|\dot{s}^q|) = 0 \qquad \text{for } p \neq q.$$

Proofs. By 3.1, $H_p(|s^q|,|c^{q-1}|) = 0$ for all q, and therefore, by the exactness of the H S. of the triad, we have ∂: $H_p(|s^q|,|\dot{s}^q|) \approx H_{p-1}(|s^{q-1}|,|\dot{s}^{q-1}|)$. Iterating this isomorphism yields $H_p(|s^q|,|\dot{s}^q|) \approx H_{p-q}(|s^0|,|\dot{s}^0|) = H_{p-q}(|s^0|)$. Since $|s^0|$ is a single point, the assertion of 3 3 follows. This also establishes 3 2.

Definition 3.4. If s^{q-1} is a $(q-1)$-face of the q-simplex $s^q (q > 0)$, the *incidence isomorphism*

$$[s^q{:}s^{q-1}]\colon\ H_q(|s^q|,|\dot{s}^q|) \approx H_{q-1}(|s^{q-1}|,|\dot{s}^{q-1}|)$$

is defined to be the isomorphism ∂ of 3.2 Explicitly, $[s^q{:}s^{q-1}]$ may be defined from the diagram

$$H_q(|s^q|,|\dot{s}^q|) \xrightarrow{\partial} H_{q-1}(|\dot{s}^q|) \xrightarrow{i_*} H_{q-1}(|\dot{s}^q|,|c^{q-1}|) \xleftarrow{j_*} H_{q-1}(|s^{q-1}|,|\dot{s}^{q-1}|)$$

where i, j are inclusions, by setting

$$[s^q{:}s^{q-1}] = j_*^{-1} i_* \partial.$$

Lemma 3 5 *If* $q > 0$, *and* $f\colon (|s^q|,|\dot{s}^q|) \to (|s_1^q|,|\dot{s}_1^q|)$ *is such that* $|s^{q-1}|$ *and* $|c^{q-1}|$ *are mapped into* $|s_1^{q-1}|$ *and* $|c_1^{q-1}|$, *respectively, then commutativity holds in the diagram*

$$\begin{array}{ccc}
H_q(|s^q|,|\dot{s}^q|) & \xrightarrow{[s^q{:}s^{q-1}]} & H_{q-1}(|s^{q-1}|,|\dot{s}^{q-1}|) \\
\downarrow{f_*} & & \downarrow{f_{1*}} \\
H_q(|s_1^q|,|\dot{s}_1^q|) & \xrightarrow{[s_1^q{:}s_1^{q-1}]} & H_{q-1}(|s_1^{q-1}|,|\dot{s}_1^{q-1}|)
\end{array}$$

where f_1 *is the map defined by* f

Proof. The incidence isomorphism is the boundary operator of the triad $(|s^q|;|s^{q-1}|,|c^{q-1}|)$, and the commutativity of this boundary operator with induced homomorphisms is contained in the statement of Theorem I,14 5.

Definition 3 6 If A is a 0-simplex (i.e. a vertex) and $g \in G$, let gA denote the element of $H_0(A)$ mapped on g by the isomorphism induced by the map $A \to P_0$ (In the notation of I,7.1, $gA = (gA)_A$.) Let s^q be an ordered q-simplex with vertices $A^0 < A^1 < \cdots < A^q$. Define s^k $(k = 0,1, \cdots, q-1)$ to be the ordered simplex with vertices

$A^{q-k} < A^{q-k+1} < \cdots < A^q$. For each $g \in G$, the element gs^q of $H_q(|s^q|,|\dot{s}^q|)$ is defined inductively by

$$gs^k = [s^k:s^{k-1}]^{-1}gs^{k-1}. \tag{1}$$

Thus

$$gs^q = [s^q:s^{q-1}]^{-1} \cdots [s^1:A^q]^{-1}gA^q.$$

THEOREM 3.7. *For every ordered q-simplex s^q, the correspondence $g \to gs^q$ is an isomorphic mapping of G onto the group $H_q(|s^q|,|\dot{s}^q|)$.*

PROOF. The proposition follows from the fact that the incidence isomorphisms are isomorphisms, and $g \to gA^q$ maps G isomorphically onto $H_0(A^q)$ (See I,7.2).

THEOREM 3.8. *If s_1 and s_2 are ordered q-simplexes, and f is a simplicial map $(s_1,\dot{s}_1) \to (s_2,\dot{s}_2)$ preserving the order, then $f_*(gs_1) = gs_2$.*

PROOF. Consider the simplexes s_i^n ($i = 1, 2$; $n = 1, \cdots, q$) as defined in 3.6. Since f is order-preserving, it defines maps

$$f_n\colon\ (s_1^n,\dot{s}_1^n) \to (s_2^n,\dot{s}_2^n).$$

Consider the diagram

$$\begin{array}{ccccccc}
H_q(|s_1^q|,|\dot{s}_1^q|) & \longrightarrow & H_{q-1}(|s_1^{q-1}|,|\dot{s}_1^{q-1}|) & \longrightarrow & \cdots & \longrightarrow & H_q(A_1^q) \\
\downarrow{\scriptstyle f_{q*}} & & \downarrow{\scriptstyle f_{q-1*}} & & & & \downarrow{\scriptstyle f_{0*}} \\
H_q(|s_2^q|,|\dot{s}_2^q|) & \longrightarrow & H_{q-1}(|s_2^{q-1}|,|\dot{s}_2^{q-1}|) & \longrightarrow & \cdots & \longrightarrow & H_q(A_2^q)
\end{array}$$

where the horizontal arrows are the incidence isomorphisms. By 3.5, commutativity holds in each square of the diagram. Since f_{0*} maps gA_1^q into gA_2^q (see I,7.2), it follows that f_{q*} maps gs_1^q into gs_2^q.

THEOREM 3 2c. *The cohomology sequence of the triad $(|s^q|,|s^{q-1}|,|c^{q-1}|)$ reduces to the isomorphism δ: $H^{q-1}(|s^{q-1}|,|\dot{s}^{q-1}|) \approx H^q(|s^q|,|\dot{s}^q|)$*

THEOREM 3.3c. $H^q(|s^q|,|\dot{s}^q|) \approx G$, $H^p(|s^q|,|\dot{s}^q|) = 0$ *for* $p \neq q$.

DEFINITION 3.4c. If s^{q-1} is a $(q-1)$-face of a q-simplex s^q, the ***incidence isomorphism***

$$[s^{q-1}:s^q] \quad H^{q-1}(|s^{q-1}|,|\dot{s}^{q-1}|) \approx H^q(|s^q|,|\dot{s}^q|)$$

is defined as the isomorphism δ of 3.2c or equivalently as

$$[s^{q-1}:s^q] = \delta i^* j^{*-1}.$$

LEMMA 3.5c. *Under the conditions of* 3.5

$$f^*[s_1^{q-1}:s_1^q] = [s^{q-1}:s^q]f_1^*$$

Definition 3.6c. If A is a vertex and $g \varepsilon G$, let $gA \varepsilon H^0(A)$ be the image of g under the isomorphism induced by the map $A \to P_0$. (In the notation of I,7.1c, $gA = g_A(A)$.) If s^q is an ordered q-simplex with vertices $A^0 < A^1 < \cdots < A^q$ and $g \varepsilon G$, define $gs^q \varepsilon H^q(|s^q|,|\dot{s}^q|)$ inductively by

$$gs^k = [s^{k-1}:s^k]gs^{k-1}. \tag{1c}$$

Thus

$$gs^q = [s^{q-1}:s^q] \cdots [A^q:s^1]gA^q.$$

Theorem 3.7c. *The correspondence $g \to gs^q$ is an isomorphism of G onto $H^q(|s^q|,|\dot{s}^q|)$.*

Theorem 3.8c. *If s_1,s_2 are ordered q-simplexes, and f: $(s_1,\dot{s}_1) \to (s_2,\dot{s}_2)$ is simplicial and order preserving, then $f^*(gs_2) = gs_1$.*

4. AUTOMORPHISMS OF A SIMPLEX

In the preceding section we introduced the symbol $gs^q \varepsilon H_q(|s^q|,|\dot{s}^q|)$, which is a function of $g \varepsilon G$ and of the order of the vertices of the q-simplex s^q. Our immediate objective is to determine the extent of the dependence on the order. The result obtained (4.3) is fundamental in connecting our axiomatic system with the classical homology theory, in particular, with that part dealing with *orientation*. As a preliminary, we study the group H_0 of a space consisting of two points (the 0-dimensional sphere) in greater detail.

Theorem 4.1. *Let A and B be the vertices of the 1-simplex s^1. We shall use the simplified notation $S^0 = |\dot{s}^1|$ (the 0-sphere). Using the notation of* I,7.1, *every element $h \varepsilon H_0(S^0)$ can be written uniquely as*

$$h = (gA)_{S^0} + (g'B)_{S^0} \qquad g,g' \varepsilon G.$$

The element h is in $\tilde{H}_0(S^0)$ if and only if $g + g' = 0$, or, equivalently,

$$h = (gA)_{S^0} - (gB)_{S^0}.$$

Proof. The first part of the theorem is a direct consequence of I,7.1, and of the direct sum theorem I,13.1. To prove the second part, consider the map f: $S^0 \to P_0$ where P_0 is the space consisting of a single point that was used to define the coefficient group G (I,6.1). Then, by I,7.2,

$$f_*h = g + g'$$

so that $f_*h = 0$ if and only if $g + g' = 0$.

DEFINITION 4.2 An *automorphism* of a simplex is any 1-1 simplicial map of the simplex on itself.

By II,4 4, the automorphisms of a simplex are in 1-1 correspondence with the elements of the group of permutations of its vertices. Any permutation is a product of simple permutations, each of which interchanges two vertices and leaves all others fixed. In general, a permutation can be expressed in many ways as a product of simple permutations; but the number of such factors is either always even or always odd. Accordingly the permutation is said to be *even* or *odd*. The same adjective is used to describe the corresponding automorphism

THEOREM 4 3. *Let* f: $(s,\dot{s}) \to (s,\dot{s})$ *be an automorphism of the q-simplex s, and* $h \in H_q(|s|,|\dot{s}|)$ *Then*

$$f_*(h) = \begin{cases} h & \text{if } f \text{ is even,} \\ -h & \text{if } f \text{ is odd.} \end{cases}$$

PROOF. For $q = 0$, the theorem is trivial. Consider the case $q = 1$. Let $S^0 = |\dot{s}^1|$ consist of the points A and B, and let f_0: $S^0 \to S^0$ be defined by f. If $f =$ identity, the result is trivial; thus we may assume that $f(A) = B$ and $f(B) = A$. Then commutativity holds in the diagram (see I,8.4).

$$\begin{array}{ccc} H_1(|s^1|,S^0) & \xrightarrow{f_*} & H_1(|s^1|,S^0) \\ \downarrow \tilde{\partial} & & \downarrow \tilde{\partial} \\ \tilde{H}_0(S^0) & \xrightarrow{f_{0*}} & \tilde{H}_0(S^0) \end{array}$$

Let $h \in H_1(|s^1|,S^0)$: then, by 4 1,

$$\tilde{\partial}h = (gA)_{S^0} - (gB)_{S^0}$$

for some $g \in G$, and, by I,7.2,

$$\tilde{\partial}f_*h = f_{0*}\tilde{\partial}h = (gf(A))_{S^0} - (gf(B))_{S^0} = (gB)_{S^0} - (gA)_{S^0} = -\tilde{\partial}h.$$

Since $H_1(|s^1|) = 0$, the kernel of $\tilde{\partial}$ is zero; hence $f_*h = -h$.

Suppose, inductively, that the theorem is true for the integer $q - 1$, $q \geqq 2$. Since every permutation is a product of simple permutations, it is sufficient to consider the case of a simple permutation f on the vertices of s. Since s has more than two vertices, there is a vertex A_0 such that $f(A_0) = A_0$. Let s' be the $(q-1)$-face of s opposite the vertex A_0. Then f maps s' onto itself and defines an automorphism

$$f'\colon\ (s',\dot{s}') \to (s',\dot{s}')$$

Moreover, by 3.5, the commutativity relation $[s{:}s']f_* = f'_*[s{:}s']$ holds. Let $h \in H_q(|s|,|\dot{s}|)$. Since f' is a simple permutation of the vertices of s', and 4.3 is assumed for the dimension $q - 1$, we have

$$f'_*[s{:}s']h = -[s{:}s']h.$$

This implies

$$[s{:}s']f_*h = [s{:}s'](-h),$$

and, since $[s{:}s']$ is an isomorphism, $f_*h = -h$

THEOREM 4.4. *Let s_1 and s_2 be two ordered simplexes both carried by the same unordered q-simplex s. Then*

$$gs_1 = \pm gs_2$$

according as the order of s_2 differs by an even or an odd permutation from the order of s_1.

PROOF. Let f be an automorphism of $(s,\dot{s})$ which maps s_2 onto s_1 in an order preserving fashion Then, on one hand, 3.8 yields

$$f_*(gs_2) = gs_1.$$

On the other hand, 4.3 implies

$$f_*(gs_2) = \pm gs_2.$$

Consequently $gs_1 = \pm gs_2$.

THEOREM 4.5. *Let s be an ordered simplex with vertices $A^0 < \cdots < A^q$, and let s_k be the face obtained by omitting the vertex A^k and not changing the order of the others. Then*

$$[s{:}s_k]gs = (-1)^k gs_k.$$

PROOF If $k = 0$, this is formula (1) of 3.6. The general case can be reduced to the case $k = 0$ as follows Let $\bar{s}$ be the new ordered simplex obtained from s by moving the vertex A^k in front of all the others Then $s_k = \bar{s}_0$ and $gs_k = g\bar{s}_0$, while, by 4.4, $gs = (-1)^k g\bar{s}$. Hence

$$[s{:}s_k]gs = [\bar{s}{:}\bar{s}_0](-1)^k g\bar{s} = (-1)^k g\bar{s}_0 = (-1)^k gs_k.$$

THEOREM 4.1c. *Every element $h \in H^0(S^0)$ is uniquely described by the elements $h(A) = g$ and $h(B) = g'$ of G (see I,7.1c), and every pair g,g' may thus be obtained The element h is in the subgroup G_{S^0} (see I,7.3c) if and only if $g = g'$.*

THEOREM 4.3c. *If f is an automorphism of the q-simplex s, and $h \in H^q(|s|,|\dot{s}|)$, then*

$$f^*(h) = \begin{cases} h & \text{if } f \text{ is even,} \\ -h & \text{if } f \text{ is odd.} \end{cases}$$

THEOREM 4.4c. *If* s_1,s_2 *are ordered q-simplexes on the same unordered q-simplex s, then* $gs_1 = \pm gs_2$ *according as the orders differ by an even or odd permutation.*

THEOREM 4.5c *If* s,s_k *are as in* 4.5, *then*

$$[s_k:s]gs_k = (-1)^k gs.$$

5. CHAINS IN A COMPLEX

DEFINITION 5.1 If K is a simplicial complex, the *q-dimensional skeleton* K^q of K is the subcomplex of K consisting of all the simplexes of K of dimensions $\leqq q$

In this and the subsequent sections, (K,L) will stand for a pair consisting of a simplicial complex K and a subcomplex L.

LEMMA 5.2 *If* $p \neq q$, *then* $H_p(|K^q \cup L|, |K^{q-1} \cup L|) = 0$.

PROOF. Let $s_1, \cdots, s_r$ be the q-simplexes of K which are not in L. Then $K^q \cup L = s_1 \cup \cdots \cup s_r \cup K^{q-1} \cup L$ and $\dot{s}_m = s_m \cap (K^{q-1} \cup L)$ $(m = 1, \cdots, r)$. Thus the direct sum theorem 2.3 can be applied. Since, by 3.3, $H_p(|s_m|,|\dot{s}_m|) = 0$ for $p \neq q$, the proposition follows.

DEFINITION 5.3. The *group of q-chains of* (K,L), denoted by $C_q(K,L)$, is defined by

$$C_q(K,L) = H_q(|K^q \cup L|, |K^{q-1} \cup L|).$$

If $f.\ (K,L) \to (K_1,L_1)$ is simplicial, the map

$$(K^q \cup L, K^{q-1} \cup L) \to (K_1^q \cup L_1, K_1^{q-1} \cup L_1),$$

that f defines, induces homomorphisms

$$f_q\colon\ C_q(K,L) \to C_q(K_1,L_1).$$

They are called the *chain homomorphisms* induced by f.

THEOREM 5.4. $C_q(K,L) = 0$ *for* $q < 0$ *and for* $q >$ dim K.

PROOF. If $q < 0$, then $C_q(K,L) = H_q(|L|,|L|)$, while, if $q > \dim K$, then $C_q(K,L) = H_q(|K|,|K|)$. In either case $C_q(K,L) = 0$ by I,8.1.

DEFINITION 5.5. Let $A^0, \cdots, A^q$ be a finite sequence of vertices of some simplex of K (with possible repetitions). For each $g \in G$ define the element $gA^0 \cdots A^q$ of $C_q(K,L)$ as follows. Let s be a simplex with ordered vertices $B^0 < \cdots < B^q$. Let $f\colon\ (s,\dot{s}) \to (K^q \cup L, K^{q-1} \cup L)$ be the simplicial map such that $f(B^i) = A^i$. Then

$$gA^0 \cdots A^q = f_*(gs).$$

It follows from 3.8 that the definition of $gA^0 \cdots A^q$ is independent of the choice of s

THEOREM 5.6. *The assignment of g to $gA^0 \cdots A^q$ defines a homomorphism $G \to C_q(K,L)$, i e*

$$(g_1 + g_2)A^0 \cdots A^q = g_1A^0 \cdots A^q + g_2A^0 \cdots A^q.$$

If $i_0, \cdots, i_q$ is a permutation of the array $0, \cdots, q$, then

$$gA^{i_0} \cdots A^{i_q} = \pm gA^0 \cdots A^q$$

according as the permutation is even or odd. If some vertex occurs at least twice in $A^0, \cdots, A^q$, then $gA^0 \cdots A^q = 0$. If $A^0, \cdots, A^q$ are in a simplex of L, then $gA^0 \cdots A^q = 0$.

PROOF. The first and the second part follow respectively from 3.7 and 4 4 If $A^0, \cdots, A^q$ involve a repetition or are in a simplex of L, then the map f: $(s,\dot{s}) \to (K^q \cup L, K^{q-1} \cup L)$ carries s into $K^{q-1} \cup L$. Thus f may be factored into maps $(s,\dot{s}) \to (K^{q-1} \cup L, K^{q-1} \cup L) \to (K^q \cup L, K^{q-1} \cup L)$ Since $H_q(K^{q-1} \cup L, K^{q-1} \cup L) = 0$ (see I,8 1), it follows that $gA^0 \cdots A^q = 0$.

THEOREM 5.7. *Let $s_1, \cdots, s_{\alpha_q}$ be the q-simplexes of K which are not in L. Suppose that for each simplex s_m an order of its vertices $A_m^0 < \cdots < A_m^q$ has been chosen Then each q-chain c of (K,L) can be written uniquely in the form*

$$c = \sum_{m=1}^{\alpha_q} g_mA_m^0 \cdots A_m^q, \qquad g_m \in G.$$

PROOF. Let i_m: $(s_m,\dot{s}_m) \subset (K^q \cup L, K^{q-1} \cup L)$ Since $K^q \cup L = s_1 \cup \cdots \cup s_{\alpha_q} \cup K^{q-1} \cup L$ and $\dot{s}_m = s_m \cap (K^{q-1} \cup L)$, the direct sum theorem 2 3 can be applied to yield a unique representation

$$c = \sum_{n=1}^{\alpha_q} i_{m*}h_m, \qquad h_m \in H_q(|s_m|,|\dot{s}_m|).$$

By 3.7, each h_m can be written uniquely as $h_m = g_ms_m$ Since $i_{m*}g_ms_m = g_mA_m^0 \cdots A_m^q$ by Definition 5 5, the conclusion of 5 7 follows.

One of the consequences of 5.7 is that the symbols $gA^0 \cdots A^q$ generate the group $C_q(K,L)$. Observe further that, using 5 6, any linear combination (with integer coefficients) of such symbols can be brought to the normal form given in 5 7 Thus the group $C_q(K,L)$ may be regarded as the group generated by symbols $gA^0 \cdots A^q$, where $g \in G$ and $A^0, \cdots, A^q$ are in a simplex of K, with relations given in 5 6.

THEOREM 5 8. *If f: $(K,L) \to (K_1,L_1)$ is simplicial and $gA^0 \cdots A^q$ is a q-chain of (K,L), then*

$$f_q(gA^0 \cdots A^q) = gf(A^0) \cdots f(A^q).$$

PROOF. Let s be an ordered q-simplex with vertices $B^0, \cdots, B^q$, and let h: $(s,s) \to (K^q \cup L, K^{q-1} \cup L)$ be a simplicial map with $h(B^i) = A^i$. By Definition 5.5, $gA^0 \cdots A^q = h_*(gs)$. Let

$$_qf\colon (K^q \cup L, K^{q-1} \cup L) \to (K_1^q \cup L_1, K_1^{q-1} \cup L_1)$$

be the map defined by f. Then $f_q = {}_qf_*$ and

$$\begin{aligned} f_q(gA^0 \cdots A^q) &= f_q h_*(gs) = {}_qf_* h_*(gs) \\ &= ({}_qfh)_*(gs) = gf(A^0) \cdots f(A^q). \end{aligned}$$

Consider the inclusion maps

$$L \xrightarrow{i} K \xrightarrow{j} (K,L).$$

The symbol $gA^0 \cdots A^q$ then represents an element of the group $C_q(K)$ as well as of the group $C_q(K,L)$. If in addition $A^0, \cdots, A^q$ are in a simplex of L, then $gA^0 \cdots A^q$ may also represent an element of $C_q(L)$. To avoid all ambiguity we may write $(gA^0 \cdots A^q)_K$ or $(gA^0 \cdots A^q)_{(K,L)}$ or $(gA^0 \cdots A^q)_L$ to indicate in which group the symbol $gA^0 \cdots A^q$ is to be taken. From 5.7 follows

$$\begin{aligned} i_q(gA^0 \cdots A^q)_L &= (gA^0 \cdots A^q)_K, \\ j_q(gA^0 \cdots A^q)_K &= (gA^0 \cdots A^q)_{(K,L)}. \end{aligned}$$

As an immediate consequence of 5.6–5.8 we have

THEOREM 5.9. *The sequence*

$$0 \to C_q(L) \xrightarrow{i_q} C_q(K) \xrightarrow{j_q} C_q(K,L) \to 0$$

is exact and the image of i_q is a direct summand of $C_q(K)$.

LEMMA 5.2c. *If $p \neq q$, then $H^p(|K^q \cup L|, |K^{q-1} \cup L|) = 0$.*

DEFINITION 5.3c. The group of q-cochains of K is denoted by $C^q(K,L)$ and is defined by

$$C^q(K,L) = H^q(|K^q \cup L|, |K^{q-1} \cup L|).$$

If f: $(K,L) \to (K_1,L_1)$ is simplicial, the map

$$(K^q \cup L, K^{q-1} \cup L) \to (K_1^q \cup L_1, K_1^{q-1} \cup L_1),$$

that f defines, induces homomorphisms

$$f^q\colon C^q(K_1,L_1) \to C^q(K,L).$$

They are called the cochain homomorphisms induced by f.

THEOREM 5.4c. *$C^q(K,L) = 0$ for $q < 0$ or $q >$ dim K.*

DEFINITION 5.5c. Let $c \in C^q(K)$ be a q-cochain of (K,L), and

$A^0, \cdots, A^q$ a finite sequence of vertices of some simplex of K (with possible repetitions). Define

$$c(A^0, \cdots, A^q) = g \, \varepsilon \, G$$

as follows: Let s be a simplex with ordered vertices $B^0 < \cdots < B^q$, and let $f \cdot (s,s) \to (K^q \cup L, K^{q-1} \cup L)$ be the simplicial map such that $f(B^i) = A^i$. Then $g \, \varepsilon \, G$ is the unique element satisfying

$$f^*c = gs.$$

By 3.8c, g is independent of the choice of s.

THEOREM 5 6c.

$$(c_1 + c_2)(A^0, \cdots, A^q) = c_1(A^0, \cdots, A^q) + c_2(A^0, \cdots, A^q).$$

If $i_0, \cdots, i_q$ is a permutation of $0, \cdots, q$, then $c(A^{i_0}, \cdots, A^{i_q}) = \pm c(A^0, \cdots, A^q)$ according as the permutation is even or odd. If some vertex occurs at least twice in $A^0, \cdots, A^q$, then $c(A^0, \cdots, A^q) = 0$ If $A^0, \cdots, A^q$ are in a simplex of L, then $c(A^0, \cdots, A^q) = 0$.

THEOREM 5 7c. *Let $s_1, \cdots, s_{\alpha_q}$ be the q-simplexes of K which are not in L Suppose that for each s_m an order of its vertices $A_m^0 < \cdots < A_m^q$ has been chosen Then, if $g_m \, \varepsilon \, G$ for $m = 1, \cdots, \alpha_q$, there exists a unique q-cochain $c \, \varepsilon \, C^q(K,L)$ such that*

$$c(A_m^0, \cdots, A_m^q) = g_m, \qquad m = 1, \cdots, \alpha_q$$

In view of 5 7c and 5 6c the group $C^q(K,L)$ may be described as the group of functions with values in G defined for arrays of vertices $A^0, \cdots, A^q$ which are in a simplex of K and subject to the conditions listed in 5 6c In view of 5 7c, we may also use the linear form notation for cochains; i.e. the symbol

$$\sum g_m A_m^0 \cdots A_m^q$$

can be interpreted to mean the cochain c which has the value g_m on $A_m^0, \cdots, A_m^q$ for each m. As the next theorem shows, the functional notation for cochains is much more convenient.

THEOREM 5 8c *If $f \cdot (K,L) \to (K_1,L_1)$ is simplicial, $c \, \varepsilon \, C^q(K_1,L_1)$ and $A^0, \cdots, A^q$ are vertices lying in a simplex of K, then*

$$(f^c c)(A^0, \cdots, A^q) = c(fA^0, \cdots, fA^q).$$

THEOREM 5.9c. *The sequence*

$$0 \to C^q(K,L) \xrightarrow{j^q} C^q(K) \xrightarrow{i^q} C^q(L) \to 0$$

is exact and the image of j^q is a direct summand of $C^q(K)$.

6. THE BOUNDARY OPERATOR

DEFINITION 6.1. The *boundary* operator for chains

$$\partial_q\colon\ C_q(K,L) \to C_{q-1}(K,L)$$

is defined to be the boundary operator of the triple $(|K^q \cup L|, |K^{q-1} \cup L|, |K^{q-2} \cup L|)$. Explicitly ∂_q is the composition of the homomorphisms

$$H_q(|K^q \cup L|, |K^{q-1} \cup L|) \xrightarrow{\partial} H_{q-1}(|K^{q-1} \cup L|) \xrightarrow{k_*} H_{q-1}(|K^{q-1} \cup L|, |K^{q-2} \cup L|)$$

where k is the indicated inclusion map

THEOREM 6 2 $\partial_q\partial_{q+1} = 0$

PROOF. If $\partial_q = k_{q-1*}\partial$ and $\partial_{q+1} = k_{q*}\partial'$, then $\partial_q\partial_{q+1} = k_{q-1*}\partial k_{q*}\partial'$. Here ∂ and k_{q*} are the homomorphisms

$$H_q(|K^q \cup L|) \xrightarrow{k_{q*}} H_q(|K^q \cup L|, |K^{q-1} \cup L|) \xrightarrow{\partial} H_{q-1}(|K^{q-1} \cup L|).$$

Since these are two consecutive homomorphisms in the H S of the pair $(|K^q \cup L|, |K^{q-1} \cup L|)$, we have $\partial k_{q*} = 0$ and thus $\partial_q\partial_{q+1} = 0$.

THEOREM 6 3 *If* $f\colon\ (K,L) \to (K_1,L_1)$ *is simplicial, then* $f_{q-1}\partial_q = \partial_q f_q$.

This is a consequence of corresponding commutativity theorem (I,10 3) for triples.

In order to give an explicit form to the boundary operator ∂_q for chains, we introduce the following notational convention.

If $A^0 \cdots A^n$ *is an ordered array of vertices, the same series of vertices with a circumflex over one or more vertices means the ordered array obtained from the first array by deleting those vertices with circumflexes, e g.*

$$A^0\hat{A}^1 \cdots \hat{A}^k \cdots A^n = A^0A^2 \cdots A^{k-1}A^{k+1} \cdots A^n.$$

THEOREM 6.4. *The boundary operator for q-chains* ∂_q *satisfies*

$$\partial_q(gA^0 \cdots A^q) = \sum_{k=0}^{q} (-1)^k gA^0 \cdots \hat{A}^k \cdots A^q.$$

PROOF. Assume first that the complex K coincides with the ordered q-simplex s with vertices $A^0 < \cdots < A^q$, and that $L = 0$ Then $K^q = K = s$ and $K^{q-1} = \dot{s}$. Let s_k denote the ordered simplex with vertices $A^0 < \cdots \hat{A}^k \cdots < A^q$. Define c_k as the subcomplex of s consisting of all faces of s except for s and s_k. The proof is based on the diagram

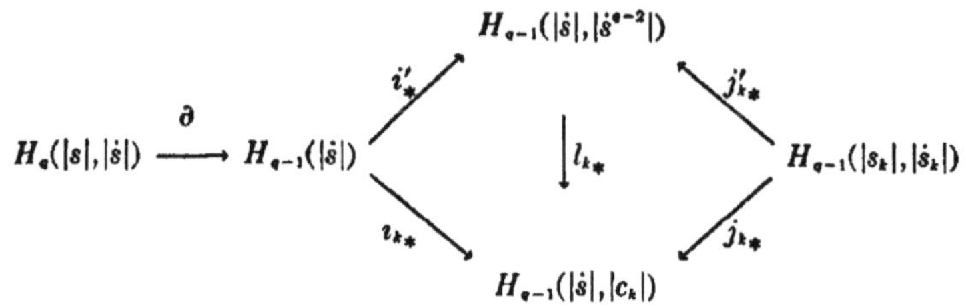

$$
\begin{array}{ccccccc}
 & & & H_{q-1}(|\dot{s}|,|\dot{s}^{q-2}|) & & & \\
 & & \nearrow i'_* & & \nwarrow j'_{k*} & & \\
H_q(|s|,|\dot{s}|) & \xrightarrow{\partial} & H_{q-1}(|\dot{s}|) & \Big\downarrow l_{k*} & & & H_{q-1}(|s_k|,|\dot{s}_k|) \\
 & & \searrow i_{k*} & & \swarrow j_{k*} & & \\
 & & & H_{q-1}(|\dot{s}|,|c_k|) & & &
\end{array}
$$

where all maps that occur are either ∂ or are induced by inclusions. Now $\partial_q(gs)$ is in $H_{q-1}(|s|,|s^{q-2}|)$. By theorem 5 8,

$$\text{(1)} \qquad \partial_q(gs) = \sum_{i=0}^{q} g_i A^0 \cdots \hat{A}^i \cdots A^q = \sum_{i=0}^{q} j'_{i*}(g_i s_i)$$

for some elements $g_i \in G$ Now apply l_{k*} to both sides of (1). If $i \neq k$, then $s_i \subset c_k$ and the map $l_k j'_i$ may be factored into inclusion maps

$$(s_i,\dot{s}_i) \rightarrow (c_k,c_k) \rightarrow (s,c_k).$$

Since, by I,8.1, $H_{q-1}(|c_k|,|c_k|) = 0$, it follows that $(l_k j'_i)_* = 0$. Therefore (1) gives

$$\text{(2)} \qquad l_{k*}\partial_q(gs) = l_{k*}j'_{k*}(gs) = j_{k*}(g_k s_k).$$

Now factor ∂_q into $i'_*\partial$, and use $l_k i' = i_k$. Then (2) reduces to

$$\text{(3)} \qquad i_{k*}\partial(gs) = j_{k*}(g_k s_k).$$

Now j_{k*} is an isomorphism, and by 3 4,

$$j_{k*}^{-1} i_{k*}\partial = [s:s_k]$$

is the incidence isomorphism. Hence (3) gives

$$[s,s_k](gs) = g_k s_k.$$

If this is compared with 4 5 and 3.6, we obtain immediately that $g_k = (-1)^k g$ Since this holds for each k, (1) reduces to the required formula for a simplex.

Returning to the general case, let $gA^0 \cdots A^q$ be a q-chain in an arbitrary pair (K,L) Let s be an ordered q-simplex with vertices $B^0 < \cdots < B^q$ and let f: $s \rightarrow (K,L)$ be a simplicial map such that $f(B^i) = A^i$, $i = 0, \cdots, q$ Then by 5.8,

$$f_q(gB^0 \cdots B^q) = gA^0 \cdots A^q,$$
$$f_q(gB^0 \cdots \hat{B}^k \cdots B^q) = gA^0 \cdots \hat{A}^k \cdots A^q.$$

Since $f_{q-1}\partial_q = \partial_q f_q$, we have

$$\partial_q(gA^0 \cdots A^q) = \partial_q f_q(gB^0 \cdots B^q) = f_{q-1}\partial_q(gB^0 \cdots B^q)$$
$$= f_{q-1}(\sum (-1)^k gB^0 \cdots \hat{B}^k \cdots B^q) = \sum (-1)^k gA^0 \cdots \hat{A}^k \cdots A^q.$$

This completes the proof.

REMARK. Since the excision map $(K^q, K^{q-1} \cup L^q) \subset (K^q \cup L, K^{q-1} \cup L)$ induces isomorphisms, an isomorphic theory could be obtained by defining $C_q(K,L) = H_q(|K^q|, |K^{q-1} \cup L^q|)$. This alternative definition has the advantage of employing only skeletons of dimension $\leqq q$ and thus yielding $C_q(K,L) = C_q(K^q,L^q)$ On the other hand the boundary operator ∂_q for chains would have to be defined as the composite of three homomorphisms, one of which is the inverse of an isomorphism induced by an excision.

DEFINITION 6.1c. *The coboundary operator.*

$$\delta^q \quad C^q(K,L) \to C^{q+1}(K,L)$$

is the coboundary operator of the triple $(|K^{q+1} \cup L|, |K^q \cup L|, |K^{q-1} \cup L|)$.

THEOREM 6.2c. $\delta^{q+1}\delta^q = 0$

THEOREM 6 3.c *If f. $(K,L) \to (K_1,L_1)$ is simplicial, then $f^{q+1}\delta^q = \delta^q f^q$.*

THEOREM 6.4c. *If $c \in C^q(K,L)$, then*

$$\delta_q c(A^0, \cdots, A^{q+1}) = \sum_{k=0}^{q+1} (-1)^k c(A^0, \cdots, \hat{A}^k, \cdots, A^{q+1}).$$

7. CYCLES AND HOMOLOGY GROUPS

DEFINITION 7.1. Let (K,L) be a pair consisting of a simplicial complex K and a subcomplex L. The kernel of ∂_q. $C_q(K,L) \to C_{q-1}(K,L)$ is called the group of *q-cycles* of (K,L) and is denoted by $Z_q(K,L)$. The image of ∂_{q+1}: $C_{q+1}(K,L) \to C_q(K,L)$ is called the group of *q-boundaries* of (K,L) and is denoted by $B_q(K,L)$. Since $\partial_q\partial_{q+1} = 0$, $B_q(K,L)$ is a subgroup of $Z_q(K,L)$; the factor group $H_q(K,L) = Z_q(K,L)/B_q(K,L)$ is called the *q-dimensional homology group* of (K,L). If f. $(K,L) \to (K_1,L_1)$ is simplicial, then f_q carries $Z_q(K,L)$ into $Z_q(K_1,L_1)$ and $B_q(K,L)$ into $B_q(K_1,L_1)$, thereby inducing homomorphisms.

$$f_* \quad H_q(K,L) \to H_q(K_1,L_1).$$

REMARK. The group $H_q(K,L)$ is not to be confused with the group $H_q(|K|,|L|)$ given by the axioms. An isomorphism of these two groups will be established in the next section, and is the main step in our uniqueness proof.

According to 5.9, the inclusion maps

$$L \xrightarrow{i} K \xrightarrow{j} (K,L)$$

induce exact sequences

$$0 \rightarrow C_q(L) \xrightarrow{i_q} C_q(K) \xrightarrow{j_q} C_q(K,L) \rightarrow 0.$$

We shall explore this fact to define homomorphisms ∂_*: $H_q(K,L) \rightarrow H_{q-1}(L)$.

LEMMA 7.2 *If we define* $\bar{Z}_q(K,L) = j_q^{-1}[Z_q(K,L)]$, $\bar{B}_q(K,L) = j_q^{-1}[B_q(K,L)]$, *and* $\bar{H}_q(K,L) = \bar{Z}_q(K,L)/\bar{B}_q(K,L)$, *then*

$$\bar{Z}_q(K,L) = \partial_q^{-1}[i_{q-1}C_{q-1}(L)], \qquad \bar{B}_q(K,L) = B_q(K) \cup i_q[C_q(L)]$$

where $\cup$ *is the operation of forming the smallest subgroup of* $C_q(K)$ *containing the two groups. Further, the homomorphism* j_q *induces isomorphisms*

$$\bar{j}: \bar{H}_q(K,L) \approx H_q(K,L).$$

PROOF. Let $c \in C_q(K)$. Then $c \in \bar{Z}_q(K,L)$ if and only if $j_qc \in Z_q(K,L)$, i.e. if and only if $\partial_q j_q c = 0$. This is equivalent to $j_{q-1}\partial_q c = 0$, and since kernel j_{q-1} = image i_{q-1}, this is equivalent to $\partial_q c \in i_{q-1}[C_{q-1}(L)]$.

Suppose $c \in \bar{B}_q(K,L)$. Then $j_q c \in B_q(K,L)$ and $j_q c = \partial_{q+1} b$ for some $b \in C_{q+1}(K,L)$. Let $d \in C_{q+1}(K)$ be such that $j_{q+1}d = b$. Then

$$j_q(c - \partial_{q+1}d) = j_q c - \partial_{q+1}j_{q+1}d = j_q c - \partial_{q+1}b = 0,$$

so that there is an $e \in C_q(L)$ with $i_q e = c - \partial_{q+1}d$. Thus $c = \partial_{q+1}d + i_q e$, and therefore $c \in B_q(K) \cup i_q[C_q(L)]$. Conversely, if $c = \partial_{q+1}d + i_q e$, then $j_q c = j_q\partial_{q+1}d + j_q i_q e = \partial_{q+1}j_{q+1}d$ and $j_q c \in B_q(K,L)$. Thus $c \in \bar{B}_q(K,L)$.

The last part of 7.2 is now a direct consequence of the Noether isomorphism theorem.

LEMMA 7.3. *The boundary homomorphism* ∂_q: $C_q(K) \rightarrow C_{q-1}(K)$ *defines homomorphisms*

$$\bar{Z}_q(K,L) \rightarrow i_{q-1}[Z_{q-1}(L)],$$
$$\bar{B}_q(K,L) \rightarrow i_{q-1}[B_{q-1}(L)].$$

Since the kernel of i_{q-1} *is zero, the composition* $i_{q-1}^{-1}\partial_q$ *defines homomorphisms*

$$\bar{Z}_q(K,L) \rightarrow Z_{q-1}(L), \qquad \bar{B}_q(K,L) \rightarrow B_{q-1}(L),$$

and thereby induces a homomorphism

$$\Delta\colon\ \bar{H}_q(K,L) \to H_{q-1}(L).$$

PROOF. Let $c \in \bar{Z}_q(K,L)$. By 7.2, we then have $\partial_q c = i_{q-1}d$ for some $d \in C_{q-1}(L)$. Then $i_{q-2}\partial_{q-1}d = \partial_{q-1}i_{q-1}d = \partial_{q-1}\partial_q c = 0$, and therefore $\partial_{q-1}d = 0$ since the kernel of i_{q-2} is zero. Thus $d \in Z_{q-1}(L)$ and $\partial_q c \in i_{q-1}[Z_{q-1}(L)]$.

If $c \in \bar{B}_q(K,L)$, then, by 7.2, $c = \partial_{q+1}d + i_q e$ for some $d \in C_{q+1}(K)$, $e \in C_q(L)$. Then $\partial_q c = \partial_q\partial_{q+1}d + \partial_q i_q e = i_{q-1}\partial_q e$, and $\partial_q c \in i_{q-1}[B_{q-1}(L)]$.

DEFINITION 7.4. The homomorphism

$$\partial_*\colon\ H_q(K,L) \to H_{q-1}(L)$$

is defined to be the composition $\partial_* = \Delta\bar{j}^{-1}$.

THEOREM 7.5. *Consider the diagram*

$$\begin{array}{ccccccc}
H_q(K,L) & \xleftarrow{\ \nu\ } & Z_q(K,L) & \xrightarrow{\ \eta\ } & C_q(K,L) & \xleftarrow{\ j_q\ } & C_q(K) \\
 & & & & & & \downarrow{\scriptstyle \partial_q} \\
H_{q-1}(L) & \xleftarrow{\ \nu_1\ } & Z_{q-1}(L) & \xrightarrow{\ \eta_1\ } & C_{q-1}(L) & \xrightarrow{\ i_{q-1}\ } & C_{q-1}(K)
\end{array}$$

in which η,η_1 are inclusions and ν,ν_1 are natural homomorphisms. Let $x \in H_q(K,L)$, $y \in H_{q-1}(L)$. Then $y = \partial_* x$ if and only if for any $c \in Z_q(K,L)$ and any $d \in C_q(K)$ such that $x = \nu c$, $\eta c = j_q d$, there is an $e \in Z_{q-1}(L)$ with

$$\partial_q d = i_{q-1}\eta_1 e, \qquad \nu_1 e = y.$$

The proof follows by inspection of the definitions of $\bar{j}$, Δ, and ∂_*.

DEFINITION 7.1c. The kernel of $\delta^q\colon\ C^q(K,L) \to C^{q+1}(K,L)$ is called the group of *q-cocycles*, and is denoted by $Z^q(K,L)$. The image of δ^{q-1}, denoted by $B^q(K,L)$, is called the group of *q-coboundaries*. The factor group $H^q(K,L) = Z^q(K,L)/B^q(K,L)$ is called the *q-dimensional cohomology group* of (K,L) If $f\colon\ (K,L) \to (K_1,L_1)$ is simplicial, then $f^*\colon\ H^q(K_1,L_1) \to H^q(K,L)$ is the homomorphism induced by f^q.

LEMMA 7.2c. *In the sequence*

$$0 \to C^q(K,L) \xrightarrow{j^q} C^q(K) \xrightarrow{i^q} C^q(L) \to 0,$$

define $\bar{Z}^q(L) = (i^q)^{-1}Z^q(L)$, $\bar{B}^q(L) = (i^q)^{-1}B^q(L)$, and $\bar{H}^q(L) = \bar{Z}^q(L)/\bar{B}^q(L)$. *Then*

$$\bar{Z}^q(L) = (\delta^q)^{-1}[j^{q+1}C^{q+1}(K,L)], \qquad \bar{B}^q(L) = B^q(K) \cup j^q[C^q(K,L)],$$

and i^q *induces isomorphisms*

$$\bar{i}.\quad \overline{H}^q(L) \approx H^q(L).$$

LEMMA 7.3c *Under* δ^q: $C^q(K) \to C^{q+1}(K)$, $\overline{Z}^q(L)$ *is carried into* $j^{q+1}Z^{q+1}(K,L)$ *and* $\overline{B}^q(L)$ *is carried into* $j^{q+1}B^{q+1}(K,L)$. *Since the kernel of* j^{q+1} *is zero, the composition* $(j^{q+1})^{-1}\delta^q$ *induces a homomorphism*

$$\Delta\colon\ \overline{H}^q(L) \to H^{q+1}(K,L).$$

DEFINITION 7.4c. The homomorphism

$$\delta^*\colon\ H^q(L) \to H^{q+1}(K,L)$$

is defined to be the composition $\delta^* = \Delta\bar{i}^{-1}$.

THEOREM 7.5c. *Consider the diagram*

$$\begin{array}{ccccccc}
H^q(K,L) & \xleftarrow{\ \nu\ } & Z^q(K,L) & \xrightarrow{\ \eta\ } & C^q(K,L) & \xrightarrow{\ j^q\ } & C^q(K) \\
 & & & & & & \uparrow{\scriptstyle \delta^{q-1}} \\
H^{q-1}(L) & \xleftarrow{\ \nu_1\ } & Z^{q-1}(L) & \xrightarrow{\ \eta_1\ } & C^{q-1}(L) & \xleftarrow{\ i^{q-1}\ } & C^{q-1}(K)
\end{array}$$

and let $x \in H^{q-1}(L)$, $y \in H^q(K,L)$ *Then* $\delta^* y = x$ *if and only if, for any* $c \in Z^{q-1}(L)$ *and any* $d \in C^{q-1}(K)$ *such that* $x = \nu_1 c$, $\eta_1 c = i^{q-1}d$, *there is an* $e \in Z^q(K,L)$ *with*

$$\delta^{q-1}d = j^q\eta e, \qquad \nu e = y.$$

8. THE MAIN ISOMORPHISM

Let (X,A) be a pair with a triangulation $T = \{t,(K,L)\}$. In this section we establish the basic result that $H_q(X,A)$ and $H_q(K,L)$ are isomorphic

LEMMA 8.1. *In the diagram*

$$H_q(|K|,|L|) \xleftarrow{\ j_*\ } H_q(|K^q \cup L|,|L|) \xrightarrow{\ l_*\ } H_q(|K^q \cup L|,|K^{q-1} \cup L|)$$

where j: $(K^q \cup L,L) \subset (K,L)$ and l: $(K^q \cup L,L) \subset (K^q \cup L, K^{q-1} \cup L)$ *are inclusion maps, the following relations hold*:

(a) *the kernel of* l_* *is zero,*

(b) *the image of* l_* *is* $Z_q(K,L)$,

(c) j_* *is onto,*

(d) *the kernel of* j_* *is* $l_*^{-1}[B_q(K,L)]$.

As an immediate consequence of this lemma we have

THEOREM 8.2. (MAIN ISOMORPHISM). *In the diagram*

$$H_q(X,A) \xleftarrow{i_*} H_q(|K|,|L|) \xleftarrow{j_*} H_q(|K^q \cup L|,|L|) \xrightarrow{l_*} C_q(K,L) \xleftarrow{\eta} Z_q(K,L) \xrightarrow{\nu} H_q(K,L)$$

where η is an inclusion and ν the natural map, the homomorphisms i_*j_* and $\nu\eta^{-1}l_*$ are onto and their kernels in $H_q(|K^q \cup L|,|L|)$ are equal. Therefore $\theta_T = \nu\eta^{-1}l_*j_*^{-1}i_*^{-1}$ is single valued and is an isomorphism

$$\theta_T\colon\ H_q(X,A) \approx H_q(K,L).$$

Explicitly, if $x \in H_q(X,A)$ and $y \in H_q(K,L)$, then $y = \theta_T x$ if and only if there exist elements $c \in Z_q(K,L)$ and $d \in H_q(|K^q \cup L|,|L|)$ with

$$x = i_*j_*d, \qquad l_*d = \eta c, \qquad \nu c = y.$$

The properties of this isomorphism will be established in the next section. We proceed now with the proof of 8.1 The following simplified notation will be used:

$$X = |K|, \qquad \hat{X}^q = |K^q \cup L|, \qquad A = \hat{X}^{-1} = |L|.$$

Of the theorems of this chapter, the proof uses only the following proposition, which is a restatement of 5 2:

LEMMA 8 3. *If $p \neq q$, then $H_q(\hat{X}^p,\hat{X}^{p-1}) = 0$*

Two lemmas will be needed

LEMMA 8 4 *The homomorphism*

$$H_q(\hat{X}^p,A) \to H_q(\hat{X}^{p+1},A)$$

induced by the inclusion map $(\hat{X}^p,A) \to (\hat{X}^{p+1},A)$ is (1) onto if $q \neq p + 1$, (2) of kernel zero if $q \neq p$, and (3) an isomorphism if $q \neq p,p + 1$.

PROOF. Consider the section

$$H_{q+1}(\hat{X}^{p+1},\hat{X}^p) \to H_q(\hat{X}^p,A) \to H_q(\hat{X}^{p+1},A) \to H_q(\hat{X}^{p+1},\hat{X}^p)$$

of the homology sequence of the triple $(\hat{X}^{p+1},\hat{X}^p,A)$. By 8 3, the left term vanishes if $q \neq p$ and the right term vanishes if $q \neq p + 1$. Thus (1) and (2) follow from the exactness of the H S of the triple (see I,10.2). Statement (3) follows from (1) and (2).

LEMMA 8 5. $H_q(\hat{X}^{q-1},A) = 0.$

PROOF. From 8.4 it follows by induction that $H_q(\hat{X}^{q-1},A) \approx H_q(\hat{X}^{q-r},A)$ for $r = 2, \cdots, q + 1$ Since $H_q(\hat{X}^{-1},A) = H_q(A,A) = 0$ by I,8 1, the proposition follows.

PROOF OF 8.1. The homomorphism j_* $H_q(\hat{X}^q,A) \to H_q(X,A)$ may be factored into homomorphisms

$$H_q(\hat{X}^q,A) \xrightarrow{j_{1*}} H_q(\hat{X}^{q+1},A) \xrightarrow{j_{2*}} \cdots \xrightarrow{j_{r*}} H_q(\hat{X}^{q+r},A) = H_q(X,A)$$

where $q + r = \dim K$, and where $j_1, j_2, \cdots$ are inclusion maps. By 8.4, j_{1*} is onto, while $j_{2*}, j_{3*}, \cdots$ are isomorphisms. This implies that j_* is onto, thus proving part (c) of 8.1. It also follows that

$$\text{(1)} \qquad \text{kernel } j_* = \text{kernel } j_{1*}.$$

The proof of the remaining three parts of 8.1 is based on the following diagram:

$$\begin{array}{ccccccc}
 & & H_{q+1}(\hat{X}^{q+1},\hat{X}^q) & & & & \\
 & & \downarrow \bar{\partial} & \searrow^{\partial_{q+1}} & & & \\
H_q(\hat{X}^{q-1},A) & \rightarrow & H_q(\hat{X}^q,A) & \xrightarrow{l_*} & H_q(\hat{X}^q,\hat{X}^{q-1}) & \xrightarrow{\bar{\partial}_1} & H_{q-1}(\hat{X}^{q-1},A) \\
 & & \downarrow j_{1*} & & & \searrow^{\partial_q} & \downarrow l_* \\
 & & H_q(\hat{X}^{q+1},A) & & & & H_{q-1}(\hat{X}^{q-1},\hat{X}^{q-2})
\end{array}$$

The commutativity conditions in the two triangles are obvious consequences of the definitions of the appropriate boundary operators.

The middle row is a portion of the H.S. of the triple $(\hat{X}^q,\hat{X}^{q-1},A)$. Since $H_q(\hat{X}^{p-1},A) = 0$ by 8.5, it follows that the kernel of l_* is zero, thus proving part (a) of 8.1. Furthermore image $l_* =$ kernel $\bar{\partial}_1$, and since the kernel of l_* is zero, we have

$$\text{kernel } \bar{\partial}_1 = \text{kernel } l_*\bar{\partial}_1 = \text{kernel } \partial_q = Z_q(K,L).$$

Thus image $l_* = Z_q(K,L)$, and part (b) of 8.1 is proved.

The vertical column is a portion of the H.S. of the triple $(\hat{X}^{q+1},\hat{X}^q,A)$; therefore, by (1) and exactness,

$$\text{kernel } j_* = \text{kernel } j_{1*} = \text{image } \bar{\partial}.$$

Thus

$$l^*[\text{kernel } j_*] = l^*[\text{image } \bar{\partial}] = \text{image } l^*\bar{\partial} = \text{image } \partial_{q+1} = B_q(K,L)$$

and therefore, since the kernel of l_* is zero, kernel $j_* = l_*^{-1}[B_q(K,L)]$. This proves (d) and concludes the proof of 8.1.

Theorem 8.6 *If (X,A) is a triangulable pair, then $H_q(X,A) = 0$ for $q < 0$.*

Proof. In view of the main isomorphism, it suffices to show that $H_q(K,L) = 0$ for $q < 0$. By 5.4, we have $C_q(K,L) = 0$ for $q < 0$. This implies $Z_q(K,L) = 0$, and thus also $H_q(K,L) = 0$.

Lemma 8.1c. *In the diagram*

$$H^q(|K|,|L|) \xrightarrow{j_*} H^q(|K^q \cup L|,|L|) \xleftarrow{l_*} H^q(|K^q \cup L|,|K^{q-1} \cup L|)$$

where j and l are inclusions, the following relations hold:

(a) l^* *is onto,*

(b) *the kernel of* l^* *is* $B^q(K,L)$

(c) *the kernel of* j^* *is zero,*

(d) *the image of* j^* *is* $l^*Z^q(K,L)$.

THEOREM 8.2c *In the diagram*

$$H^q(X,A) \xrightarrow{t^*} H^q(|K|,|L|) \xrightarrow{j^*} H^q(|K^q \cup L|,|L|) \xleftarrow{l^*} C^q(K,L)$$
$$\xleftarrow{\eta} Z^q(K,L) \xrightarrow{\nu} H^q(K,L)$$

t^* is an isomorphism, the kernel of j^* is zero, the images of j^* and $l^*\eta$ coincide, and the kernels of $l^*\eta$ and ν coincide. Therefore $\theta_T = \nu\eta^{-1}l^{*-1}j^*t^*$ is a single-valued isomorphism

$$\theta_T\colon\ H^q(X,A) \approx H^q(K,L).$$

Explicitly, if $x \in H^q(X,A)$ and $y \in H^q(K,L)$, then $y = \theta_T x$ if and only if there is a $c \in Z^q(K,L)$ such that

$$j^*t^*x = l^*\eta c, \qquad \nu c = y.$$

LEMMA 8.3c. *If $p \neq q$, then $H^q(\hat{X}^p,\hat{X}^{p-1}) = 0$.*

LEMMA 8.4c. *The homomorphism $H^q(\hat{X}^{p+1},A) \to H^q(\hat{X}^p,A)$ induced by the inclusion map is* (1) *of kernel zero if $q \neq p+1$,* (2) *a homomorphism onto if $q \neq p$, and* (3) *an isomorphism if $q \neq p,p+1$.*

LEMMA 8.5c. $H^q(\hat{X}^{q-1},A) = 0$.

THEOREM 8.6. *If (X,A) is triangulable, then $H^q(X,A) = 0$ for $q < 0$.*

9. PROPERTIES OF THE MAIN ISOMORPHISM

The following three important properties of the isomorphism θ_T of the last section will now be established.

THEOREM 9.1. *In the diagram*

$$\begin{array}{ccc} H_q(X,A) & \xrightarrow{\theta_T} & H_q(K,L) \\ \downarrow \partial & & \downarrow \partial_* \\ H_{q-1}(A) & \xrightarrow{\theta_{T'}} & H_{q-1}(L) \end{array}$$

where $T' = \{t',L\}$ is the triangulation of A defined by T, the commutativity relation $\partial_*\theta_T = \theta_{T'}\partial$ holds.

THEOREM 9.2. *If $(X,A),(X_1,A_1)$ are pairs with triangulations $T =$*

$\{t,(K,L)\}$, $T_1 = \{t_1,(K_1,L_1)\}$ respectively, and if $f: (X,A) \to (X_1,A_1)$ is simplicial, then, in the diagram

$$\begin{array}{ccc} H_q(X,A) & \xrightarrow{\theta_T} & H_q(K,L) \\ \downarrow f_* & & \downarrow g_* \\ H_q(X_1,A_1) & \xrightarrow{\theta_{T_1}} & H_q(K_1,L_1) \end{array}$$

where $g: (K,L) \to (K_1,L_1)$ is the simplicial map $g = t_1^{-1}ft$, the commutativity relation $g_*\theta_T = \theta_{T_1}f_*$ holds.

Let P_0 be the space consisting of the single reference point. Regarding P_0 as a simplicial complex K_0, we have $Z_0(K_0) = C_0(K_0) = G$ and $B_0(P_0) = 0$ so that $H_0(K_0) = G$.

THEOREM 9.3. *θ: $H_0(P_0) \to H_0(K_0)$ is the identity map $G \to G$.*

The last proposition is an immediate consequence of the definition of θ. Theorem 9 2 follows from standard commutativity relations in each square of the following diagram

$$\begin{array}{ccccccccc} H_q(X,A) & \leftarrow & H_q(|K|,|L|) & \leftarrow & H_q(|K^q|,|L^{q-1}|) & \to & Z_q(K,L) & \to & H_q(K,L) \\ \downarrow f_* & & \downarrow g_* & & \downarrow h_* & & \downarrow g_q & & \downarrow g_* \\ H_q(X_1,A_1) & \leftarrow & H_q(|K_1|,|L_1|) & \leftarrow & H_q(|K_1^q|,|L_1^{q-1}|) & \to & Z_q(K_1,L_1) & \to & H_q(K_1,L_1) \end{array}$$

where h is defined by g.

The proof of 9 1 requires some preparation

LEMMA 9 4. *The homomorphism*

$$k_*: H_q(|K^q|,|L^{q-1}|) \to H_q(|K^q \cup L|,|L|)$$

induced by inclusion, is a homomorphism onto

PROOF The homomorphism k_* may be factored into homomorphisms

$$H_q(|K^q|,|L^{q-1}|) \xrightarrow{k_{1*}} H_q(|K^q|,|L^q|) \xrightarrow{k_{2*}} H_q(|K^q \cup L|,|L|)$$

induced by appropriate inclusion maps. Since $L^q = K^q \cap L$, the homomorphism k_{2*} is an isomorphism by the excision theorem 2.1 To examine k_{1*} consider the portion

$$H_q(|K^q|,|L^{q-1}|) \xrightarrow{k_{1*}} H_q(|K^q|,|L^q|) \to H_{q-1}(|L^q|,|L^{q-1}|)$$

of the H.S. of the triple (K^q,L^q,L^{q-1}) By 8 3, the end term is zero. It follows from exactness that k_{1*} is onto. Thus k_* is onto

PROOF OF 9.1. Consider the following diagram

$$
\begin{array}{ccccccccc}
 & & & & & & & & H_q(K,L) \\
 & & & & & & & & \uparrow \nu \\
 & & & & & & & & Z_q(K,L) \\
 & & & j_* & & & l_* & & \downarrow \eta \\
 & & \swarrow & & H_q(|K^q \cup L|,|L|) & & \rightarrow & & C_q(K,L) \\
 & t_* & & g_* & \uparrow k_* & & f_* & & \uparrow j_q \\
H_q(X,A) & \leftarrow & H_q(|K|,|L|) & \leftarrow & H_q(|K^q|,|L^{q-1}|) & & \rightarrow & & C_q(K) \\
\downarrow \partial & t'_* & \downarrow \partial' & j_{1*} & \downarrow \partial_1 & & f_{1*} & & \downarrow \partial_q \\
H_{q-1}(A) & \leftarrow & H_{q-1}(|L|) & \leftarrow & H_{q-1}(|L^{q-1}|) & & \rightarrow & & C_{q-1}(K) \\
 & & & & & & & & \uparrow i_{q-1} \\
 & & & & & \searrow & l_{1*} & \rightarrow & C_{q-1}(L) \\
 & & & & & & & & \uparrow \eta_1 \\
 & & & & & & & & Z_{q-1}(L) \\
 & & & & & & & & \downarrow \nu_1 \\
 & & & & & & & & H_{q-1}(L)
\end{array}
$$

In this diagram η, η_1 are inclusions, ν, ν_1 are natural homomorphisms, and all other homomorphisms are either boundary operators, or are induced by t and t', or by appropriate inclusion maps. The commutativity relations in each square are easy to verify

Let $x \in H_q(X,A)$. By definition of θ_T there exist elements

$$c \in Z_q(K,L), \qquad b \in H_q(|K^q \cup L|,|L|)$$

with

$$x = t_* j_* b, \qquad l_* b = \eta c, \qquad \nu c = \theta_T x$$

By 9 4, k_* is onto, and therefore there is an $a \in H_q(|K^q|,|L^{q-1}|)$ with $k_* a = b$. Define $d = f_* a$ Then

$$j_q d = l_* k_* a = l_* b = \eta c.$$

Thus, by 7.5, there is an $e \in Z_{q-1}(L)$ with

$$\partial_q d = i_{q-1} \eta_1 e, \qquad \nu_1 e = \partial_* \theta_T x.$$

Since

$$t'_* j_{1*} \partial_1 a = \partial t_* g_* a = \partial t_* j_* b = \partial x,$$
$$i_{q-1} l_{1*} \partial_1 a = \partial_q f_* a = \partial_q d = i_{q-1} \eta_1 e,$$

and since the kernel of i_{q-1} is zero, it follows that

$$\partial x = t'_* j_{1*} \partial_1 a, \qquad l_{1*} \partial_1 a = \eta_1 e.$$

Thus, by the definition of $\theta_{T'}$, it follows that $\theta_{T'} \partial x = \nu_1 e = \partial_* \theta_T x$.

THEOREM 9.1c. *In the diagram*

$$\begin{array}{ccc} H^q(X,A) & \xrightarrow{\ \theta_T\ } & H^q(K,L) \\ \uparrow\delta & & \uparrow\delta^* \\ H^{q-1}(A) & \xrightarrow{\ \theta_{T'}\ } & H^{q-1}(L) \end{array}$$

the commutativity relation $\theta_T\delta = \delta^*\theta_{T'}$ *holds.*

THEOREM 9.2c *If* f: $(X,A) \to (X_1,A_1)$ *is simplicial relative to triangulations* $T = \{t,(K,L)\}$, $T_1 = \{t_1,(K_1,L_1)\}$, *then in the diagram*

$$\begin{array}{ccc} H^q(X,A) & \xrightarrow{\ \theta_T\ } & H^q(K,L) \\ \uparrow f^* & & \uparrow g^* \\ H^q(X_1,A_1) & \xrightarrow{\ \theta_{T_1}\ } & H^q(K_1,L_1) \end{array}$$

the commutativity relation $g^*\theta_{T_1} = \theta_T f^*$ *holds, where* $g = t_1^{-1}ft$.

THEOREM 9.3c θ: $H_0(P_0) \to H_0(K_0)$ *is the identity map* $G \to G$, *where* K_0 *is the simplicial complex consisting of the single vertex* P_0.

LEMMA 9.4c. *The homomorphism*

$$k^* \quad H^q(|K^q \cup L|,|L|) \to H^q(|K^q|,|L^{q-1}|),$$

induced by the inclusion map, has kernel zero

The proof of 9.1c is not quite dual to that of 9.1, and will therefore be given here. Consider the diagram

$$\begin{array}{ccccccccc}
 & & & & & & & & H^q(K,L) \\
 & & & & & & & & \uparrow\nu \\
 & & & & & & & & Z^q(K,L) \\
 & & & j^* & & & l^* & & \downarrow\eta \\
 & & \ulcorner & \longrightarrow & H^q(|K^q \cup L|,|L|) & & \leftarrow & & C^q(K,L) \\
 & t^* & & g^* & \downarrow k^* & & f^* & & \downarrow j^q \\
H^q(X,A) & \to & H^q(|K|,|L|) & \to & H^q(|K^q|,|L^{q-1}|) & & \leftarrow & & C^q(K) \\
\uparrow\delta & t'^* & \uparrow\delta' & j_1^* & \uparrow\delta_1 & & f_1^* & & \uparrow\delta^q \\
H^{q-1}(A) & \to & H^{q-1}(|L|) & \to & H^{q-1}(|L^{q-1}|) & & \leftarrow & & C^{q-1}(K) \\
 & & & & \uparrow & & l_1^* & & \downarrow i^q \\
 & & & & \llcorner & \text{---} & \text{---} & & C^{q-1}(L) \\
 & & & & & & & & \uparrow\eta_1 \\
 & & & & & & & & Z^{q-1}(L) \\
 & & & & & & & & \downarrow\nu_1 \\
 & & & & & & & & H^{q-1}(L)
\end{array}$$

Let $x \in H^{q-1}(A)$. By the definition of $\theta_{T'}$, there exists a $c \in Z^{q-1}(L)$ with

$$j_1^* i'^* x = l_1^* \eta_1 c, \qquad \nu_1 c = \theta_{T'} x.$$

Since i^q is onto, we may select $d \in C^{q-1}(K)$ so that $i^q d = \eta_1 c$. Then, by 7.5c, there is an $e \in Z^q(K,L)$ satisfying

$$\delta^q d = j^q \eta e, \qquad \nu e = \delta^* \theta_{T'} x.$$

Then

$$\begin{aligned} k^* j^* i^* \delta x &= \delta_1 j_1^* i'^* x = \delta_1 l_1^* \eta_1 c = \delta_1 l_1^* i^q d = f^* \delta^q d \\ &= f^* g^q \eta e = k^* l^* \eta e. \end{aligned}$$

Since, by 9.4c, the kernel of k^* is zero, it follows that

$$j^* i^* \delta x = l^* \eta e.$$

Thus the definition of θ_T implies $\theta_T \delta x = \nu e = \delta^* \theta_{T'} x$.

10. THE UNIQUENESS THEOREM

Let H and $\overline{H}$ be two homology theories defined on admissible categories containing all triangulable pairs and their maps. The main result of this chapter is that every isomorphism of the coefficient groups G and $\overline{G}$ yields an isomorphism of the theories H and $\overline{H}$ over the class of triangulable spaces and their maps.

THEOREM 10.1 (UNIQUENESS THEOREM) *Given two homology theories, H and $\overline{H}$, and given a homomorphism*

$$h_0\colon\ G \to \overline{G}$$

of their coefficient groups, there exists a unique set of homomorphisms

$$h(q,X,A)\colon\ H_q(X,A) \to \overline{H}_q(X,A)$$

defined for each triangulable pair (X,A) and each integer q, such that

(1) $$h(0,P_0) = h_0$$

(2) *if $f\colon\ (X,A) \to (X_1,A_1)$ is a map, and (X,A) and (X_1,A_1) are triangulable, then the commutativity relation $\overline{f}_* h(q,X,A) = h(q,X_1,A_1) f_*$ holds in the diagram*

$$\begin{array}{ccc} H_q(X,A) & \xrightarrow{\ h\ } & \overline{H}_q(X,A) \\ \downarrow f_* & & \downarrow \overline{f}_* \\ H_q(X_1,A_1) & \xrightarrow{\ h\ } & \overline{H}_q(X_1,A_1) \end{array}$$

(3) *the commutativity relation* $\bar{\partial} h(q,X,A) = h(q-1,A)\partial$ *holds in the diagram*

$$\begin{array}{ccc} H_q(X,A) & \xrightarrow{h} & \bar{H}_q(X,A) \\ \downarrow \partial & & \downarrow \bar{\partial} \\ H_{q-1}(A) & \xrightarrow{h} & \bar{H}_{q-1}(A) \end{array}$$

If h_0: $G \approx \bar{G}$ *is an isomorphism, then each* $h(q,X,A)$ *is also an isomorphism*

PROOF. We shall construct homomorphisms

$$\zeta(q,K,L) \quad H_q(K,L) \to \bar{H}_q(K,L)$$

for each simplicial pair (K,L) and each integer q satisfying the following conditions:

$$\zeta(0,P_0) = h_0, \tag{4}$$

(5) if f. $(K,L) \to (K_1,L_1)$ is simplicial, then the commutativity relation $\bar{f}_* \zeta(q,K,L) = \zeta(q,K_1,L_1) f_*$ holds in the diagram

$$\begin{array}{ccc} H_q(K,L) & \xrightarrow{\zeta} & \bar{H}_q(K,L) \\ \downarrow f_* & & \downarrow \bar{f}_* \\ H_q(K_1,L_1) & \xrightarrow{\zeta} & \bar{H}_q(K_1,L_1) \end{array}$$

(6) the commutativity relation $\bar{\partial}_* \zeta(q,K,L) = \zeta(q-1,L)\partial_*$ holds in the diagram

$$\begin{array}{ccc} H_q(K,L) & \xrightarrow{\zeta} & \bar{H}_q(K,L) \\ \downarrow \partial_* & & \downarrow \bar{\partial}_* \\ H_{q-1}(L) & \xrightarrow{\zeta} & \bar{H}_{q-1}(L) \end{array}$$

Let $g \in G$ and let $gA^0 \cdots A^q$ be a q-chain of K Define

$$\eta(gA^0 \cdots A^q) = h_0(g)A^0 \cdots A^q$$

It follows from 5.4-5.7 that η yields a homomorphism

$$\eta. \quad C_q(K) \to \bar{C}_q(K).$$

From 5.8 and 6.4 it follows that η commutes properly with f_q and ∂_q. Consequently η defines homomorphisms

$$Z_q(K,L) \to \bar{Z}_q(K,L), \qquad B_q(K,L) \to \bar{B}_q(K,L),$$

and thus induces homomorphisms of their quotient groups

$$\zeta\colon\ H_q(K,L) \to \bar{H}_q(K,L)$$

satisfying (4)-(6).

Now, given a pair (X,A) with a triangulation $T = \{t,(K,L)\}$, consider the diagram

$$H_q(X,A) \xrightarrow{\theta_T} H_q(K,L) \xrightarrow{\zeta} \bar{H}_q(K,L) \xleftarrow{\bar{\theta}_T} \bar{H}_q(X,A)$$

where θ_T and $\bar{\theta}_T$ are defined for H and $\bar{H}$ in 8.2. Define

$$h_T(q,X,A) = \bar{\theta}_T^{-1}\zeta\theta_T,$$

then

$$h_T(q,X,A)\colon\ H_q(X,A) \to \bar{H}_q(X,A).$$

We will show that h_T is independent of the triangulation T and satisfies (1)-(3).

Observe that, if $(X,A) = P_0$, then $\theta_T\colon\ G \approx G$ and $\bar{\theta}_T\colon\ \bar{G} \approx \bar{G}$ are identity maps, so that $h_T(0,P_0) = \zeta(0,P_0) = h_0$ as desired.

Now let $(X,A),(X_1,A_1)$ be pairs with triangulations $T = \{t,(K,L)\}$, $T_1 = \{t_1,(K_1,L_1)\}$, and let $f\colon\ (X,A) \to (X_1,A_1)$ be a map (not necessarily simplicial). In addition, we consider the identity map $i\colon$ $(X,A) \to (X,A)$. Theorems II,8.4 and II,8.6 imply the existence of an integer n and of maps

$$f_n\colon\ (X,A) \to (X_1,A_1), \qquad i_n\colon\ (X,A) \to (X,A)$$

with the following properties: f_n is simplicial relative to the triangulations ${}^nT = \{{}^nt,({}^nK,{}^nL)\}$, $T_1 = \{t_1,(K_1,L_1)\}$, and is homotopic to f, while i_n is simplicial relative to the triangulations ${}^nT = \{{}^nt,({}^nK,{}^nL)\}$, $T = \{t,(K,L)\}$, and is homotopic to i. Define the simplicial maps

$$g\colon\ ({}^nK,{}^nL) \to (K_1,L_1), \qquad k\colon\ ({}^nK,{}^nL) \to (K,L)$$

by

$$g = t_1^{-1}f_n({}^nt), \qquad k = t^{-1}i_n({}^nt).$$

We now have the diagram

$$\begin{array}{ccccccc}
H_q(X,A) & \xrightarrow{\theta} & H_q(K,L) & \xrightarrow{\zeta} & \bar{H}_q(K,L) & \xleftarrow{\bar{\theta}} & \bar{H}_q(X,A) \\
\uparrow{\scriptstyle i_{n*}} & & \uparrow{\scriptstyle k_*} & & \uparrow{\scriptstyle \bar{k}_*} & & \uparrow{\scriptstyle \bar{i}_{n*}} \\
H_q(X,A) & \xrightarrow{\theta} & H_q({}^nK,{}^nL) & \xrightarrow{\zeta} & \bar{H}_q({}^nK,{}^nL) & \xleftarrow{\bar{\theta}} & \bar{H}_q(X,A) \\
\downarrow{\scriptstyle f_{n*}} & & \downarrow{\scriptstyle g_*} & & \downarrow{\scriptstyle \bar{g}_*} & & \downarrow{\scriptstyle \bar{f}_{n*}} \\
H_q(X_1,A_1) & \xrightarrow{\theta} & H_q(K_1,L_1) & \xrightarrow{\zeta} & \bar{H}_q(K_1,L_1) & \xleftarrow{\bar{\theta}} & \bar{H}_q(X_1,A_1)
\end{array}$$

The commutativity relations involving ζ follow from (5) since k and g are simplicial. The commutativity relations involving θ and $\bar{\theta}$ follow from 9.2 since i_n and f_n are simplicial. Since i_n is homotopic to i, it follows that i_{n*} and $\bar{i}_{n*}$ are identity maps. Since f_n is homotopic to f, it follows that $f_{n*} = f_*$ and $\bar{f}_{n*} = \bar{f}_*$. Thus we obtain

$$\bar{f}_* h_T(q,X,A) = h_{T_1}(q,X_1,A_1) f_* \tag{7}$$

Applying this relation to the special case $(X,A) = (X_1,A_1)$, and f = identity, we find $h_T(q,X,A) = h_{T_1}(q,X,A)$ so that indeed $h_T(q,X,A)$ is independent of the triangulation T, and may be written as $h(q,X,A)$. Consequently (7) yields (2).

To prove (3) consider (X,A) with a triangulation $T = \{t,(K,L)\}$ and denote by T' the induced triangulation of A. In the diagram

$$\begin{array}{ccccccc}
H_q(X,A) & \xrightarrow{\theta} & H_q(K,L) & \xrightarrow{\zeta} & \bar{H}_q(K,L) & \xleftarrow{\bar{\theta}} & \bar{H}_q(X,A) \\
\downarrow{\scriptstyle \partial} & & \downarrow{\scriptstyle \partial_*} & & \downarrow{\scriptstyle \bar{\partial}_*} & & \downarrow{\scriptstyle \bar{\partial}} \\
H_{q-1}(A) & \xrightarrow{\theta} & H_{q-1}(L) & \xrightarrow{\zeta} & \bar{H}_{q-1}(L) & \xleftarrow{\bar{\theta}} & \bar{H}_{q-1}(A)
\end{array}$$

commutativity in the center square is asserted by (6) while commutativity in the outside squares follows from 9.1. Thus $\bar{\partial} h(q,X,A) = h(q-1,A)\partial$ as desired.

We now prove that the homomorphisms $h(q,X,A)$ satisfying (1)-(3) are unique. Two auxiliary propositions will be established first.

If s^q is an ordered q-simplex and $g \in G$, then

$$h(q,|s^q|,|\dot{s}^q|)(gs^q) = h_0(g)s^q \tag{8}$$

If $q = 0$, the proposition is an immediate consequence of (1). Assume then that $q > 0$ and that (8) holds with $q - 1$ in place of q.

Let $A^0 < \cdots < A^q$ be the vertices of s^q and let s^{q-1} be the face of s^q with vertices $A^1 < \cdots < A^q$. Then, by 3.7,

$$[s^q : s^{q-1}]gs^q = gs^{q-1}.$$

Since $[s^q:s^{q-1}] = j_*^{-1}i_*\partial$, it follows from (2) and (3) that

$$h(q-1, |s^{q-1}|, |\dot{s}^{q-1}|)[s^q : s^{q-1}] = \overline{[s^q : s^{q-1}]}h(q, |s^q|, |\dot{s}^q|).$$

This implies

$$\begin{aligned}\overline{[s^q:s^{q-1}]}h(q,|s^q|,|\dot{s}^q|)(gs^q) &= h(q-1,|s^{q-1}|,|\dot{s}^{q-1}|)(gs^{q-1}) \\ &= h_0(g)s^{q-1} = \overline{[s^q \cdot s^{q-1}]}h(g)s^q\end{aligned}$$

Since $\overline{[s^q : s^{q-1}]}$ is an isomorphism, the proposition follows.

Let $gA^0 \cdots A^q \in C_q(K,L)$. Then

(9) $\quad h(q,|K^q \cup L|, |K^{q-1} \cup L|)(gA^0 \cdots A^q) = h_0(g)A^0 \cdots A^q.$

By 5.5, we have $gA^0 \cdots A^q = f_*(g)$ where s is an ordered q-simplex, and f is a suitable map $f : (s,\dot{s}) \to (K^q \cup L, K^{q-1} \cup L)$. Then, by (2) and (8),

$$h(gA^0 \cdots A^q) = h(f_*(g)) = \bar{f}_*h(gs) = \bar{f}_*h_0(g)s = h_0(g)A^0 \cdots A^q$$

as required.

Now let h and h' be two families of homomorphisms satisfying (1)-(3). Let (K,L) be a simplicial pair. We introduce abbreviations:

$$h_1 = h(q,|K^q \cup L|,|K^{q-1} \cup L|), \qquad h_2 = h(q,|K^q \cup L|,|L|),$$

and similarly define h_1' and h_2'. By (9) we have $h_1 = h_1'$. Let $l : (K^q \cup L, L) \to (K^q \cup L, K^{q-1} \cup L)$ be the inclusion map. Then by (2)

$$\bar{l}_*h_2 = h_1 l_* = h_1' l_* = \bar{l}_*h_2'.$$

Since, by 8.1, the kernel of $\bar{l}_*$ is zero, it follows that $h_2 = h_2'$. Now consider the inclusion map $j : (K^q \cup L, L) \to (K,L)$. Then

$$h(q,|K|,|L|)j_* = \bar{j}_*h_2 = \bar{j}_*h_2' = h'(q,|K|,|L|)j_*.$$

Since, by 8.1, j_* is onto, it follows that $h(q,|K|,|L|) = h'(q,|K|,|L|)$.

Let (X,A) be a triangulable pair and $T = \{t,(K,L)\}$ a triangulation of (X,A). Then

$$\bar{t}_*h(q,X,A) = h(q,|K|,|L|)t_* = h'(q,|K|,|L|)t_* = \bar{t}_*h'(q,X,A).$$

Since $\bar{t}_*$ is an isomorphism, it follows that $h(q,X,A) = h'(q,X,A)$. Thus $h = h'$ and uniqueness has been proved.

Now let h_0: $G \approx \bar{G}$ be an isomorphism, and let $\bar{h}_0$: $\bar{G} = G$ be the inverse isomorphism. Let $\bar{h}(q,X,A)$. $\bar{H}_q(X,A) \to H_q(X,A)$ be the homomorphisms satisfying (1)-(3) relative to $\bar{h}_0$. Then

$$\bar{h}(q,X,A)h(q,X,A): \quad H_q(X,A) \to H_q(X,A)$$

are homomorphisms satisfying (1)-(3) relative to the identity map $G \to G$. Therefore, by the uniqueness property, we have $\bar{h}(q,X,A)h(q,X,A) =$ identity. Similarly $h(q,X,A)\bar{h}(q,X,A) =$ identity Thus $h(q,X,A)$ is an isomorphism with $\bar{h}(q,X,A)$ as inverse This concludes the proof of the theorem.

THEOREM 10.1c. *Given two cohomology theories H and $\bar{H}$ and a homomorphism*

$$h_0: \quad G \to \bar{G}$$

of their coefficient groups, there exists a unique set of homomorphisms

$$h(q,X,A). \quad H^q(X,A) \to \bar{H}^q(X,A)$$

defined for each triangulable pair (X,A) and each integer q, such that

(1) $$h(0,P_0) = h_0,$$
(2) $$\bar{\delta}h(q,A) = h(q+1,X,A)\delta.$$
(3) *If f $(X,A) \to (X_1,A_1)$, then* $f^*h(q,X_1,A_1) = h(q,X,A)f^*$.

If h_0 $G \approx \bar{G}$, then each $h(q,X,A)$ is an isomorphism.

EXERCISES

A. INCIDENCE ISOMORPHISMS.

1. Let s_1^{q-1},s_2^{q-1} be $(q-1)$-faces of a q-simplex s^q and let s^{q-2} be their common $(q-2)$-face. Verify that

$$[s_1^{q-1}{:}s^{q-2}][s^q{:}s_1^{q-1}] = -[s_2^{q-1}{:}s^{q-2}][s^q{:}s_2^{q-1}].$$

Hint: use 4 5

B. CELLS AND SPHERES.

1. Using the notations of I,16, consider the map h_i: $(E^n,S^{n-1}) \to (E^n,S^{n-1})$ defined by $h_i(x_1, \cdots, x_i, \cdots, x_n) = (x_1, \cdots, -x_i, \cdots, x_n)$. Prove that $h_{i*}z = -z$ for $z \in H_n(E^n,S^{n-1})$.

2 Consider the map f_i. $S^{n-1} \to S^{n-1}$ defined by h_i in problem 1. Prove that $f_{i*}z = -z$ for $z \in H_{n-1}(S^{n-1})$

3. Consider the map h: $(E^n,S^{n-1}) \to (E^n,S^{n-1})$ defined in vector notation by $h(x) = -x$. Prove that $h_*z = (-1)^n z$ for $z \in H_n(E^n,S^{n-1})$.

4. Consider the map f: $S^{n-1} \to S^{n-1}$ defined by h in problem 3. Prove that $f_* z = (-1)^n z$ for $z \in H_{n-1}(S^{n-1})$.

C. THE EULER CHARACTERISTIC.

Assume that the values of the homology theory are D-modules (see exercises I,H and I,I), and assume that the coefficient group has rank 1.

1. Let (X,A) be a pair with a triangulation $T = \{t,(K,L)\}$. Let α_q be the number of q-simplexes of K which are not in L Show that the Euler characteristic $\chi(X,A)$ exists and satisfies

$$\chi(X,A) = \sum_{q=0}^{n} (-1)^q \alpha_q$$

where $n = \dim K$. (Hint: prove $r[C_q(K,L)] = \alpha_q$)

D. n-CIRCUITS

DEFINITION A simplicial pair (K,L) is called a *simple n-circuit* if (1) each point of K is a point of some n-simplex of K, (2) each $(n-1)$-simplex of $K - L$ is a face of two n-simplexes of $K - L$, and (3) if σ and τ are two n-simplexes of $K - L$, there exists an ordered set

$$\sigma = \sigma_1, \sigma_2, \cdots, \sigma_k = \tau$$

of alternately n and $(n-1)$-simplexes of $K - L$ such that each $(n-1)$-simplex is a face of the neighboring n-simplexes

1 Assume that the coefficient group G of the cohomology theory is infinite cyclic (i.e. isomorphic to the integers) Let $A^0, \cdots, A^n$ be the vertices of an n-simplex of $K - L$ where (K,L) is a simple n-circuit If $g \in G$, show that $gA^0 \cdots A^n$ is an n-cocycle of K mod L (see Remark after 5.8c), and show that any n-cocycle of K mod L is cohomologous to one of this form Show that $H^n(K,L)$ is either infinite cyclic or cyclic of order 2. In the first case (K,L) is called an *orientable* n-circuit, in the second case, *nonorientable.*

2. Let (K,L) be a simple n-circuit and let the coefficient group be infinite cyclic If (K,L) is orientable, show that $H_n(K,L)$ is infinite cyclic, and $H_{n-1}(K,L)$ has no elements of finite order. If (K,L) is nonorientable, show that $H_n(K,L) = 0$, and the subgroup of $H_{n-1}(K,L)$ of elements of finite order is cyclic of order 2.

3 Define real projective n-space P^n to be the space obtained from an n-sphere by identifying each point with its antipode Show that P^n is triangulable as an n-circuit which is orientable or not according as n is odd or even

4. Show that the cartesian product of any finite number of spheres can be triangulated so as to be an orientable simple n-circuit

5. Let K be a 2-circuit. Show that the numbers α_0, α_1, α_2 of simplexes of dimensions 0, 1, and 2 satisfy the conditions

$$\begin{aligned} 3\alpha_2 &= 2\alpha_1, \\ \alpha_1 &= 3(\alpha_0 - \chi(K)), \\ \alpha_0 &\geqq \tfrac{1}{2}\left(7 + \sqrt{49 - 24\chi(K)}\right). \end{aligned}$$

6. Show that 2-sphere S^2, projective plane P^2, and the torus T^2 can be triangulated as simple 2-circuits Their Euler characteristics are respectively 2, 1, and 0 Show that, for any such triangulation, the following inequalities hold:

$$\begin{array}{llll} S^2. & \alpha_0 \geqq 4, & \alpha_1 \geqq 6, & \alpha_2 \geqq 4 \\ P^2. & \alpha_0 \geqq 6, & \alpha_1 \geqq 15, & \alpha_2 \geqq 10. \\ T^2: & \alpha_0 \geqq 7, & \alpha_1 \geqq 21, & \alpha_2 \geqq 14 \end{array}$$

Find triangulations in which the minimal values are attained.

CHAPTER IV

Categories and functors

1. INTRODUCTION

The first objective of this chapter is to introduce and illustrate the concepts of *category, functor*, and related notions These are needed in subsequent chapters to facilitate the statements of uniqueness and existence theorems Only as much of the subject is included as is used in the sequel. A thorough treatment can be found in a paper of Eilenberg and MacLane [Trans Amer. Math Soc 58 (1945), 231-294]

The ideas of category and functor inspired in part the axiomatic treatment of homology theory given in this book In addition, the point of view that these ideas engender has controlled its development at every stage.

The second part of the chapter is devoted to homology theories defined on abstract h-categories An admissible category for homology theory, as defined in I,1, has much more structure than is needed for the statement of the axioms for a homology theory. Abstracting the essential elements leads to the notion of an h-category. There is a corresponding concept of h-functor It is shown that the composition of an h-functor and a homology theory, when defined, is again a homology theory. We thereby obtain a rule for deriving one homology theory from another. It is used frequently in subsequent chapters.

2. CATEGORIES

The definitions below arise from the consideration of the common properties of collections such as (1) topological spaces and their continuous mappings, (2) groups and their homomorphisms, and (3) simplicial complexes and their simplicial maps. An examination of the properties of continuous maps, homomorphisms, and simplicial maps leads to the following definition.

DEFINITION 2 1. A set $\mathcal{C}$ of elements $\{\gamma\}$ is called a *multiplicative system* if, for some pairs $\gamma_1,\gamma_2 \varepsilon \mathcal{C}$, a *product* $\gamma_2\gamma_1 \varepsilon \mathcal{C}$ is defined. An element $\epsilon \varepsilon \mathcal{C}$ is called an *identity* (or a *unit*) if $\epsilon\gamma_1 = \gamma_1$ and $\gamma_2\epsilon = \gamma_2$ whenever $\epsilon\gamma_1$ and $\gamma_2\epsilon$ are defined. The multiplicative system is called an *abstract category* if the following axioms are satisfied:

(1) The triple product $\gamma_3(\gamma_2\gamma_1)$ is defined if and only if $(\gamma_3\gamma_2)\gamma_1$ is defined. When either is defined the associative law

$$\gamma_3(\gamma_2\gamma_1) = (\gamma_3\gamma_2)\gamma_1$$

holds This triple product will be written as $\gamma_3\gamma_2\gamma_1$.

(2) The triple product $\gamma_3\gamma_2\gamma_1$ is defined whenever both products $\gamma_3\gamma_2$ and $\gamma_2\gamma_1$ are defined.

(3) For each $\gamma \in \mathfrak{C}$ there exist identities $\epsilon_1, \epsilon_2 \in \mathfrak{C}$ such that $\gamma\epsilon_1$ and $\epsilon_2\gamma$ are defined.

LEMMA 2 2. *For each $\gamma \in \mathfrak{C}$, the identities ϵ_1 and ϵ_2 such that $\gamma\epsilon_1$ and $\epsilon_2\gamma$ are defined, are unique.*

PROOF. Suppose $\gamma\epsilon_1$ and $\gamma\epsilon_1'$ are both defined with ϵ_1, ϵ_1' both identities Then $(\gamma\epsilon_1)\epsilon_1' = \gamma\epsilon_1'$ is defined. Hence $\epsilon_1\epsilon_1'$ is defined and $\epsilon_1 = \epsilon_1\epsilon_1' = \epsilon_1'$. Similarly ϵ_2 is unique.

DEFINITION 2.3. A *category* $\mathfrak{C}$ consists of a collection $\{C\}$ of elements called *objects* and a collection $\{\gamma\}$ of elements called *mappings* The mappings form an *abstract category* in the sense of 2.1 The objects are in a 1-1 correspondence $C \to i_C$ with the set of identities of the abstract category. Thus to each mapping γ there correspond unique objects C_1 and C_2 such that γi_{C_1} and $i_{C_2}\gamma$ are defined The objects C_1, C_2 are called the *domain* and the *range* of γ respectively; notations $C_1 = D(\gamma)$, $C_2 = R(\gamma)$, γ: $C_1 \to C_2$.

LEMMA 2 4. *The product $\gamma_2\gamma_1$ is defined if and only if $R(\gamma_1) = D(\gamma_2)$. If $\gamma_2\gamma_1$ is defined, then $R(\gamma_2\gamma_1) = R(\gamma_2)$, $D(\gamma_2\gamma_1) = D(\gamma_1)$.*

In other words the lemma states that, if γ_1: $C_1 \to C_2$, γ_2: $C_2' \to C_3$, then $\gamma_2\gamma_1$ is defined if and only if $C_2 = C_2'$ and then $\gamma_2\gamma_1$: $C_1 \to C_3$.

PROOF. Let $\epsilon = i_{R(\gamma_1)}$ If $\gamma_2\gamma_1$ is defined, then $\gamma_2(\epsilon\gamma_1) = \gamma_2\gamma_1$ is defined. Thus $\gamma_2\epsilon$ is defined and $R(\gamma_1) = D(\gamma_2)$. Conversely, if $R(\gamma_1) = D(\gamma_2)$, then $\gamma_2\epsilon$ and $\epsilon\gamma_1$ are both defined for $\epsilon = i_{R(\gamma_1)}$ and therefore $(\gamma_2\epsilon)\gamma_1 = \gamma_2\gamma_1$ is defined. If $\gamma_2\gamma_1$ is defined and $\epsilon = i_{D(\gamma_1)}$, then $\gamma_1\epsilon$ is defined. Hence $\gamma_2\gamma_1\epsilon = (\gamma_2\gamma_1)\epsilon$ is defined so that $D(\gamma_1) = D(\gamma_2\gamma_1)$. Similarly $R(\gamma_2\gamma_1) = R(\gamma_2)$.

DEFINITION 2.5 A map γ: $C_1 \to C_2$ in $\mathfrak{C}$ is called an *equivalence* if there exists a map γ': $C_2 \to C_1$ in $\mathfrak{C}$ such that $\gamma'\gamma = i_{C_1}$, $\gamma\gamma' = i_{C_2}$.

LEMMA 2.6. *The map γ' of 2 5 is unique and is called the inverse of γ, notation: γ^{-1}. γ^{-1} is also an equivalence with $(\gamma^{-1})^{-1} = \gamma$. If γ_1: $C_1 \to C_2$, γ_2: $C_2 \to C_3$ are equivalences, then $\gamma_2\gamma_1$ is an equivalence with $(\gamma_2\gamma_1)^{-1} = \gamma_1^{-1}\gamma_2^{-1}$. Each identity ϵ is an equivalence with $\epsilon^{-1} = \epsilon$.*

PROOF. Suppose that γ' and γ'' satisfy $\gamma'\gamma = \gamma''\gamma = \epsilon_{C_1}$, $\gamma\gamma' = \gamma\gamma'' = \epsilon_{C_2}$. Then

$$\gamma' = \gamma'\epsilon_{C_2} = \gamma'\gamma\gamma'' = \epsilon_{C_1}\gamma'' = \gamma''.$$

The fact that γ^{-1} is an equivalence and that $(\gamma^{-1})^{-1} = \gamma$ follows directly from the definition. If γ_2,γ_1 are equivalences and $\gamma_2\gamma_1$ is defined, then $(\gamma_2\gamma_1)(\gamma_1^{-1}\gamma_2^{-1})$ is defined and equal to ϵ_{c_2}, similarly $(\gamma_1^{-1}\gamma_2^{-1})(\gamma_2\gamma_1) = \epsilon_{c_1}$. Thus $(\gamma_2\gamma_1)^{-1} = \gamma_1^{-1}\gamma_2^{-1}$. If ϵ is an identity and $\epsilon' = \epsilon_{D(\epsilon)}$, then $\epsilon\epsilon'$ is defined and $\epsilon' = \epsilon\epsilon' = \epsilon$. Thus $\epsilon = \epsilon\epsilon$ so that $\epsilon = i_{D(\epsilon)} = i_{R(\epsilon)}$ and $\epsilon^{-1} = \epsilon$.

DEFINITION 2.7. A *subcategory* $\mathcal{C}_0$ of $\mathcal{C}$ is a subaggregate of $\mathcal{C}$ such that

1°. if $C \in \mathcal{C}_0$, then $i_C \in \mathcal{C}_0$,

2°. if $\gamma_2,\gamma_1 \in \mathcal{C}_0$ and $\gamma_2\gamma_1$ is defined, then $\gamma_2\gamma_1 \in \mathcal{C}_0$, and

3°. if $\gamma \in \mathcal{C}_0$, then $D(\gamma)$ and $R(\gamma)$ are in $\mathcal{C}_0$.

A subcategory $\mathcal{C}_0$ of $\mathcal{C}$ is called *full* if, for each $\gamma \in \mathcal{C}$, conditions $D(\gamma) \in \mathcal{C}_0$, $R(\gamma) \in \mathcal{C}_0$ imply $\gamma \in \mathcal{C}_0$

It is easy to see that a subcategory is itself a category However a map $\gamma \in \mathcal{C}_0$ may be an equivalence in $\mathcal{C}$ without being an equivalence in $\mathcal{C}_0$.

The process of obtaining a subcategory may be broken up in two steps. First select a subset of the objects of $\mathcal{C}$ and consider the full subcategory determined by these objects. Then without further limiting the objects select a subaggregate of maps satisfying 1° and 2°.

3. EXAMPLES OF CATEGORIES

The first example of a category is composed of topological spaces and continuous maps The objects are topological spaces, the maps are continuous maps of one topological space into another. The composition (product) of two maps is the usual function of a function The equivalences in this category are the homeomorphic maps of one space onto another

The category of prime importance in our axiomatic treatment has, as its objects, pairs (X,A) where X is a topological space and $A \subset X$; and has, as its mappings, continuous maps $f. (X,A) \to (Y,B)$ This category will be denoted by $\mathfrak{A}_1$ The admissible categories discussed in I,1 are all subcategories of $\mathfrak{A}_1$. In an admissible category $\mathfrak{A}$ the equivalences are precisely the homeomorphisms as defined in I,5.

In I,2 we encountered the categories $\mathcal{G}_C$ and $\mathcal{G}_R$, the objects in these categories are, respectively, compact abelian groups, and modules over a ring R. The mappings in these categories are the continuous homomorphisms in the first case, and linear maps in the second. When R is the ring of integers, $\mathcal{G}_R$ is just the category $\mathcal{G}$ of ordinary abelian groups and their homomorphisms

For the categories $\mathcal{G}_C,\mathcal{G}_R$ we may consider the categories $\mathcal{S}_i\mathcal{G}_C,\mathcal{S}_i\mathcal{G}_R$

whose objects are lower sequences of groups in $\mathcal{G}_C$ or $\mathcal{G}_R$. The mappings are the homomorphisms of one such lower sequence into another The exact lower sequences form subcategories $\mathcal{E}_l\mathcal{G}_C, \mathcal{E}_l\mathcal{G}_R$. The similarly defined categories of upper sequences and exact upper sequences will be denoted by $\mathcal{S}_u\mathcal{G}_C, \mathcal{E}_u\mathcal{G}_C$, etc.

The categories studied in Chapter II are denoted by $\mathcal{K}_l, \mathcal{K}_s$; the objects in both these categories are pairs (K,L) where K is a simplicial complex and L is a subcomplex The mappings in $\mathcal{K}_l$ are linear mappings, while in $\mathcal{K}_s$ they are simplicial. Clearly $\mathcal{K}_s$ is a subcategory of $\mathcal{K}_l$ Theorem II,4 8 states that every equivalence in $\mathcal{K}_l$ is in $\mathcal{K}_s$.

In III,1 the important category $\mathfrak{J}$ of triangulable pairs is introduced. The objects are triangulable pairs (X,A), the mappings are continuous maps of one such pair into another. Clearly $\mathfrak{J}$ is a full subcategory of $\mathcal{A}_1$ and is an admissible category for homology theory.

4. FUNCTORS

Let $\mathcal{C}$ and $\mathcal{D}$ be categories and let T be a function which maps the objects of $\mathcal{C}$ into the objects of $\mathcal{D}$ and, in addition, assigns to each map $f \varepsilon \mathcal{C}$ a map $T(f) \varepsilon \mathcal{D}$ The map T is called a *covariant functor* (from $\mathcal{C}$ to $\mathcal{D}$) if it satisfies the following conditions

1° If f $C_1 \to C_2$, then $T(f)$ $T(C_1) \to T(C_2)$.

2° $T(i_C) = i_{T(C)}$

3°. If f_2f_1 is defined, then $T(f_2f_1) = T(f_2)T(f_1)$.

The map T is called a *contravariant functor* if these conditions are replaced by

1′ If f $C_1 \to C_2$, then $T(f)$ $T(C_2) \to T(C_1)$.

2′ $T(i_C) = i_{T(C)}$.

3′. If f_2f_1 is defined, then $T(f_2f_1) = T(f_1)T(f_2)$.

The condition 1° can be rewritten $T(Df) = DT(f)$ and $T(Rf) = RT(f)$ Thus T is a covariant functor if it commutes with the operations of the category

In view of condition 2° a functor T is completely determined by the function $T(f)$ defined for maps f only Thus a covariant functor T is essentially a homomorphism of the abstract category associated with $\mathcal{C}$ into the abstract category of $\mathcal{D}$ subject to the condition that identities be mapped into identities A contravariant functor yields an anti-homomorphism of the abstract categories

If T is a functor from $\mathcal{C}$ to $\mathcal{D}$, and T' is a functor from $\mathcal{D}$ to $\mathcal{E}$, they can be composed in the obvious way to form a functor $T'T$ from $\mathcal{C}$ to $\mathcal{E}$. If TT' have the same (opposite) variance, then $T'T$ is covariant (contravariant).

5. EXAMPLES OF FUNCTORS

Let $\mathcal{A}$ be an admissible category on which a homology theory is given. Let q be a fixed integer, and define for an admissible map $f\colon (X,A) \to (Y,B)$

$$H_q(f) = f_*\colon \ H_q(X,A) \to H_q(Y,B).$$

Then Axioms 1 and 2 for a homology theory assert that the pair of functions $H_q(X,A), H_q(f)$ is a covariant functor H_q on the category $\mathcal{A}$ with values in the category $\mathcal{G}_R$ or $\mathcal{G}_C$.

Instead of using the category $\mathcal{G}_R$ we may use the category $\mathcal{E}_l\mathcal{G}_R$ of exact lower sequences in $\mathcal{G}_R$. We then define $H(X,A)$ to be the homology sequence of (X,A), and $H(f)$ to be the homomorphism f_{**} of the H.S. of (X,A) into that of (Y,B) induced by f (see I,4.1). Then Axioms 1, 2, 3, and 4 insure that H is a covariant functor on $\mathcal{A}$ to $\mathcal{E}_l\mathcal{G}_R$ or $\mathcal{E}_l\mathcal{G}_C$. This functor is referred to briefly as the *homology functor*.

In a similar way, the *cohomology functor* is a contravariant functor on $\mathcal{A}$ to $\mathcal{E}_u\mathcal{G}_R$ or $\mathcal{E}_u\mathcal{G}_C$.

Another covariant functor on $\mathcal{A}$, this time with values also in $\mathcal{A}$, is obtained by setting

$$T(X,A) = A, \qquad T(f) = f|A$$

where $f|A$ is the map $A \to B$ defined by a map $f\colon (X,A) \to (Y,B)$.

A covariant functor on $\mathcal{K}_s$ to $\mathcal{A}_C$ (compact pairs) is obtained by setting

$$T(K,L) = (|K|,|L|), \qquad T(f) = f.$$

6. TRANSFORMATIONS OF FUNCTORS

Let T and S be two covariant functors from $\mathcal{C}$ to $\mathcal{D}$. A function Γ which assigns to each object $C \in \mathcal{C}$ a map $\Gamma(C) \in \mathcal{D}$ such that

1°. $\Gamma(C)\colon \ T(C) \to S(C)$,

2°. if $f\colon \ C_1 \to C_2$, then $\Gamma(C_2)T(f) = S(f)\Gamma(C_1)$,

is called a *natural transformation* of the functor T into the functor S. In case T and S are contravariant functors, the condition 2° on Γ is replaced by

2′. if $f\colon \ C_1 \to C_2$, then $\Gamma(C_1)T(f) = S(f)\Gamma(C_2)$.

If the map $\Gamma(C)$ is an equivalence for each $C \in \mathcal{C}$, then Γ is called a *natural equivalence* of the functors T and S (notation: $\Gamma\colon \ T \rightleftarrows S$). In this case, 2° can be written

$$\Gamma(C_2)T(f)\Gamma(C_1)^{-1} = S(f).$$

The condition 1° is equivalent to the condition that the compositions in 2° are always defined. Condition 2° asserts that commutativity holds in the following diagram

$$\begin{array}{ccc} T(C_1) & \xrightarrow{T(f)} & T(C_2) \\ \downarrow{\scriptstyle \Gamma(C_1)} & & \downarrow{\scriptstyle \Gamma(C_2)} \\ S(C_1) & \xrightarrow{S(f)} & S(C_2) \end{array}$$

It is easily verified that the composition of two natural transformations is also natural. If Γ: $T \to S$ is a natural equivalence, then Γ^{-1}: $S \to T$ defined by $\Gamma^{-1}(C) = [\Gamma(C)]^{-1}$ is also a natural equivalence, and $\Gamma_0 = \Gamma\Gamma^{-1} = \Gamma^{-1}\Gamma$ has the property $\Gamma_0(C) = \iota_{T(C)}$. Thus Γ_0: $T \rightleftarrows T$. It follows that the concept of natural equivalence is reflexive, symmetric, and transitive.

7. EXAMPLES OF TRANSFORMATIONS OF FUNCTORS

Let $\mathcal{A}$ be an admissible category on which a homology functor H is given, and let T be the functor from $\mathcal{A}$ to $\mathcal{A}$ defined by

$$T(X,A) = A, \qquad T(f) = f|A.$$

Let q be a positive integer and let H_q, H_{q-1} be the homology functors from $\mathcal{A}$ to the category $\mathcal{G}_R$ (or $\mathcal{G}_C$). Then Axiom 3: $\partial f_* = (f|A)_*\partial$ is precisely the condition 2° that the homomorphism

$$\partial\colon\ H_q(X,A) \to H_{q-1}(A) = H_{q-1}(T(X,A))$$

shall be a natural transformation of the functor H_q into the composite functor $H_{q-1}T$.

In Chapter III it was proved that ∂ is an equivalence ∂: $H_q \rightleftarrows H_{q-1}T$ over the subcategory of pairs $(|s|,|\dot{s}|)$ where s is a simplex and $\dot{s}$ is its boundary.

The group of q-chains $C_q(K)$ is a covariant functor on $\mathcal{K}_s$ to $\mathcal{G}_R$ (or $\mathcal{G}_C$), provided we define $C_q(f) = f_q$ for each simplicial map f: $(K,L) \to (K_1,L_1)$. In the same fashion the groups $Z_q(K,L)$ of the q-cycles mod L and $B_q(K,L)$ of the bounding q-cycles mod L are functors on $\mathcal{K}_s$ to $\mathcal{G}_R$ (or $\mathcal{G}_C$). Since $B_q(K,L) \subset Z_q(K,L)$, the former is called a *sub-functor* of the latter. The *quotient functor* $H_q(K,L) = Z_q(K,L)/B_q(K,L)$ has the same domain and range. The map $\nu_q(K,L)$: $Z_q(K,L) \to H_q(K,L)$ which assigns to each cycle its homology class is a natural transformation of Z_q into H_q.

In order to distinguish this q-functor from the qth homology functor defined on $\mathcal{A}$ to $\mathcal{G}_R$ we shall write $\bar{H}_q$ for the qth homology functor on $\mathcal{K}_s$.

For each simplicial pair (K,L) we may regard $(|K|,|L|)$ as a triangulated pair with the triangulation $T = \{t,(K,L)\}$ where t is the identity map of $(|K|,|L|)$. The isomorphisms $\theta = \theta_T \quad H_q(|K|,|L|) \to H_q(K,L)$, defined in III,8, yield a natural equivalence of two functors

$$\theta: \quad H_q T \rightleftarrows \bar{H}_q$$

where T is the covariant functor on $\mathcal{K}_s$ to $\mathcal{A}$ defined by $T(K,L) = (|K|,|L|)$, $T(f) = f$. Theorem II,8.4 is the statement of the naturality of θ.

8. c-CATEGORIES AND ∂-FUNCTORS

In order to achieve uniformity and avoid repetition in the chapters ahead, it is useful to introduce concepts similar to the homology and cohomology theories of Chapter I but defined on more general categories than admissible categories for homology. The analogy with homology and cohomology theories is complete as far as the first four axioms are concerned

DEFINITION 8.1. A *category with couples* (briefly: a *c*-category) is a category $\mathcal{C}$ in which certain pairs (α,β) of maps, called *couples*, are distinguished, subject to the sole condition that the composition $\beta\alpha$ is defined If α. $A \to B$, β: $B \to C$ is such a couple, we write

$$(\alpha,\beta): \quad A \to B \to C.$$

A covariant [contravariant] functor T. $\mathcal{C} \to \mathfrak{D}$ of a c-category into a c-category, is called a *c-functor* if for each couple (α,β) in $\mathcal{C}$, the pair of maps $(T\alpha,T\beta)$ $[(T\beta,T\alpha)]$ is a couple in $\mathfrak{D}$.

As an example of a c-category consider an admissible category $\mathcal{A}$ in the sense of I,1. For each pair (X,A) in $\mathcal{A}$ consider the inclusion maps i: $A \subset X$, j. $X \subset (X,A)$ as forming a couple

$$(i,j): \quad A \to X \to (X,A).$$

The same admissible category $\mathcal{A}$ gives rise to another c-category in which the couples

$$(i,j): \quad (A,B) \to (X,B) \to (X,A)$$

are the appropriate inclusion maps of an admissible triple (X,A,B).

Another example of a c-category is obtained from the category $\mathcal{G}_R$ (or $\mathcal{G}_C$) of groups by defining couples

$$(\phi,\psi): \quad G_1 \to G_2 \to G_3$$

whenever ϕ: $G_1 \to G_2$ has kernel zero, ψ: $G_2 \to G_3$ is onto, and kernel ψ = image ϕ, i e whenever the sequence

$$0 \to G_1 \xrightarrow{\phi} G_2 \xrightarrow{\psi} G_3 \to 0$$

is exact.

DEFINITION 8 2. Let (α,β): $A \to B \to C$ and (α_1,β_1): $A_1 \to B_1 \to C_1$ be couples in a c-category $\mathfrak{C}$ We define a *map* $(\alpha,\beta) \to (\alpha_1,\beta_1)$ to be a triple of maps

$$\gamma_1\colon A \to A_1, \qquad \gamma_2 \quad B \to B_1, \qquad \gamma_3\cdot \quad C \to C_1$$

in $\mathfrak{C}$ such that commutativity holds in the two squares of the diagram

$$\begin{array}{ccccc} A & \xrightarrow{\alpha} & B & \xrightarrow{\beta} & C \\ \downarrow{\scriptstyle\gamma_1} & & \downarrow{\scriptstyle\gamma_2} & & \downarrow{\scriptstyle\gamma_3} \\ A_1 & \xrightarrow{\alpha_1} & B_1 & \xrightarrow{\beta_1} & C_1 \end{array}$$

It is easy to verify that, with the maps thus defined, the couples (α,β) in C form a category of their own A c-functor T: $\mathfrak{C} \to \mathfrak{D}$ induces a functor on the category of couples of $\mathfrak{C}$ into that of $\mathfrak{D}$.

We now come to the main objective of this section Let $\mathfrak{C}$ be a c-category We shall consider systems $H = \{H_q(A),\alpha_*,\partial_{(\alpha\ \beta)}\}$ where (1) for each object $A \in \mathfrak{C}$ and each integer q, $H_q(A)$ is a group, (2) for each map α: $A \to B$ in $\mathfrak{C}$ and each integer q, α_*: $H_q(A) \to H_q(B)$ is a homomorphism, (3) for each couple (α,β) $A \to B \to C$ and each integer q, $\partial_{(\alpha\ \beta)}$. $H_q(C) \to H_{q-1}(A)$ is a homomorphism. The groups and homomorphisms belong to just one of the categories $\mathcal{G}_R$ or $\mathcal{G}_C$.

Such a system H will be called a *covariant ∂-functor* on the c-category $\mathfrak{C}$ provided the following four axioms hold:

AXIOM 1. If α = identity, then α_* = identity.

AXIOM 2. $(\beta\alpha)_* = \beta_*\alpha_*$.

AXIOM 3. If $\gamma_1,\gamma_2,\gamma_3$ form a map of the couple (α,β): $A \to B \to C$ into the couple (α_1,β_1): $A_1 \to B_1 \to C_1$, then commutativity holds in the diagram

$$\begin{array}{ccc} H_q(C) & \xrightarrow{\gamma_{3*}} & H_q(C_1) \\ \downarrow{\scriptstyle\partial_{(\alpha,\beta)}} & & \downarrow{\scriptstyle\partial_{(\alpha_1\ \beta_1)}} \\ H_{q-1}(A) & \xrightarrow{\gamma_{1*}} & H_{q-1}(A_1) \end{array}$$

AXIOM 4. For every couple (α,β): $A \to B \to C$ the sequence

$$\cdots \leftarrow H_{q-1}(A) \overset{\partial}{\leftarrow} H_q(C) \overset{\beta_*}{\leftarrow} H_q(B) \overset{\alpha_*}{\leftarrow} H_q(A) \leftarrow \cdots$$

is exact

These four axioms are precise replicas of the first four axioms for a homology theory.

In an analogous fashion we define a *contravariant δ-functor* $H = \{H^q(A),\alpha^*,\delta_{(\alpha,\beta)}\}$ where α^*. $H^q(B) \to H^q(A)$ for α: $A \to B$ and $\delta_{(\alpha,\beta)}$. $H^q(A) \to H^{q+1}(C)$ for (α,β): $A \to B \to C$. The axioms are replicas of the first four axioms of cohomology

Let H be a covariant ∂-functor on a c-category $\mathfrak{D}$ and let T. $\mathfrak{C} \to \mathfrak{D}$ be a covariant c-functor. The composition

$$HT = \{H_q(TA),(T\alpha)_*,\partial_{(T\alpha,T\beta)}\}$$

is then clearly a covariant ∂-functor on $\mathfrak{C}$ Similarly if H is a contravariant δ-functor on $\mathfrak{D}$ and T. $\mathfrak{C} \to \mathfrak{D}$ is covariant, then

$$HT = \{H^q(TA),(T\alpha)^*,\delta_{(\alpha,T\beta)}\}$$

still is a contravariant δ-functor on $\mathfrak{C}$. Thus the composition HT is well defined whenever T is covariant.

If T is contravariant, HT may still be defined formally as above but the result is neither a covariant ∂-functor nor a contravariant δ-functor This suggests the definition of two additional "mixed" types of functors, namely of covariant δ-functors and of contravariant ∂-functors. In a covariant δ-functor $H = \{H^q(A),\alpha_*,\delta_{(\alpha,\beta)}\}$ we have α_*: $H^q(A) \to H^q(B)$ for α: $A \to B$ and $\delta_{(\alpha,\beta)}$: $H^q(C) \to H^{q+1}(A)$ for (α,β): $A \to B \to C$. In a contravariant ∂-functor $H = \{H_q(A), \alpha^*,\partial_{(\alpha,\beta)}\}$ we have α^*: $H_q(B) \to H_q(A)$ for α. $A \to B$ and $\partial_{(\alpha,\beta)}$. $H_q(A) \to H_{q-1}(C)$ for (α,β) $A \to B \to C$ The axioms required are the obvious reformulations of the four axioms listed above.

With the mixed theories at hand, the composition HT is always defined The situation is summarized in the following theorem, the proof of which is immediate.

THEOREM 8 3 *Let T: $\mathfrak{C} \to \mathfrak{D}$ be a c-functor and H a ∂-functor [δ-functor] on $\mathfrak{D}$ The composition HT is then a ∂-functor [δ-functor] on $\mathfrak{C}$. Further HT is covariant if H and T have the same variance, and HT is contravariant if H and T have opposite variances.*

Each homology theory on an admissible category $\mathfrak{a}$ furnishes an example of a covariant ∂-functor on either of the c-categories associated with $\mathfrak{a}$ Similarly each cohomology theory on $\mathfrak{a}$ furnishes an example of a contravariant δ-functor Examples of mixed functors will occur in Chapter v

The connection between the mixed functors and the unmixed ones can be further illuminated by a procedure that will be referred to as "the sign-changing trick" Given a covariant ∂-functor $H = \{H_q(A),\alpha_*,\partial_{(\alpha,\beta)}\}$, define a covariant δ-functor $\bar{H} = \{\bar{H}^q(A),\bar{\alpha}_*,\bar{\delta}_{(\alpha,\beta)}\}$ as follows:

$$\bar{H}^q(A) = H_{-q}(A), \qquad \bar{\alpha}_* = \alpha_*, \qquad \bar{\delta}_{(\alpha,\beta)} = \partial_{(\alpha,\beta)}$$

(precisely, $\bar{\alpha}_*$: $\bar{H}^q(A) \to \bar{H}^q(B)$ is defined as α_*: $H_{-q}(A) \to H_{-q}(B)$ etc) Conversely applying the sign-changing trick to $\bar{H}$ we obtain H Thus the correspondence $H \to \bar{H}$ establishes a 1-1 correspondence between covariant ∂-functors and covariant δ-functors A similar discussion applies to contravariant functors

9. h-CATEGORIES AND h-FUNCTORS

The introduction of the concept of a c-category was a step toward the consideration of homology theories defined on categories more general and abstract than on admissible categories of spaces We saw that the first four axioms can be formulated with ease in this more general setting As for the remaining axioms, it is clear that they cannot be stated since, in a c-category, we lack the concepts of "homotopy," "excision," and "point" To complete the picture, we make the following definition.

DEFINITION 9 1. An *h-category* $\mathcal{C}$ is a c-category in which (i) a binary relation $\alpha \simeq \beta$(α *homotopic* to β) is given for maps α,β $A \to B$ in $\mathcal{C}$, (ii) certain maps α. $A \to B$ in $\mathcal{C}$ are singled out and are called *excisions*, and (iii) certain objects of $\mathcal{C}$ are singled out and are called *points*. A covariant ∂-functor [contravariant δ-functor] on $\mathcal{C}$ which satisfies the analogs of the Homotopy, Excision, and Dimension axioms will be called a *homology* [*cohomology*] *theory* on the h-category $\mathcal{C}$.

It should be noted that since no properties of homotopies, excisions, and points are required, the above definition has a purely formal character, and is made only to supply a language convenient for later uses.

DEFINITION 9 2 Let α. $A \to B$, β. $B \to A$ be two maps in an h-category $\mathcal{C}$. If $\beta\alpha$ $A \to A$ and $\alpha\beta$ $B \to B$ are both homotopic to identity maps, then α and β are both called *homotopy equivalences*, β is called a *homotopy inverse* of α and vice versa A map α $A \to B$ in $\mathcal{C}$ which is a composition of a finite number of excisions and homotopy equivalences is called a *generalized excision*.

Theorem 9.3. *If H is a homology [cohomology] theory on an h-category $\mathfrak{C}$ and α: $A \to B$ is a generalized excision, then*

$$\alpha_*: H_q(A) \approx H_q(B) \qquad [\alpha^*: H^q(B) \approx H^q(A)].$$

Proof. Let $\alpha = \alpha^1 \cdots \alpha^n$ be a representation of α as a composition of excisions and homotopy equivalences. Since $\alpha_* = \alpha^1_* \cdots \alpha^n_*$, it suffices to show that each α^i_* is an isomorphism. If α^i is an excision, this follows from the Excision axiom. If α^i is a homotopy equivalence with β^i as homotopy inverse, then $\alpha^i_*\beta^i_* = (\alpha^i\beta^i)_* =$ identity and similarly $\beta^i_*\alpha^i_* =$ identity. Thus α^i_* is an isomorphism.

Definition 9.4. A covariant c-functor T: $\mathfrak{C} \to \mathfrak{D}$ on the h-category $\mathfrak{C}$ with values in the h-category $\mathfrak{D}$ is called an *h-functor* if T preserves homotopies, generalized excisions, and points. Explicitly: if $\alpha \simeq \beta$ in $\mathfrak{C}$, then $T\alpha \simeq T\beta$ in $\mathfrak{D}$; if α is a generalized excision in $\mathfrak{C}$, then $T\alpha$ is a generalized excision in $\mathfrak{D}$; and if A is a point in $\mathfrak{C}$, then TA is a point in $\mathfrak{D}$.

Theorem 9.5. *Let T: $\mathfrak{C} \to \mathfrak{D}$ be a covariant h-functor and H a homology [cohomology] theory on $\mathfrak{D}$. Then the composition HT is a homology [cohomology] theory on $\mathfrak{C}$.*

Proof. By 8.3, HT is a covariant ∂-functor [contravariant δ-functor]. There remains to prove that HT satisfies the Homotopy, Excision, and Dimension axioms. This however is an immediate consequence of 9.4 and 9.3.

As in 8.3, Definition 9.4 and Theorem 9.5 could also be formulated for contravariant functors T provided we consider theories H of a mixed character, i.e. covariant δ-functors and contravariant ∂-functors satisfying the analogs of the last three axioms. These mixed theories do not have any geometric analog, but will arise occasionally in later chapters. In particular the following theorem will be used in v,12.

Theorem 9.5c. *Let T: $\mathfrak{C} \to \mathfrak{D}$ be a contravariant h-functor and H a covariant δ-functor [contravariant ∂-functor] on $\mathfrak{D}$ satisfying the Homotopy, Excision, and Dimension axioms. Then the composition HT is a cohomology [homology] theory on $\mathfrak{C}$.*

Remark. In proving that a c-functor T is an h-functor, it suffices to show that T preserves homotopies and points and carries excisions into generalized excisions. Then it follows immediately that T carries generalized excisions into generalized excisions.

10. COMPARISON OF ∂-FUNCTORS AND OF HOMOLOGY THEORIES

In this section the discussion will be limited to covariant ∂-functors with values in the category $\mathcal{G}_R$, but all that will be said applies equally well to the contravariant case as well as to δ-functors with values in $\mathcal{G}_R$ or $\mathcal{G}_C$.

DEFINITION 10.1. Let $H = \{H_q(A),\alpha_*,\partial_{(\alpha,\beta)}\}$ and $\overline{H} = \{\overline{H}_q(A), \overline{\alpha}_*,\overline{\partial}_{(\alpha,\beta)}\}$ be two covariant ∂-functors on the c-category $\mathcal{C}$ with values in $\mathcal{G}_R$. A *homomorphism*

$$h: \ H \to \overline{H}$$

is a family of homomorphisms

$$h(q,A): \ H_q(A) \to \overline{H}_q(A)$$

defined for each $A \varepsilon \mathcal{C}$ and each integer q, subject to the following two conditions. If $\alpha: \ A \to B$ is a mapping in $\mathcal{C}$, then the commutativity relation $\overline{\alpha}_* h(q,A) = h(q,B)\alpha_*$ holds in the diagram

$$\begin{array}{ccc} H_q(A) & \xrightarrow{h} & \overline{H}_q(A) \\ \downarrow{\scriptstyle \alpha_*} & & \downarrow{\scriptstyle \overline{\alpha}_*} \\ H_q(B) & \xrightarrow{h} & \overline{H}_q(B) \end{array}$$

If $(\alpha,\beta): \ A \to B \to C$ is a couple in $\mathcal{C}$, then the commutativity relation $\overline{\partial}_{(\alpha,\beta)} h(q,C) = h(q-1,A)\partial_{(\alpha,\beta)}$ holds in the diagram

$$\begin{array}{ccc} H_q(C) & \xrightarrow{h} & \overline{H}_q(C) \\ \downarrow{\scriptstyle \partial} & & \downarrow{\scriptstyle \overline{\partial}} \\ H_{q-1}(A) & \xrightarrow{h} & \overline{H}_{q-1}(A) \end{array}$$

If each $h(q,A)$ is an isomorphism $H_q(A) \approx \overline{H}_q(A)$, then we say that h is an *isomorphism* and write $h: \ H \approx \overline{H}$.

The above definition applies automatically to homology theories since a homology theory is a ∂-functor defined on an admissible category (or an h-category) and satisfying additional axioms. Theorems III,10.1 and 10.1c may now be restated as follows.

THEOREM 10.2. *Let H and $\overline{H}$ be two homology [cohomology] theories defined on the category $\mathcal{T}$ of triangulable pairs. Every homomorphism*

$$h_0: \ G \to \overline{G}$$

of their coefficient groups admits a unique extension to a homomorphism

$$h: \ H \to \overline{H} \text{ on } \mathcal{T}.$$

If h_0 is an isomorphism, then so is h.

DEFINITION 10 3. An admissible category $\mathcal{A}$ is called a *uniqueness category* for homology [cohomology] if, for any two homology [cohomology] theories defined over $\mathcal{A}$, each homomorphism of the coefficient groups

$$h_0\colon\ G \to \overline{G}$$

can be extended uniquely to a homomorphism,

$$h\quad H \to \overline{H} \text{ on } \mathcal{A}.$$

Whenever h_0 is an isomorphism, so also is h This is shown by extending h_0^{-1} to h', and observing that $h'h$ and hh' are extensions of identities $G \to G$, $\overline{G} \to \overline{G}$. Then uniqueness insures that hh' and $h'h$ are identities.

In this terminology 10 2 may be restated as

THEOREM 10.4. *The category $\mathcal{J}$ of triangulable pairs and their maps is a uniqueness category both for homology and cohomology*

In Chapter XII a considerably larger uniqueness category $\mathcal{U}$ will be constructed

With the uniqueness theorem at hand, one can see why the mixed types of homology theories are not considered for categories of spaces. Suppose, for example, that H is a covariant δ-functor defined on the category $\mathcal{J}$ satisfying the Homotopy, Excision, and Dimension axioms. Apply the sign-changing trick of §8 to H There results a homology theory $\overline{H}$ on $\mathcal{J}$. By the uniqueness theorem, $\overline{H}_q(X,A) = 0$ for $q < 0$ It follows that $H_q(X,A) = 0$ for $q > 0$ Thus, for mixed theories, the positive dimensional groups are trivial.

NOTES

Logical foundations The categories of "all topological spaces" and "all groups," if not handled carefully, lead to the antinomies usually associated with a "set of all sets." To avoid these, one must either restrict the notion of a category and not speak of "the category of *all* groups," or else place oneself in an axiomatic system in which the "category of all groups" is a legitimate object not leading to antinomies. Best suited for the latter purpose seems to be the system of von Neumann-Bernays-Godel [P Bernays, *A system of axiomatic set theory*, Journal of Symbolic Logic 2 (1937), 65-77, 6 (1941), 1-17; 7 (1942), 65-89, 133-145; 8 (1943), 89-106; 13 (1948) 65-79 K. Gödel, *The consistency of the continuum hypothesis*, Annals of Math Studies 3, Princeton 1940] This system contains the notions of "class" and "set," each set being also a class, but not vice versa. Then the various categories introduced in Chapters I and IV can be proved to exist as classes.

One must take care not to perform on these categories certain operations (such as forming the set of all subsets) which can be performed on sets but not on classes.

The transitivity of homotopies In 9 1, no conditions were imposed on the binary relation $\alpha \simeq \beta$ (α is homotopic to β) In most of the h-categories that will be considered this relation will be reflexive, symmetric, and transitive It will also be compositive in the sense that, if $\alpha,\beta\colon A \to B$, α',β' $B \to C$, $\alpha \simeq \beta$, and $\alpha' \simeq \beta'$, then $\alpha'\alpha \simeq \beta'\beta$.

One may always expand the relation of homotopy in an h-category $\mathcal{C}$ so as to obtain these properties. This is done as follows. As a first step, define $\alpha_1 \simeq' \alpha_2$ if $\alpha_1 = \beta_1\gamma_1$, $\alpha_2 = \beta_2\gamma_2$ with $\beta_1 \simeq \beta_2$ and $\gamma_1 \simeq \gamma_2$ The next step is to define $\alpha \simeq'' \beta$ if there exists a finite sequence $\alpha = \alpha_1, \alpha_2, \cdots, \alpha_n = \beta$ ($n > 0$) such that, for each $i < n$, either $\alpha_i \simeq' \alpha_{i+1}$ or $\alpha_{i+1} \simeq' \alpha_i$ The new relation $\simeq''$ is reflexive, symmetric, transitive and compositive. For any homology theory on $\mathcal{C}$, it is easy to prove that $\alpha \simeq'' \beta$ implies $\alpha_* = \beta_*$ Thus we may always replace the relation $\simeq$ by the relation $\simeq''$.

If we use this broadened concept of homotopy, the relation of homotopy equivalence defined in 9 2 becomes reflexive, symmetric, and transitive.

EXERCISES

A. FUNCTORS OF SEVERAL VARIABLES

1 Given categories $\mathcal{C}_1, \cdots, \mathcal{C}_m, \mathfrak{D}_1, \cdots, \mathfrak{D}_n, \mathcal{E}$, define the concept of a functor T covariant in $\mathcal{C}_1, \cdots, \mathcal{C}_m$ contravariant in $\mathfrak{D}_1, \cdots, \mathfrak{D}_n$ with values in $\mathcal{E}$ Generalize the concepts of natural transformation and natural equivalence to such functors

2 Exhibit the commutativity and associativity laws of the (external) direct sum

$$G_1 + G_2 \approx G_2 + G_1$$
$$G_1 + (G_2 + G_3) \approx (G_1 + G_2) + G_3$$

as natural equivalences of functors.

B FUNCTORS OF GROUPS.

1 Consider the category of all groups (including nonabelian groups) and homomorphisms. Consider the functions which assign to each group (1) its commutator subgroup, (2) its center, and (3) its group of automorphisms Convert these into functors on appropriate subcategories

Definition. A functor T on one of the categories $\mathcal{G}_R$, $\mathcal{G}_C$ with values in the same or another of these categories is said to be *additive* if $T(\phi_1 + \phi_2) = T\phi_1 + T\phi_2$ whenever ϕ_1,ϕ_2: $G \to G'$ The functor T is said to *preserve exactness* if it transforms each exact sequence into an exact sequence.

2. Let T be additive. If $\phi = 0$, then $T\phi = 0$. If $G = 0$, then $TG = 0$

3. Let T be a covariant additive functor If i_α: $G_\alpha \to G$ $(\alpha = 1, \cdots, n)$ is an injective representation of G as a direct sum, then Ti_α. $TG_\alpha \to TG$ is an injective representation of TG as a direct sum.

4 Let T be a covariant additive functor. If

$$0 \to G_1 \xrightarrow{\phi} G \xrightarrow{\psi} G_2 \to 0$$

is exact, and the image of ϕ is a direct summand of G, then the same is true of

$$0 \longrightarrow TG_1 \xrightarrow{T\phi} TG \xrightarrow{T\psi} TG_2 \longrightarrow 0.$$

5 Prove that a functor preserves exactness if and only if it carries each exact sequence of the form $0 \to G' \to G \to G'' \to 0$ into another such exact sequence.

C. ∂-functors as functors

1 Given a covariant [contravariant] ∂ or δ-functor H on a c-category $\mathcal{C}$, assign to each couple (α,β) of $\mathcal{C}$ the exact sequence given by Axiom 4 Show that if $(\gamma_1,\gamma_2,\gamma_3)$ maps (α,β) into another couple (α_1,β_1), then $(\gamma_{1*},\gamma_{2*},\gamma_{3*})$ $[(\gamma_1^*,\gamma_2^*,\gamma_3^*)]$ yields a homomorphism of the corresponding exact sequences Show that one obtains thus a new covariant [contravariant] functor H^* defined on the category of pairs of $\mathcal{C}$ with values in the category of exact lower or upper sequences Formulate the Axioms 1-4 in terms of the functor H^* If H and $\bar{H}$ are two ∂-functors and h: $H \to \bar{H}$ is a homomorphism, then h induces a natural transformation h^*. $H^* \to \bar{H}^*$ What conditions must a natural transformation Γ· $H^* \to \bar{H}^*$ satisfy in order that it be of the form $\Gamma = h^*$?

2. Show that the operation of replacing superscripts by subscripts and changing their sign at the same time carries a δ-functor into a ∂-functor of the same variance

3. If H is a covariant ∂-functor and T is a covariant [contravariant] functor on groups to groups which preserves exactness, then TH is a covariant ∂-functor [contravariant δ-functor] Examine the situation for the remaining three cases for H Show that the result also applies

to homology and cohomology theories on an admissible category, or on an h-category.

D. MODULES AND VECTOR SPACES.

Let G be a module over an integral domain D (see I, Exer. H). Consider pairs (d,g), $d \varepsilon D$, $d \neq 0$, $g \varepsilon G$ and define equivalence $(d_1,g_1) \sim (d_2,g_2)$ to mean that $d'd_2g_1 = d'd_1g_2$ for some $d' \neq 0$

1. Show that this is a proper equivalence relation Denote the equivalence class of (d,g) by $[d,g]$ and the set of all these equivalence classes by $\hat{G}$.

2 Show that $\hat{G}$ becomes a D-module under the operations

$$[d_1,g_1] + [d_2,g_2] = [d_1d_2,d_2g_1 + d_1g_2], \qquad d'[d,g] = [d,d'g].$$

3 Show that the map $g \rightarrow [1,g]$ is a linear homomorphism $G \rightarrow \hat{G}$, and examine its kernel.

4. Consider the case $D = G$ (i e D regarded as a D-module with the obvious composition law). Show that $\hat{D}$ is a field, namely the field of quotients of D

5. Show that $\hat{G}$ is a vector space over $\hat{D}$ under the operation

$$[d_1,d_2][d,g] = [d_1d,d_2g].$$

6. Show that the rank of G over D is the same as the rank of $\hat{G}$ over $\hat{D}$

7 Given a homomorphism ϕ. $G_1 \rightarrow G_2$ of the D-modules G_1 and G_2, define $\hat{\phi}[d,g] = [d,\phi g]$ and show that $\hat{\phi}$ $\hat{G}_1 \rightarrow \hat{G}_2$ is linear over $\hat{D}$. Show that the pair of operations $\hat{G},\hat{\phi}$ is a covariant functor on the category $\mathcal{G}_D$ of D-modules with values in the category $\mathcal{G}_{\hat{D}}$ of vector spaces over $\hat{D}$. Show that this functor preserves exactness.

8 Given a homology [cohomology] theory H with values in the category $\mathcal{G}_D$ of D-modules, define $\hat{H}$ as the composition of H with the functor of 7 Show that $\hat{H}$ again is a homology [cohomology] theory with values in $\mathcal{G}_{\hat{D}}$ Show that the Betti numbers are the same in H and $\hat{H}$.

CHAPTER V

Chain complexes

1. INTRODUCTION

The primary aim of this and the next four chapters is the construction of homology theories with prescribed coefficient groups The construction divides roughly into three steps, namely. (i) space → complex (ii) complex → chain complex, and (iii) chain complex → homology groups This chapter is devoted to the third step and is purely algebraic in character

A chain complex (called a group system by W. Mayer) is a lower sequence of groups in which the composition of any two successive homomorphisms is zero The constructions on chain complexes leading to their homology groups are suggested by the results of Chapter III Analogs of the Axioms of Chapter I are proved for these derived groups. In this way we obtain a "Homology theory" for chain complexes

Cochain complexes are also defined However, they bear only a formal difference from chain complexes

In order to introduce homology groups over a general coefficient group, two methods of constructing groups out of groups are discussed These are the tensor product, $C \otimes G$, of two groups, and the group of homomorphisms $\mathrm{Hom}(C,G)$ of C into G Using these, we assign a new chain complex and a new cochain complex to any given chain complex and a coefficient group G; and we obtain thereby the homology and cohomology groups of the chain complex over G

2. CHAIN COMPLEXES

DEFINITION 2 1 A *chain complex* K is a lower sequence $\{C_q(K),\partial_q\}$ of groups and homomorphisms ∂_q $C_q(K) \to C_{q-1}(K)$ such that $\partial_{q-1}\partial_q = 0$ for each integer q $C_q(K)$ is called the group of *q-chains* of K, and ∂_q is called the *boundary* homomorphism A map f $K \to K'$ of one chain complex into another is (as in I,2) a sequence of homomorphisms f_q $C_q(K) \to C_q(K')$ defined for each q and such that $f_{q-1}\partial_q = \partial'_q f_q$.

DEFINITION 2 2. Let $K = \{C_q(K),\partial_q\}$ be a chain complex. The kernel, $Z_q(K)$, of ∂_q is called the group of *q-cycles* of K. The image, $B_q(K)$ of ∂_{q+1} is called the group of *q-boundaries* of K. Since $\partial_q\partial_{q+1} = 0$, $B_q(K)$ is a subgroup of $Z_q(K)$, and the factor group $H_q(K) = Z_q(K)/B_q(K)$ is called the *q-dimensional homology group* of K. If $f: K \to K'$ is a map, then f_q sends $Z_q(K)$ into $Z_q(K')$ and $B_q(K)$ into $B_q(K')$, thereby inducing homomorphisms $f_*: H_q(K) \to H_q(K')$.

The following theorem is an immediate consequence of the definitions.

THEOREM 2.3. *If $f: K \to K$ is the identity map, then f_* is the identity homomorphism. If $f: K \to K'$ and $g: K' \to K''$, then $(gf)_* = g_*f_*$.*

Chain complexes K and their maps f constitute a category denoted by $\partial\mathcal{G}_R$ or $\partial\mathcal{G}_C$ according as the groups $C_q(K)$ are in the category $\mathcal{G}_R$ or $\mathcal{G}_C$. Then 2.3 asserts that the pair $H_q(K),f_*$ is a covariant functor from $\partial\mathcal{G}_R$ to $\mathcal{G}_R$ [or from $\partial\mathcal{G}_C$ to $\mathcal{G}_C$].

The concept of a cochain complex $K = \{C^q(K),\delta^q\}$ differs from a chain complex in that $\delta^q: C^q(K) \to C^{q+1}(K)$ and that the dimension is written as a superscript. Thus an application of the "sign-changing trick" (IV,8) converts a chain complex into a cochain complex and vice versa. The discussion of cochain complexes therefore differs from the discussion of chain complexes in terminology alone. δ^q is called the coboundary operator, $Z^q(K)$ is the group of q-cocycles, $H^q(K)$ is the q-dimensional cohomology group, etc. The pair $H^q(K),f^*$ constitutes a covariant functor on the category $\delta\mathcal{G}_R$ [or $\delta\mathcal{G}_C$] of cochain complexes with values in the category $\mathcal{G}_R$ [or $\mathcal{G}_C$].

3. COUPLES

In this section we shall convert the categories of chain complexes into c-categories (IV,8) and we shall extend $H_q(K),f_*$ to a covariant ∂-functor.

We begin with the observation that if K is a chain complex and L is a subsequence of K (in the sense of I,2), then both L and K/L are again chain complexes called the *subcomplex* and *factor complex* respectively. Moreover the inclusion map $i: L \to K$, and the natural map $\eta: K \to K/L$ yield an exact sequence

$$0 \to L \xrightarrow{i} K \xrightarrow{\eta} K/L \to 0. \tag{1}$$

This suggests the following definition.

DEFINITION 3.1. Let L,K,M be chain complexes. The maps

ϕ: $L \to K$, ψ. $K \to M$ are said to form a *couple* (ϕ,ψ): $L \to K \to M$, provided the sequence

$$\text{(ii)} \qquad 0 \to L \xrightarrow{\phi} K \xrightarrow{\psi} M \to 0$$

is exact, i e. if the kernel of ϕ is zero, ψ is onto, and kernel ψ = image ϕ.

If, further, the image of ϕ is a direct summand of K (i e. if $\phi(C_q(L))$ is a direct summand of $C_q(K)$ for all q), then the couple (ϕ,ψ) is called *direct*

It is clear that, with couples defined as above, the categories $\partial\mathcal{G}_R$ and $\partial\mathcal{G}_C$ become c-categories. One obtains different c-categories by taking direct couples only. Circumstances under which it is necessary to consider direct couples rather than all couples will appear in §§11 and 12.

In the subsequent lemmas and definitions, we assume that an exact sequence (ii) is given

LEMMA 3.2. *Let* $\overline{Z}_q(M) = \psi^{-1}(Z_q(M))$, $\overline{B}_q(M) = \psi^{-1}(B_q(M))$, *and* $\overline{H}_q(M) = \overline{Z}_q(M)/\overline{B}_q(M)$. *Then*

$$\overline{Z}_q(M) = \partial^{-1}(\phi C_{q-1}(L)), \qquad \overline{B}_q(M) = B_q(K) \cup \phi(C_q(L)),$$

and ψ *induces isomorphisms*

$$\overline{\psi}\colon\ \overline{H}_q(M) \approx H_q(M).$$

PROOF. Let $c \in C_q(K)$ Then $c \in \overline{Z}_q(M)$ if and only if $\psi c \in Z_q(M)$, i.e. if and only if $\partial\psi c = 0$ This is equivalent to $\psi\partial c = 0$, and since kernel ψ = image ϕ, this is equivalent to $\partial c \in \phi(C_{q-1}(L))$

Suppose $c \in \overline{B}_q(M)$ Then $\psi c \in B_q(M)$ and $\psi c = \partial b$ for some $b \in C_{q+1}(M)$. Let $d \in C_{q+1}(K)$ be such that $\psi d = b$. Then $\psi(c - \partial d) = \psi c - \partial\psi d = \psi c - \partial b = 0$, so that there is an $e \in C_q(L)$ with $\phi e = c - \partial d$. Thus $c = \partial d + \phi e$, and therefore $c \in B_q(K) \cup \phi(C_q(L))$ Conversely, if $c = \partial d + \phi e$, then $\psi c = \psi\partial d + \psi\phi e = \partial\psi d$, and $\psi c \in B_q(M)$ Thus $c \in \overline{B}_q(M)$

The last part of 3 2 is a direct consequence of the Noether isomorphism theorem.

LEMMA 3 3 *The boundary homomorphism of the chain complex* K *defines homomorphisms*

$$\overline{Z}_q(M) \to \phi[Z_{q-1}(L)],$$
$$\overline{B}_q(M) \to \phi[B_{q-1}(L)].$$

Since the kernel of ϕ *is zero,* $\phi^{-1}\partial$ *defines homomorphisms*

$$\overline{Z}_q(M) \to Z_{q-1}(L), \qquad \overline{B}_q(M) \to B_{q-1}(L),$$

and induces a homomorphism

$$\Delta\colon\ \overline{H}_q(M) \to H_{q-1}(L).$$

PROOF. Let $c \in \overline{Z}_q(M)$. By 3.1, we have $\partial c = \phi d$ for some $d \in C_{q-1}(L)$. Then $\phi\partial d = \partial\phi d = \partial\partial c = 0$, and therefore $\partial d = 0$. Thus $d \in Z_{q-1}(L)$ and $\partial c \in \phi[Z_{q-1}(L)]$.

If $c \in \overline{B}_q(M)$, then, by 3.2, $c = \partial d + \phi e$ for some $d \in C_{q+1}(K)$ and $e \in C_q(L)$. Then $\partial c = \partial\partial d + \partial\phi e = \phi\partial e$, and $\partial c \in \phi B_{q-1}(L)$.

DEFINITION 3.4. The homomorphism

$$\partial_*: H_q(M) \rightarrow H_{q-1}(L)$$

defined as the composition $\partial_* = \Delta\overline{\psi}^{-1}$ is called the *boundary homomorphism* of the couple (ϕ,ψ): $L \rightarrow K \rightarrow M$.

It will be useful to have a more direct description of the homomorphism ∂_*. Suppose $h \in H_q(M)$. Choose $z \in Z_q(M)$ belonging to the coset h, and choose $c \in C_q(K)$ with $\psi c = z$. Then ∂c is in the image of ϕ and $\phi^{-1}\partial c \in Z_{q-1}(L)$ is a cycle in the coset $\partial_* h$. Another description of ∂_* is obtained by stating the analog of III,7.5.

The star in ∂_* has been inserted to distinguish it from the boundary operator within the chain complexes. However, in later uses we shall omit the star.

THEOREM 3.5. *The system $H = \{H_q(K), f_*, \partial_*\}$ is a covariant ∂-functor on the c-category $\partial\mathcal{G}_R$ $[\partial\mathcal{G}_C]$ of chain complexes with values in the category $\mathcal{G}_R$ $[\mathcal{G}_C]$.*

Axioms 1 and 2 of IV,8 are contained in 2.3, while Axioms 3 and 4 correspond to Theorems 3.6 and 3.7 that follow.

THEOREM 3.6. *Let (g,f,h): $(\phi,\psi) \rightarrow (\phi',\psi')$ be a map of the couple (ϕ,ψ): $L \rightarrow K \rightarrow M$ into the couple (ϕ',ψ'): $L' \rightarrow K' \rightarrow M'$, then commutativity holds in the diagram*

$$\begin{array}{ccc} H_q(M) & \xrightarrow{\ \partial_*\ } & H_{q-1}(L) \\ \Big\downarrow{\scriptstyle h_*} & & \Big\downarrow{\scriptstyle g_*} \\ H_q(M') & \xrightarrow{\ \partial_*\ } & H_{q-1}(L') \end{array}$$

PROOF. This is a consequence of the commutativity relations in the diagram

$$\begin{array}{ccccc} H_q(M) & \xleftarrow{\ \overline{\psi}\ } & \overline{H}_q(M) & \xrightarrow{\ \Delta\ } & H_{q-1}(L) \\ \Big\downarrow{\scriptstyle h_*} & & \Big\downarrow{\scriptstyle \overline{f}_*} & & \Big\downarrow{\scriptstyle g_*} \\ H_q(M') & \xleftarrow{\ \overline{\psi}\ } & \overline{H}_q(M') & \xrightarrow{\ \Delta\ } & H_{q-1}(L) \end{array}$$

where $\overline{f}_*$ is induced by f. The commutativity relations in this diagram are immediate consequences of the definition of ψ and Δ.

THEOREM 3 7. *If* (ϕ,ψ). $L \to K \to M$ *is a couple, then the sequence*

$$\cdots \leftarrow H_{q-1}(L) \overset{\partial_*}{\leftarrow} H_q(M) \overset{\psi_*}{\leftarrow} H_q(K) \overset{\phi_*}{\leftarrow} H_q(L) \leftarrow \cdots$$

is exact

PROOF. By definition, we must prove three propositions.

(1) $\text{kernel } \psi_* = \text{image } \phi_*,$

(2) $\text{kernel } \partial_* = \text{image } \psi_*,$

(3) $\text{kernel } \phi_* = \text{image } \partial_*$

Suppose that $h \in H_q(L)$ and $z \in Z_q(L)$ lies in the coset h Then $\psi\phi z = 0$, and therefore $\psi_*\phi_* h = 0$ Thus, image $\phi_* \subset$ kernel ψ_* Suppose now that $h \in H_q(K)$ and $\psi_* h = 0$. Then, if $z \in Z_q(K)$ lies in the coset h, it follows that $\psi z \in B_q(M)$, i e $\psi z = \partial c$ for some $c \in C_{q+1}(M)$. Let $b \in C_{q+1}(K)$ be such that $\psi b = c$ Then $\psi(z - \partial b) = \psi z - \partial\psi b = \partial c - \partial c = 0$, and therefore there is an $a \in C_q(L)$ with $\phi a = z - \partial b$ Since $\phi\partial a = \partial\phi a = \partial z - \partial\partial b = 0$, it follows that $\partial a = 0$ and $a \in Z_q(L)$. If $h' \in H_q(L)$ is the coset of a, then, since ϕa and z are in the same coset, it follows that $\phi_* h' = h$

To prove (2) suppose that $h \in H_q(K)$ and let $z \in Z_q(K)$ be in the coset h Then ψz is in the coset $\psi_* h$ of $H_q(M)$ Since $\partial z = 0$, it follows from the definition of ∂_* that $\partial_*\psi_* h = 0$ Therefore image $\psi_* \subset$ kernel ∂_* Suppose now that $h \in H_q(M)$ and that $\partial_* h = 0$ Let $z \in Z_q(M)$ be in the coset h, and let $c \in C_q(K)$ be such that $\psi c = z$ Then $\phi^{-1}\partial c$ is in $Z_{q-1}(L)$, and is in the coset $\partial_* h$ Since $\partial_* h = 0$, there is a $b \in C_q(L)$ such that $\partial b = \phi^{-1}\partial c$ or $\phi\partial b = \partial c$ Then $\partial(c - \phi b) = \partial c - \phi\partial b = 0$, so that $c - \phi b \in Z_q(K)$, and $\psi(c - \phi b) = \psi c = z$ Thus if h' is the coset of $c - \phi b$ in $H_q(K)$, it follows that $\psi_* h' = h$

To prove (3) suppose that $h \in H_q(M)$, and let $z \in Z_q(M)$ lie in the coset h. Select $c \in C_q(K)$ with $\psi c = z$. Then there is a $b \in Z_{q-1}(L)$ with $\phi b = \partial c$, and b lies in the coset $\partial_* h$ of $H_{q-1}(L)$ Since $\phi b \in B_{q-1}(K)$, it follows that $\phi_*\partial_* h = 0$ Thus image $\partial_* \subset$ kernel ϕ_* Suppose now that $h \in H_q(L)$ and $\phi_* h = 0$ Let $z \in Z_q(L)$ be in the coset h Then $\phi z \in B_q(K)$, and $\phi z = \partial c$ for some $c \in C_{q+1}(K)$ Then $\partial\psi c = \psi\partial c = \psi\phi z = 0$. Thus $\psi c \in Z_{q+1}(M)$. If $h' \in H_{q+1}(M)$ is the coset of ψc, then it follows from the definition of ∂_* that $\partial_* h' = h$ This completes the proof of 3 7.

The definitions and theorems for cochain complexes are quite similar and are obtainable by the sign-changing trick The analog of 3 5 is

THEOREM 3 5c. *The system* $H = \{H^q(K), f_*, \delta^*\}$ *is a covariant* δ-

functor on the c-category $\delta\mathcal{G}_R$ [$\delta\mathcal{G}_C$] of cochain complexes with values in the category $\mathcal{G}_R$ [$\mathcal{G}_C$]

We return to the remarks made at the beginning of this section and observe that whenever possible we use the operation of factorization in groups and chain complexes to pass from pairs to single objects Thus instead of the pair (K,L) where L is a subcomplex of K we consider the exact sequence $0 \to L \to K \to K/L \to 0$ Similarly instead of the triple (K,L,N) we consider the exact sequence $0 \to L/N \to K/N \to K/L \to 0$ This point of view will be followed systematically in the future

4. HOMOTOPIES, EXCISIONS, POINTS

In this section we define homotopies, excisions, and points in the c-categories $\partial\mathcal{G}_R$ and $\partial\mathcal{G}_C$, thereby converting them into h-categories. The system $H = \{H_q(K),f_*,\partial_*\}$ becomes then a homology theory The particular definitions of "homotopy," "excision," and "point" adopted are motivated by the applications of the next two chapters

DEFINITION 4.1. Let K and K' be chain complexes and let f,g be two maps of K into K' A *chain homotopy* D of f into g (notation $D: f \simeq g$) is a sequence of homomorphisms

$$D_q: C_q(K) \to C_{q+1}(K')$$

such that

$$\partial_{q+1}D_q + D_{q-1}\partial_q = g_q - f_q$$

If such a homotopy D exists, f and g are called *homotopic* and we write $f \simeq g$. If $c \in C_q(K)$, then D_qc is called the *deformation chain* of c

LEMMA 4 2 *The relation $f \simeq g$ is reflexive, symmetric, and transitive.*

PROOF. If $D = 0$, then $D: f \simeq f$ If $D. f \simeq g$, then $-D$: $g \simeq f$ If $D: f \simeq g$ and $D'. g \simeq h$, then $D + D'. f \simeq h$.

LEMMA 4.3. *If $f,g: K \to L$, $f',g': L \to M$, $f \simeq g$ and $f' \simeq g'$, then $f'f \simeq g'g$.*

PROOF. Let $D. f \simeq g$ and $D'. f' \simeq g'$ Define $D''_q: C_q(K) \to C_{q+1}(M)$ by $D''_q = f'_{q+1}D_q + D'_qg_q$; then a short computation gives $D'': f'f \simeq g'g$.

THEOREM 4 4 *If $f,g: K \to K'$ are chain homotopic, then their induced homomorphisms*

$$f_*,g_*: H_q(K) \to H_q(K')$$

coincide: $f_* = g_*$.

PROOF. Let D: $f \simeq g$, and let $z \in Z_q(K)$ Then $\partial Dz = gz - fz$. Thus $gz - fz \in B_q(K')$ and $f_* = g_*$

DEFINITION 4 5. A map $f\colon K \to L$ is called an *excision* if and only if maps K isomorphically onto L.

THEOREM 4 6 *If $f\colon K \to L$ is an excision, then f_* maps $H_q(K)$ isomorphically onto $H_q(L)$*

The proof of this "Excision axiom" is of course trivial due to the very narrow definition of excision that has been adopted.

DEFINITION 4 7 A chain complex $K = \{C_q(K),\partial\}$ is called *pointlike* if $\partial_q\colon C_q(K) \to C_{q-1}(K)$ is an isomorphism for q even and >0, and also for q odd and <0

THEOREM 4 8. *If K is pointlike, then $H_q(K) = 0$ for $q \neq 0$, and $H_0(K) = C_0(K)$.*

PROOF. Let q be even and >0 Since ∂_q: $C_q(K) \approx C_{q-1}(K)$, it follows that $Z_q(K) = 0$ and $B_{q-1}(K) = C_{q-1}(K)$ Thus $H_q(K) = 0$ and $H_{q-1}(K) = 0$ Similarly $H_q(K) = 0$, $H_{q-1}(K) = 0$ for q odd and <0. Since $\partial_1\partial_2 = 0$ and $\partial_{-1}\partial_0 = 0$ and ∂_2 and ∂_{-1} are isomorphisms, it follows that $\partial_1 = 0$, $\partial_0 = 0$. Thus $Z_0(K) = C_0(K)$ and $B_0(K) = 0$ Hence $H_0(K) = C_0(K)$.

Theorems 4.4, 4 6, and 4 8 combined with 3 5 yield

THEOREM 4.9 *The system $H = \{H_q(K),f_*,\partial_*\}$ is a homology theory on the h-category $\partial\mathcal{G}_R$ $[\partial\mathcal{G}_C]$ of chain complexes, with values in the category $\mathcal{G}_R$ $[\mathcal{G}_C]$.*

It should be kept in mind that actually $\partial\mathcal{G}_R$ $[\partial\mathcal{G}_C]$ represents two h-categories, one with all couples and the other with direct couples only. Theorem 4 9 holds with either meaning for $\partial\mathcal{G}_R$ $[\partial\mathcal{G}_C]$

In the h-category $\partial\mathcal{G}_R$ $[\partial\mathcal{G}_C]$, we have the concepts of homotopy equivalence and generalized excision as defined in IV,9 2. Since every excision in $\partial\mathcal{G}_R$ $[\partial\mathcal{G}_C]$ is also a homotopy equivalence, and since the composition of homotopy equivalences is, by 4 3, again a homotopy equivalence, it follows that in the category $\partial\mathcal{G}_R$ $[\partial\mathcal{G}_C]$ homotopy equivalences and generalized excisions coincide.

The definitions of homotopies, excisions, and points in the c-categories $\delta\mathcal{G}_R,\delta\mathcal{G}_C$ of cochain complexes are quite similar and are obtainable from those for chain complexes by the sign-changing trick The results for cochain complexes may be summed up in the following analog of 4 9:

THEOREM 4.9c. *The system $H = \{H^q(K),f_*,\delta^*\}$ is a covariant δ-functor on the category $\delta\mathcal{G}_R$ $[\delta\mathcal{G}_C]$ of cochain complexes, satisfying the Homotopy, Excision, and Dimension axioms.*

5. DIRECT SUMS AND PRODUCTS

This section reviews the definitions and basic properties of the cartesian product of spaces, and the direct sum and direct product of groups.

DEFINITION 5.1. Let $\{X_\alpha\}$ be a collection of sets indexed by a set M, i.e. for each $\alpha \in M$, X_α is a set of the collection. The *product*

$$\prod_{\alpha \in M} X_\alpha \tag{1}$$

of the collection is the totality of functions $x = \{x_\alpha\}$ defined for $\alpha \in M$ and such that x_α—the value of x on α—is an element of X_α. The element x_α is called the *α-coordinate* of x. For each $\beta \in M$, define the *projection*

$$p_\beta : \prod_{\alpha \in M} X_\alpha \to X_\beta \quad \text{by} \quad p_\beta(x) = x_\beta. \tag{2}$$

In case the sets X_α all coincide with a set X, then the product (1) is denoted by X^M, and is simply the set of all functions from M to X.

DEFINITION 5.2. If each X_α is a topological space, a topology is introduced in (1) as follows. If a finite number of X_α's are replaced by open subsets $U_\alpha \subset X_\alpha$, the product of the resulting collection is a subset of (1) and is called a rectangular open set of (1). Any union of rectangular open sets is called an open set of the product. The product with this topology is called the *cartesian product*. It is immediate that the projections (2) are continuous.

LEMMA 5.3. *A function*

$$f : Y \to \prod_{\alpha \in M} X_\alpha$$

defined on a space Y with values in a cartesian product is continuous if and only if, for each $\alpha \in M$, the function

$$p_\alpha f : Y \to X_\alpha$$

is continuous.

The proof is left as an exercise to the reader.

The following classical result is due to Tychonoff. For a proof, see Lefschetz [*Algebraic Topology*, Colloq. Pub. Amer. Math. Soc., 1942, p. 19].

THEOREM 5.4. *The cartesian product of a collection of compact spaces is compact.*

We recall that the term "compact space" is used here to denote a Hausdorff space in which the Borel covering theorem holds.

DEFINITION 5.5. If each X_α is an abelian group, then an *addition* is defined in (1) by the usual method of adding functional values:

$$(x + x')_\alpha = x_\alpha + x'_\alpha. \tag{3}$$

In this way (1) becomes an abelian group and is called the *direct product* of the groups $\{X_\alpha\}$. If each X_α is a compact abelian group, then the product with the topology of 5.2 and the group operation just defined is a compact abelian group called the *direct product*

If each X_α is an R-module over the same ring R, define addition in (1) by means of (3) and multiplication by a scalar $r \in R$ by

$$(4) \qquad (rx)_\alpha = r \cdot x_\alpha.$$

Then (1) becomes an R-module called the *direct product.*

In all cases the projections (2) are homomorphisms.

The direct product is thus defined for a collection of groups from either of the categories $\mathcal{G}_R, \mathcal{G}_C$, and the product belongs to the same category. The same symbolism (1) will be used to denote each of these products; the particular product in question can always be determined from the nature of the X_α's.

In case all the groups $X_\alpha, \alpha \in M$ coincide with a single group G, the direct product (1) is denoted by G^M and is simply the group of functions on M with values in G.

DEFINITION 5.6. Let $\{G_\alpha\}$ be an indexed collection of R-modules. Their *direct sum*

$$\sum_{\alpha \in M} G_\alpha$$

is the subgroup of their direct product $\prod G_\alpha$ consisting of those elements having all but a finite number of coordinates equal to zero, i.e. $g_\alpha = 0 \in G_\alpha$ for all but a finite number of $\alpha \in M$. For each $\beta \in M$, define the *injection*

$$(5) \quad i_\beta: \ G_\beta \to \sum G_\alpha \qquad \text{by} \qquad (i_\beta(g))_\alpha = \begin{cases} g & \text{for} \quad \alpha = \beta, \\ 0 & \text{for} \quad \alpha \neq \beta. \end{cases}$$

It is clear that i_β is an isomorphism of G_β onto a subgroup of $\sum G_\alpha$

In case M is a finite collection, then of course the direct sum and direct product coincide.

In case each G_α is a compact group, $\sum G_\alpha$, as a subset of $\prod G_\alpha$, has a topology which makes of it a topological group. In addition, each p_β is continuous As a subspace of the direct product, $\sum G_\alpha$ is not a closed subspace (indeed, it is everywhere dense), and is therefore not compact. It is for this reason that the direct sum is not a useful operation to apply to compact groups.

DEFINITION 5.7. A *projective representation* of a group G as a direct product consists of a family of groups $\{G_\alpha\}$ indexed by a set M, and a homomorphism η_α of G onto G_α for each α, such that the homomorphism η: $G \to \prod G_\alpha$ defined by $(\eta g)_\alpha = \eta_\alpha g$ is an isomorphism. If, in addition, each G_α is given as a factor group of G by a subgroup, and η_α is the natural homomorphism, then G is said to *decompose* into the direct product of the factor groups $\{G_\alpha\}$.

DEFINITION 5.8. Let G be an R-module An *injective representation* of G as a direct sum consists of a family of R-modules $\{G_\alpha\}$ indexed by a set M, and a homomorphism ξ_α of G_α into G for each α, such that the homomorphism ξ. $\sum G_\alpha \to G$, defined by assigning to each element of $\sum G_\alpha$ the sum $\sum \xi_\alpha g_\alpha$ of the images of its nonzero coordinates in G, is an isomorphism. If, in addition, each G_α is a subgroup of G and ξ_α is the inclusion map, then G is said to *decompose* into the direct sum of the subgroups $\{G_\alpha\}$. This is the case if and only if each $g \in G$ can be expressed uniquely in the form

$$g = g_{\alpha_1} + \cdots + g_{\alpha_m}, \qquad g_{\alpha_i} \in G_{\alpha_i}$$

where $\alpha_1, \cdots, \alpha_m$ are distinct elements of M.

If $\{G_\alpha\}$ is any collection of groups, it is clear that $\prod G_\alpha$ and $\sum G_\alpha$ are represented as the direct product and sum, respectively, of the groups $\{G_\alpha\}$ by means of $\{p_\alpha\}$ and $\{i_\alpha\}$.

6. FREE MODULES AND THEIR FACTOR GROUPS

DEFINITION 6.1. Let G be an R-module and X a subset of G The set X is said to be *linearly independent* if, for any distinct elements $x_1, \cdots, x_n$ of X, the relation $r_1x_1 + \cdots + r_nx_n = 0$, $r_i \in R$, implies $r_1 = \cdots = r_n = 0$. The set X is said to *generate* G if no proper submodule of G contains X. If X is linearly independent and generates G, then X is said to be a *base* of G and G is called *free* In this case every element $g \in G$ can be represented uniquely as $g = r_1x_1 + \cdots + r_nx_n$ where $r_i \in R$, $x_i \in X$, and $x_1, \cdots, x_n$ are distinct.

If X is a base of G and G' is any R-module, then it is clear that every function ϕ: $X \to G'$ can be extended uniquely to a homomorphism ϕ $G \to G'$.

It is well known that, if $R = F$ is a field, then every vector space G over F is free.

LEMMA 6.2 *If G,G' and H are R-modules, ϕ: $G \to G'$ and ψ: $H \to G'$ are homomorphisms, ϕ is onto and H is free, then there exists a homomorphism θ $H \to G$ such that $\phi\theta = \psi$.*

PROOF. Let X be a base of H. For each $x \in X$ select $\theta(x) \in G$ so that $\phi\theta(x) = \psi(x)$. This is possible since ϕ is onto The function θ. $X \to G$ extends to a homomorphism θ $H \to G$ which then satisfies $\phi\theta = \psi$

LEMMA 6.3. *If H is a submodule of G and G/H is free, then H is a direct summand of G.*

Proof. Consider the natural homomorphism η: $G \to G/H$ and the identity map i: $G/H \to G/H$. By 6.2 there is a θ. $G/H \to G$ such that $\eta\theta = i$ This implies that G decomposes into the direct sum of the image of θ and the kernel of η which is H.

Since every vector space is free, we have

Corollary 6.4 *Every subspace of a vector space is a direct summand.*

Definition 6.5. For any set X, let R_X be the set of all functions f $X \to R$ such that $fx \neq 0$ for at most a finite number of $x \in X$. With addition and multiplication defined by

$$(f_1 + f_2)x = f_1x + f_2x, \qquad (rf)x = r(fx),$$

R_X is an R-module. As is customary, we identify each $x \in X$ with the function which is 1 on x and zero elsewhere. Then, if f has the nonzero values $r_1, \cdots, r_n$ on $x_1, \cdots, x_n$ respectively, and is zero elsewhere, we have $f = \sum_1^n r_ix_i$. It follows that R_X is a free R-module having X as a base. It is called the *free R-module generated by X.*

If X' is a subset of X, the inclusion map $X' \subset X$ extends uniquely to an isomorphism of $R_{X'}$ with a submodule of R_X. If X is represented as a union $\cup X_\alpha$ of disjoint subsets, the inclusion maps $X_\alpha \subset X$ induce an injective representation of R_X as a direct sum. $R_X \approx \sum_\alpha R_{X_\alpha}$

Definition 6 6. Let G be an R-module and $X \subset G$ a set generating G. The inclusion map $X \to G$ can then be extended to a homomorphism θ: $R_X \to G$ which maps R_X onto G. Let Y be any set generating the kernel of θ. Then we say that G is *represented* by generators X and relations Y. Note that each element of Y can be written as a formal finite linear combination of elements of X with coefficients in R, and, when evaluated in G, this linear combination yields zero

Lemma 6.7. *Let G be the R-module given by generators X and relations Y and let G′ be given by generators X′ and relations Y′. A function* f. $X \to X'$ *can be extended to a homomorphism* ϕ: $G \to G'$ *if and only if for every element* $r_1x_1 + \cdots + r_nx_n \in Y$, $r_i \in R$, $x_i \in X$, *the element* $r_1f(x_1) + \cdots + r_nf(x_n) \in R_{X'}$, *is a linear combination of elements of* Y'. *If this condition is satisfied, the homomorphism* ϕ *extending* f *is unique.*

Proof: Let ψ: $R_X \to R_{X'}$ be the homomorphism extending f and let N and N' be the kernels of the natural maps $R_X \to G$, $R_{X'} \to G'$. Then Y and Y' generate N and N' and the condition of the lemma is equivalent to $\psi(N) \subset N'$. The homomorphism ϕ is then induced by ψ.

Lemma 6.8 *If R is a principal ideal ring (i.e. a domain of integrity in which each ideal has the form Rr for some* $r \in R$*), then every submodule of a free R-module is also a free R-module.*

Proof. Let G be a free R-module, let X be a base for G, and let H be a submodule of G. We assume the choice axiom, and may therefore suppose the elements of X are well ordered: $x_1, x_2, \cdots, x_\alpha, \cdots$. For each ordinal α, let A_α be the submodule of those elements of H which are linear combinations of the x_β with $\beta \leqq \alpha$, and let B_α be the submodule of those elements of H which are linear combinations of the x_β with $\beta < \alpha$. Then each $a \in A_\alpha$ is of the form $a = b + rx_\alpha$ where $b \in B_\alpha$. The coefficient r is a function of a and provides a homomorphism of A_α onto an ideal I_α of R. The hypothesis on R implies that I_α is a free R-module. By 6.3, A_α is the direct sum of B_α and a submodule C_α isomorphic with I_α. We will show that H is the direct sum of the submodules C_α. If $\alpha < \beta$, then $C_\alpha \subset A_\alpha \subset B_\beta$, hence $C_\alpha \cap C_\beta = 0$. Let $a_0 \in H$, then $a_0 \in A_{\alpha_0}$ for some α_0. Thus $a_0 = a_1 + c_0$ where $c_0 \in C_{\alpha_0}$ and $a_1 \in B_{\alpha_0}$. Then $a_1 \in A_{\alpha_1}$ for some $\alpha_1 < \alpha_0$. Iterating this procedure, we obtain a sequence $\alpha_n < \alpha_{n-1} < \cdots < \alpha_0$ and elements $c_i \in C_{\alpha_i}$ for $i = 0,1, \cdots, n$ such that $a_0 = a_{n+1} + c_n + \cdots + c_0$ where $a_{n+1} \in B_{\alpha_n}$. Since each decreasing sequence of ordinals is finite, it follows that $a_{n+1} = 0$ for sufficiently large n. Thus H is the direct sum of the modules C_α. Since each C_α is free, it follows that H is free.

When R is the ring of integers, an R-module is an ordinary abelian group. Hence 6.8 implies

COROLLARY 6.9. *Any subgroup of a free abelian group is free.*

7. FINITELY GENERATED GROUPS

In a great many of the applications of homology theory the spaces involved are triangulable and the coefficient group is the group of integers. In such a case, the homology groups encountered are generated by a finite number of elements. The structure of such a group is easily described: it is a direct sum of cyclic groups. Its structure can be characterized by numerical invariants. Moreover, starting with any finite set of generators and relations, there is an algorithm for obtaining a second set of generators which gives the direct sum decomposition, and, at the same time, the numerical invariants. Consequently the finitely generated groups are of the utmost importance in applications of homology theory.

The object of this section is to review this algebraic material and state the applications to chain complexes.

We recall that a matrix $A = (a_{ij})$ with integer elements is said to be *unimodular* if it has the same number of rows as columns, and its determinant is $+1$ or -1. In this case the inverse of A, denoted by $A^{-1} = (\bar{a}_{ij})$, has integer elements and is unimodular. Under ordinary matrix multiplication, the unimodular matrices, of a fixed order, form a group.

Let F be a free group with a finite base $x_1, \cdots, x_n$, and let (a_{ij}) be a unimodular matrix of integers of order n. Then the elements $y_i = \sum_{j=1}^{n} a_{ij}x_j$ $(i = 1, \cdots, n)$ also form a base for F. To prove this, observe, first, that the y's generate F; for $x_j = \sum \bar{a}_{jk}y_k$. In addition, the y's are independent For suppose $\sum_{i=1}^{n} c_i y_i = 0$. Then $\sum_{i,j} c_i a_{ij} x_j = 0$. Since the x's are independent, this implies $\sum_i c_i a_{ij} = 0$ for each j. Therefore $0 = \sum_{ij} c_i a_{ij}\bar{a}_{jk} = \sum c_i \delta_{ik} = c_k$

Suppose now that F,F' are free groups with finite bases $x_1, \cdots, x_n$ and $x_1', \cdots, x_m'$ respectively (see 6.1 with R = integers). If f: $F \to F'$ is a homomorphism, then, for each i, $f(x_i) = \sum_{j=1}^{m} b_{ij}x_j'$ for some integers b_{ij}, and f is completely determined by the matrix $B = (b_{ij})$. Conversely, any $m \times n$-matrix B corresponds to a homomorphism If the bases $(x),(x')$ are transformed into bases $(y),(y')$ by unimodular matrices of integers A and A', then the matrix of f in terms of $(y),(y')$ becomes ABA'^{-1}. Thus, to the homomorphism f corresponds an equivalence class of matrices of integers. At this point we state for future reference the classical theorem on the reduction of such matrices to canonical form by an equivalence.

THEOREM 7 1. *If B is an $m \times n$-matrix of integers, then there exist unimodular matrices of integers A and A' of orders n and m, respectively, such that* (1) *the matrix $C = ABA'^{-1}$ is in diagonal form, and* (2) *if $d_1,d_2, \cdots$ are the diagonal elements of C, then each $d_i \geqq 0$ and d_i is a divisor of d_{i+1} for $i = 1, \cdots, Min\ (m,n) - 1$. Furthermore C is unique in the sense that the array of diagonal elements is the same for all equivalent diagonal matrices satisfying* (2).

In terms of the homomorphism f: $F \to F'$, the theorem asserts that there exist bases $(y),(y')$ of F,F' such that

$$(1) \qquad f(y_i) = d_i y_i', \qquad (i = 1, \cdots, \rho).$$

The number ρ is the rank of B and is the number of nonzero diagonal elements of C. This result has many consequences

Suppose, for example, that f is an isomorphism. Then each $d_i = 1$ and $m = n$ Therefore C is unimodular, and $B = A^{-1}A'$ is also unimodular. Thus, in terms of any bases $(x),(x')$, an isomorphism corresponds to a unimodular matrix.

If this last result is applied to the case $F = F'$ and f = identity, we obtain that any two bases for F are related by a unimodular matrix. The number n of elements in any base is the same for all bases. It is easily seen that n is the rank of F (see Exercise I,H for definition of rank).

THEOREM 7.2 *Suppose the group G is isomorphic to the factor group of a free group F' of finite rank m by the homomorphic image of a free* *group F of finite rank Then G can be decomposed into the direct sum of $r \leqq m$ infinite cyclic groups and $\tau \leqq m - r$ finite cyclic groups of orders $\theta_1, \cdots, \theta_\tau$ where each $\theta_i > 1$ and θ_i divides θ_{i+1} for $i = 1, \cdots, \tau - 1$. The numbers $r, \theta_1, \cdots, \theta_\tau$ are invariants of G Precisely, if G is decomposed in two ways as a direct sum of cyclic groups in which the orders of the finite cyclic groups divide one another successively, then the numbers $r, \theta_1, \cdots, \theta_\tau$ are the same for the two decompositions.*

PROOF. As shown above we may choose bases (y), (y') in F,F' so that the homomorphism $f\colon F \to F'$ has the form (1) Let J'_i ($i = 1, \cdots, m$) be the subgroup of F' generated by y'_i, and J_i ($i = 1, \cdots, \rho$) the subgroup generated by $d_i y'_i$. Then F' decomposes into the direct sum $\sum_1^m J'_i$, and the image of f decomposes into $\sum_1^\rho J_i$ This yields

$$G \approx \sum_1^\rho \frac{J'_i}{J_i} + \sum_{\rho+1}^m J'_i.$$

If we define $\theta_1, \cdots, \theta_\tau$ to be those d's which exceed 1, then the asserted decomposition has been demonstrated

Since r is just the rank of G, its invariance is obvious Let G' be the subgroup of elements of finite order in G. For any integer n, $g \to ng$ defines a homomorphism $G' \to G'$; let nG' denote the image group, and $\phi(n)$ the number of elements in nG'. Given any decomposition of G as described above, the numbers θ_i are easily shown to satisfy

$$\phi(n) = \frac{\theta_1}{(\theta_1,n)} \cdot \frac{\theta_2}{(\theta_2,n)} \cdot \quad \cdots \cdot \frac{\theta_\tau}{(\theta_\tau,n)}$$

where (θ,n) denotes the greatest common divisor of θ and n. For any $\sigma = 1, \cdots, \tau$, the equation

$$\frac{\theta_1}{(\theta_1,n)} \cdots \frac{\theta_\sigma}{(\theta_\sigma,n)} = 1$$

holds if we set $n = \theta_\sigma$, and is false for any positive n less than θ_σ. This follows since θ_i divides θ_{i+1} Therefore θ_τ is the least positive value of n satisfying $\phi(n) = 1$; and θ_σ $(\sigma < \tau)$ is the least positive value of n satisfying

$$\phi(n) = \frac{\theta_{\sigma+1}}{(\theta_{\sigma+1},n)} \cdots \frac{\theta_\tau}{(\theta_\tau,n)} \cdot$$

This provides an inductive definition of $\theta_\tau, \theta_{\tau-1}, \cdots, \theta_1$ in terms of the function ϕ Since ϕ is invariantly defined, the θ's are likewise invariants. This completes the proof.

THEOREM 7.3. *The conclusions of the preceding theorem hold for any group G having a finite number m of generators*

Let $x_1, \cdots, x_m$ be a set of generators of G, and let F' be the free group they generate (apply 6 5 with R = the integers). Then the inclusion $\{x_i\} \subset G$ extends to a unique homomorphism h of F' onto G. It follows that G is isomorphic to F'/F where F is the kernel of h By 6.9, F is a free group. As the rank of F cannot exceed that of F', it follows that F has a *finite* base Let f be the inclusion map $F \subset F'$. Then we have shown that G satisfies the hypotheses of 7.2, and the proof is complete

REMARK. An analysis of the proof of 7 3 reveals several nonconstructive steps. One is the nonconstructive nature of the proof that the integers form a principal ideal ring (see 6.8), and another is the use of the choice axiom in the proof of 6 2. In contrast, the proof of 7 2 (including the reduction of the matrix to diagonal form) is entirely constructive; the decomposition of G can be found in a finite number of steps starting with bases in F,F' and the matrix for f In applications of this section to finite chain complexes, we shall use 7.2 rather than 7.3, and thereby remain within the realm of effective procedures.

8. CANONICAL BASES IN A FINITE COMPLEX

DEFINITION 8.1 A chain complex K is said to be *finite*, if, for each q, $C_q(K)$ is a free group on a finite base. Then $H_q(K)$ also has a finite set of generators. The rank R_q of $H_q(K)$ is called the qth *Betti number* of K The invariants $\theta_1^q, \cdots, \theta_{\tau_q}^q$ of $H_q(K)$, described in 7 2, are called the qth *torsion numbers* of K The ranks of $C_q(K)$ and $B_q(K)$ are denoted by α_q and β_q

THEOREM 8 2 *If K is a finite chain complex, then*

$$R_q = \alpha_q - \beta_q - \beta_{q-1}$$

Furthermore, there exists a set of bases, one for each $C_q(K)$, with the following properties. For each q, the base for $C_q(K)$ is composed of five types of elements,

$$\begin{array}{ll} a_q^i, & i = 1, \cdots, \beta_q - \tau_q, \\ b_q^j, & j = 1, \cdots, \tau_q, \\ c_q^k, & k = 1, \cdots, R_q, \\ d_q^l, & l = 1, \cdots, \tau_{q-1}, \\ e_q^m, & m = 1, \cdots, \beta_{q-1} - \tau_{q-1}. \end{array}$$

For each q, the homomorphism ∂_q is given by

$$\partial_q a_q^i = 0, \qquad \partial_q b_q^j = 0, \qquad \partial_q c_q^k = 0$$
$$\partial_q d_q^l = \theta_l^{q-1} b_{q-1}^l, \qquad \partial_q e_q^m = a_{q-1}^m.$$

Such a set of bases is called canonical, and the boundary operator is said to be in diagonal form

PROOF. Since $C_{q-1}(K)$ is free, we have, by 6 3, that $Z_q(K)$, the kernel of ∂_q, is a direct summand of $C_q(K)$. For each q, we choose a decomposition of $C_q(K)$ into a direct sum $C_q(K) = Z_q(K) + W_q(K)$. Then ∂_q defines an isomorphism of $W_q(K)$ into $Z_{q-1}(K)$. Since both are free groups on finite bases, there exist bases $(y),(y')$ in these groups (see 7.1) such that ∂_q has the form

$$\partial_q y_i = d_i y'_i, \qquad i = 1, \cdots, \beta_{q-1}.$$

Since the kernel of ∂_q is $Z_q(K)$, no d_i is zero. Those y_m for which $d_m = 1$ are relabeled e^m_q, the corresponding y'_m are denoted by a^m_{q-1}. The remaining y's are denoted by d^l_q, and the corresponding y''s by b^l_{q-1}. The remaining y''s are denoted by c^k_{q-1} With these choices of base elements for each q, the boundary operator has diagonal form Then $H_q(K)$ decomposes into the direct sum of cyclic groups generated by the cosets of the base elements b^j_q and c^k_q This decomposition has the form described in 7 2 Therefore the number of generators c^k_q is R_q. Since the torsion numbers are invariants, the d's associated with the b_q's must be the torsion numbers of $H_q(K)$ This establishes the canonical form of the boundary operator and the ranges of the indices j, k, and l Since the elements a^i_q and $\theta^q_j b^j_q$ form a base for $B_q(K)$ of rank β_q, the range of i, and therefore of m, must be as indicated The relation on R_q follows from the fact that the number of generators of all five types is α_q

It is important from the computational standpoint to observe that the reduction to the canonical form of 8 2 can be carried out in a finite number of steps, for any finite number of q's, starting with bases in each group and the matrix of integers describing each ∂_q in terms of these bases This follows from the fact that the reduction to diagonal form described in 7 1 is a finite process One would begin with the least q and reduce ∂_q to some diagonal form This provides a new base for C_q including a base for Z_q, and a new form for the matrix of ∂_{q+1} which now involves only the base elements of Z_q The matrix of ∂_{q+1}. $C_{q+1} \to Z_q$ is then reduced to diagonal form This provides a new base for C_{q+1} including a base for Z_{q+1}, and a new matrix for ∂_{q+2}. $C_{q+2} \to Z_{q+1}$, etc.

The analogs of 8 1 and 8 2 for finite cochain complexes are left to the reader.

*

9. THE TENSOR PRODUCT

DEFINITION 9.1.* The *tensor product* $C \otimes G$ of two R-modules C and G is the R-module generated by the set of all pairs (c,g), $c \in C$, $g \in G$ with relations

$$(1) \qquad (c_1 + c_2, g) - (c_1, g) - (c_2, g) = 0,$$
$$(c, g_1 + g_2) - (c, g_1) - (c, g_2) = 0,$$
$$(2) \qquad (rc, g) - r(c,g) = 0, \qquad (c, rg) - r(c,g) = 0.$$

Following 6.5, $C \otimes G$ is obtained then as follows: Let $R(C,G)$ be the free R-module generated by the set of pairs (c,g) and let $Y(C,G)$ be the least subgroup of $R(C,G)$ containing all the elements of the form

$$(c_1 + c_2, g) - (c_1, g) - (c_2, g), \qquad (c, g_1 + g_2) - (c, g_1) - (c, g_2),$$
$$(rc, g) - r(c,g), \qquad (c, rg) - r(c,g),$$

then

$$C \otimes G = R(C,G)/Y(C,G)$$

The element of $C \otimes G$ which is the image of the generator (c,g) of $R(C,G)$ will be denoted by $c \otimes g$. These elements generate the group $C \otimes G$ and the relations are

$$(1') \qquad (c_1 + c_2) \otimes g = c_1 \otimes g + c_2 \otimes g,$$
$$c \otimes (g_1 + g_2) = c \otimes g_1 + c \otimes g_2,$$
$$(2') \qquad (rc) \otimes g = r(c \otimes g) = c \otimes (rg)$$

If g is a fixed element of G, the correspondence $c \to c \otimes g$ is a homomorphism $C \to C \otimes G$. Similarly, if c is fixed, $g \to c \otimes g$ is a homomorphism $G \to C \otimes G$. Thus (1') implies

$$(3) \qquad (\sum c_i) \otimes g = \sum (c_i \otimes g), \qquad c \otimes \sum g_i = \sum (c \otimes g_i),$$
$$(4) \qquad 0 \otimes g = 0, \qquad c \otimes 0 = 0.$$

LEMMA 9.2. *If G is an R-module, the correspondence f: $R \otimes G \to G$ defined by $f(r \otimes g) = rg$ is an isomorphism. Similarly, $G \otimes R = G$. In the sequel both $R \otimes G$ and $G \otimes R$ will be identified with G by these isomorphisms.*

PROOF. Consider first the map f': $R(R,G) \to G$ defined by $f'(r,g) = rg$. Since $R(R,G)$ is free, and the pairs (r,g) form a base, f' is a homomorphism. It is easily seen that f' has the value 0 on each of the four relations of (1) and (2). Therefore f' maps $Y(R,G)$ into 0, and thereby induces a homomorphism f of the factor group $R \otimes G$ into

*Assume throughout that R is commutative. Tensor products over non-commutative rings are treated in the book of H. Cartan and S. Eilenberg, Homological Algebra (Princeton Press) 1956.

G such that $f(\tau \otimes g) = \tau g$. Since $f(1 \otimes g) = g$, f is onto. Applying (2′) and (3), we have

$$\sum_{i=1}^{k} \tau_i \otimes g_i = \sum_{i=1}^{k} 1 \otimes \tau_i g_i = 1 \otimes \sum_{i=1}^{k} \tau_i g_i.$$

Therefore each element of $R \otimes G$ can be written in the form $1 \otimes g$. If $f(1 \otimes g) = 0$, then $g = 0$. This implies $1 \otimes g = 0$. Hence f is an isomorphism.

DEFINITION 9.3. If $f\colon C \to C'$ and $h\colon G \to G'$ are homomorphisms, the correspondence $c \otimes g \to (fc) \otimes (hg)$ defines a homomorphism

$$f \otimes h\colon \quad C \otimes G \to C' \otimes G'$$

called the homomorphism of the tensor product *induced* by the homomorphisms f and h. Precisely, the map $(c,g) \to (fc,hg)$ defines a homomorphism $R(C,G) \to R(C',G')$ which carries $Y(C,G)$ into $Y(C'G')$, and thus induces a homomorphism $f \otimes h$ of the tensor products.

In case $G = G'$ and h is the identity, we shall speak of $f \otimes h$ as the homomorphism $C \otimes G \to C' \otimes G$ induced by f and will denote it by a symbol such as f'.

The proof of the following proposition is immediate.

THEOREM 9.4 *If $\imath\colon C \to C$, $\jmath\colon G \to G$ are identity maps, then $\imath \otimes \jmath\colon C \otimes G \to C \otimes G$ is the identity. If $f\colon C \to C'$, $f'\colon C' \to C''$, $h\colon G \to G'$, $h'\colon G' \to G''$, then $(f'f) \otimes (h'h) = (f' \otimes h')(f \otimes h)$*

This theorem states that $\otimes$ is a covariant functor of two variables, in the category $\mathcal{G}_R$ with values in $\mathcal{G}_R$.

THEOREM 9.5. *Let C and G be represented as the direct sums (see 5 8)*

$$C = \sum_{\alpha \in M} C_\alpha, \qquad G = \sum_{\beta \in N} G_\beta$$

by means of the injections

$$i_\alpha\colon \ C_\alpha \to C, \qquad j_\beta\colon \ G_\beta \to G.$$

Then, for $\alpha \in M$ and $\beta \in N$, the homomorphisms

$$\imath_\alpha \otimes \jmath_\beta\colon \ C_\alpha \otimes G_\beta \to C \otimes G$$

have zero kernels, and they provide a direct sum representation

$$C \otimes G = \sum_{(\alpha,\beta)} C_\alpha \otimes G_\beta.$$

PROOF Let $p_\alpha\colon C \to C_\alpha$, $r_\beta\colon G \to G_\beta$ be maps such that $p_\alpha \imath_\alpha =$ identity, $p_{\alpha'} \imath_\alpha = 0$ for $\alpha' \neq \alpha$, $r_\beta \jmath_\beta =$ identity, and $r_{\beta'} j_\beta = 0$ for $\beta' \neq \beta$. It follows that $(p_\alpha \otimes r_\beta)(\imath_\alpha \otimes \jmath_\beta) =$ identity and $(p_{\alpha'} \otimes r_{\beta'})$

$(i_\alpha \otimes j_\beta) = 0$ for $(\alpha,\beta) \neq (\alpha',\beta')$. Any element h of $\sum C_\alpha \otimes G_\beta$ can be written uniquely as a finite sum $h = \sum h_{\alpha\beta}$ of its nonzero components. Define

$$f\colon\ \sum C_\alpha \otimes G_\beta \to C \otimes G \qquad \text{by} \qquad f(h) = \sum (i_\alpha \otimes j_\beta) h_{\alpha\beta}.$$

Then $(p_\alpha \otimes r_\beta)f$ projects h into its component $h_{\alpha\beta}$. It follows that f has kernel zero. Let $c \otimes g$ be any generator of $C \otimes G$ Then $c = \sum i_\alpha c_\alpha$ and $g = \sum j_\beta g_\beta$, and $f(\sum c_\alpha \otimes g_\beta) = c \otimes g$ Therefore f is onto, and the proof is complete

LEMMA 9 6 *If C is a free R-module with base X, then $C \otimes G$ is generated by elements $x \otimes g$ with relations $x \otimes (g_1 + g_2) = x \otimes g_1 + x \otimes g_2$ and $r(x \otimes g) = x \otimes rg$. If G is also a free R-module with base Y, then $C \otimes G$ is a free R-module with base $\{x \otimes y\}$, $x \in X$, $y \in Y$*

This is an easy consequence of 9 5 and 9 2

LEMMA 9 7 *If f is a homomorphism of B onto C, then the induced homomorphism $B \otimes G \to C \otimes G$ is also onto*

PROOF. An element of $C \otimes G$ is a finite sum $\sum c_i \otimes g_i$ For each c_i select a $b_i \in B$ such that $f(b_i) = c_i$. Then $f(\sum b_i \otimes g_i) = \sum c_i \otimes g_i$

LEMMA 9 8 *If*

$$0 \to A \xrightarrow{f} B \xrightarrow{h} C \to 0$$

is an exact sequence of R-modules and homomorphisms, then the induced sequence

$$A \otimes G \xrightarrow{f'} B \otimes G \xrightarrow{h'} C \otimes G \to 0$$

is also exact (Note: no statement is made about the kernel of f') If, further, the image of f is a direct summand of B, then the sequence

$$0 \to A \otimes G \xrightarrow{f'} B \otimes G \xrightarrow{h'} C \otimes G \to 0$$

is exact, and the image of f' is a direct summand of $B \otimes G$.

PROOF. The second part of the lemma follows from the first and from 9 5 By 9 7, h' is onto. It therefore remains to show that image f' = kernel h' Let Γ denote the image of f'. If $a \in A$, $g \in G$, then $h'f'(a \otimes g) = (hfa) \otimes g = 0 \otimes g = 0$ Therefore Γ is contained in the kernel of h'. Hence h' induces a homomorphism h''. $B \otimes G/\Gamma \to C \otimes G$ such that the composition of the natural map η: $B \otimes G \to B \otimes G/\Gamma$ followed by h'' is h'.

Since h is onto, there is a function ϕ: $C \to B$ (not necessarily a homomorphism) such that $h\phi$ is the identity Define

$$\phi'\colon\ R(C,G) \to B \otimes G/\Gamma \quad \text{by} \quad \phi'(c,g) = \eta((\phi c) \otimes g).$$

Since $R(C,G)$ is free, ϕ' is a homomorphism It is obvious that ϕ' maps $(c,g_1 + g_2) - (c,g_1) - (c,g_2)$, $r(c,g) - (c,rg)$, and $r(c,g) - (rc,g)$ into zero Suppose $c_1,c_2 \varepsilon C$ and $g \varepsilon G$ Since

$$h(\phi(c_1 + c_2) - \phi c_1 - \phi c_2) = h\phi(c_1 + c_2) - h\phi c_1 - h\phi c_2 = (c_1 + c_2) - c_1 - c_2 = 0,$$

exactness implies there is an $a \varepsilon A$ such that $f(a) = \phi(c_1 + c_2) - \phi c_1 - \phi c_2$ Hence

$$\begin{aligned}\phi'((c_1 + c_2,g) - (c_1,g) - (c_2,g)) &= \eta(\phi(c_1 + c_2) \otimes g - (\phi c_1) \otimes g - (\phi c_2) \otimes g) \\ &= \eta((\phi(c_1 + c_2) - \phi c_1 - \phi c_2) \otimes g) = \eta(f(a) \otimes g) = \eta f'(a \otimes g) = 0.\end{aligned}$$

Thus ϕ' maps $Y(C,G)$ into zero Therefore ϕ' induces a homomorphism ϕ'': $C \otimes G \to B \otimes G/\Gamma$ Furthermore

$$\phi''h''\eta(b \otimes g) = \phi''h'(b \otimes g) = \phi'((hb) \otimes g) = \eta((\phi hb) \otimes g).$$

But $\phi hb - b$ is in $f(A)$; hence $\eta((\phi hb) \otimes g) = \eta(b \otimes g)$ This implies $\phi''h''$ = identity, and, therefore, kernel $h'' = 0$ Since $h' = h''\eta$, it follows that Γ is the kernel of h' This proves 9 8

Let G be an ordinary abelian group, or equivalently, a J-module where J is the ring of integers. If p is an integer, the operation $g \to pg$ is a homomorphism of G into G whose image is denoted by pG

LEMMA 9 9 *Let G be an abelian group, p an integer, and let η $G \to G/pG$ be the natural homomorphism Then the correspondence $n \otimes g \to \eta(ng)$ defines an isomorphism*

$$(J/pJ) \otimes G \approx G/pG.$$

PROOF. Define f: $J \to J$ by $f(n) = pn$, and let h: $J \to J/pJ$ be the natural homomorphism Then f,h satisfy the hypothesis of 9 8 Therefore

$$J \otimes G \xrightarrow{f'} J \otimes G \xrightarrow{h'} (J/pJ) \otimes G \to 0$$

is exact. By 9.2, we may identify $J \otimes G$ with G. Then f' becomes the homomorphism $g \to pg$; hence kernel h' = image $f' = pG$ Since h' is onto, the stated isomorphism is proved.

LEMMA 9.10. *If p and q are integers, and r is their greatest common divisor, then*

$$(J/pJ) \otimes (J/qJ) \approx J/rJ.$$

This follows from 9 9 in view of the well-known proposition

$$(J/qJ)/p(J/qJ) = J/rJ.$$

As an example, let G be a cyclic group of order 4, and let $f\colon J \to J$ be defined by $f(n) = 2n$. Then $J \otimes G$ is cyclic of order 4, and $f'\colon J \otimes G \to J \otimes G$ is given by $f'(n \otimes g) = 2(n \otimes g)$ Thus the kernel of f is zero, while that of f' is not This example shows that the concept "subgroup" is not preserved in general under tensor products. This means that the symbol $c \otimes g$ is ambiguous in a situation where $c \in C \subset C'$ or $g \in G \subset G'$.

So far we have considered $\otimes$ as a functor (of two variables) defined on $\mathcal{G}_R$ and $\mathcal{G}_R$ with values in $\mathcal{G}_R$. However we may also consider $\otimes$ as a functor defined on $\mathcal{G}$ (= the category of ordinary abelian groups) and $\mathcal{G}_R$ with values in $\mathcal{G}_R$. Indeed let C be an abelian group and G an R-module. Define $C \otimes G$ by regarding G as an abelian group. Then convert $C \otimes G$ into an R-module by setting

$$r(c \otimes g) = c \otimes rg$$

We must verify that, if $f\colon C \to C'$, $h\colon G \to G'$ are homomorphisms, then $f \otimes h$ is a homomorphism of R-modules. Indeed we have

$$\begin{aligned} r(f \otimes h)(c \otimes g) &= r[(fc) \otimes (hg)] = (fc) \otimes r(hg) \\ &= (fc) \otimes h(rg) = (f \otimes h)(c \otimes rg) \\ &= (f \otimes h)r(c \otimes g). \end{aligned}$$

With this verified all the previously stated results remain valid for this modified $\otimes$ product.

We shall also generalize the tensor product to a functor on $\mathcal{G}'$ and $\mathcal{G}_C$ with values in $\mathcal{G}_C$, where $\mathcal{G}'$ is the subcategory of $\mathcal{G}$ consisting of finitely generated abelian groups and their homomorphisms. This requires some preliminaries.

Lemma 9.11. *Let C be a free group on a finite base $c_1, \cdots, c_n$ Each element of $C \otimes G$ can be written uniquely in the form $\sum_1^n c_i \otimes g_i$ The function defined by*

$$f\left(\sum_1^n c_i \otimes g_i\right) = (g_1, \cdots, g_n)$$

is an isomorphism of $C \otimes G$ with the direct sum G_n (= the direct product G^n) of n-factors equal to G.

Proof. If C_k is the subgroup generated by c_k, then C decomposes into $\sum C_k$. Applying 9.5, we obtain an isomorphism $C \otimes G \approx \sum C_k \otimes G$ Then 9.2 provides an isomorphism $C_k \otimes G \approx G$. Combining these yields the isomorphism f

Definition 9.12. Let G be a compact group and C a free group on the base $c_1, \cdots, c_n$ The direct product G^n of n-factors G is a compact group. The isomorphism $f\colon C \otimes G \approx G^n$ of 9.11 is now used to carry over the topology of G^n into a *topology* for $C \otimes G$. Then $C \otimes G$ is a compact group and f is continuous

LEMMA 9.13. *The topology of $C \otimes G$ is independent of the choice of the base in C.*

LEMMA 9.14 *Let C,D be free groups on finite bases, and f: $C \to D$ a homomorphism. Let G,H be compact groups and h: $G \to H$ a homomorphism. Then $f \otimes h$. $C \otimes G \to D \otimes H$ is continuous*

PROOF. We shall prove 9.13 and 9.14 simultaneously. Let $c_1, \cdots, c_n$ and $d_1, \cdots, d_m$ be bases in C and D used to define the topologies in $C \otimes G$ and $D \otimes H$. Then f is given by $f(c_i) = \sum_{j=1}^{m} a_{ij}d_j$ $(i = 1, \cdots, n)$ where (a_{ij}) is a matrix of integers. It follows that

$$(f \otimes h)\left(\sum_{i=1}^{n} c_i \otimes g_i \right) = \sum_{i=1}^{n} \left(\sum_{j=1}^{m} a_{ij}d_j \right) \otimes hg_i$$
$$= \sum_{j=1}^{m} d_j \otimes \left(\sum_{i=1}^{n} a_{ij}hg_i \right).$$

Using the isomorphisms $C \otimes G \approx G^n$, $D \otimes H \approx H^m$ of 9 11, we see that $f \otimes h$ corresponds to a map ϕ $G^n \to H^m$ given by

$$\phi(g_1, \cdots, g_n) = \left(\sum_{i=1}^{n} a_{i1}hg_i, \cdots, \sum_{i=1}^{n} a_{im}hg_i \right).$$

Since h is continuous and the a's are integers, it follows that ϕ is continuous. Hence $f \otimes h$ is continuous; and 9.14 is proved Now let $C = D$ and $G = G'$, and let f,h be the identities. Then $f \otimes h$ is the identity and is continuous This proves 9.13

DEFINITION 9.15. Let C be a group with finite set of generators, and let

$$R \xrightarrow{i} F \xrightarrow{\eta} C$$

be a representation of C as a factor group of a free group F on a finite base (i is the inclusion map, and η is the natural homomorphism). Let G be a compact group. In the induced diagram

$$R \otimes G \xrightarrow{i'} F \otimes G \xrightarrow{\eta'} C \otimes G$$

$R \otimes G$ and $F \otimes G$ are compact, i' is continuous, and by 9.8, η' is onto and kernel η' = image i'. In this way $C \otimes G$ is isomorphic to the compact group $F \otimes G/\text{image } i'$. Using this isomorphism we carry over the topology of the factor group to provide a *topology* in $C \otimes G$. This is equivalent to defining a set $U \subset C \otimes G$ to be *open* if $\eta'^{-1}(U)$ is open in $F \otimes G$.

LEMMA 9.16 *If C has a finite set of generators and G is compact, the topology of $C \otimes G$ is independent of the representation $C = F/R$.*

LEMMA 9.17. *If C,D have finite sets of generators, and G,H are compact, and f: $C \to D$, h: $G \to H$ are homomorphisms, then $f \otimes h$: $C \otimes G \to D \otimes H$ is continuous.*

PROOF. The two lemmas are proved simultaneously. Let

$$R \xrightarrow{i} F \xrightarrow{\eta} C, \qquad R_1 \xrightarrow{i_1} F_1 \xrightarrow{\eta_1} D$$

be the representations used in defining the topologies of $C \otimes G$ and $D \otimes H$. By 6.2, there is a homomorphism f_1: $F \to F_1$ such that $f\eta = \eta_1 f_1$. Therefore commutativity holds in the diagram

$$\begin{array}{ccc} F \otimes G & \xrightarrow{\eta'} & C \otimes G \\ \downarrow{\scriptstyle f_1 \otimes h} & & \downarrow{\scriptstyle f \otimes h} \\ F_1 \otimes H & \xrightarrow{\eta_1'} & D \otimes H \end{array}$$

By 9.15, η' and η_1' are continuous. By 9.14, $f_1 \otimes h$ is continuous. Then $\eta_1'(f_1 \otimes h)$ is continuous. This implies that $(f \otimes h)\eta'$ is continuous. Thus, if U is open in $D \otimes H$, it follows that $\eta'^{-1}(f \otimes h)^{-1}(U)$ is open in $F \otimes G$. By the remark at the end of 9.15, this last implies that $(f \otimes h)^{-1}(U)$ is open. Hence $f \otimes h$ is continuous, and 9.17 holds. If we specialize f and h to be identities, we obtain 9.16.

It is necessary to review previous results under the assumption that $C \otimes G$ has a topology and observe the effect on the conclusions. In 9.2, the function f is continuous since it is the one used in defining the topology. The same remark applies to the function f of 9.11, and to the isomorphism of 9.9. If the hypotheses of 9.4,7,8 are strengthened by assuming the continuity of the appropriate functions, the corresponding functions in the conclusions are also continuous by 9.16. In 9.5, if G is to be compact and C to be finitely generated, the index ranges M and N must be finite, each C_α must be finitely generated, and each G_β compact. Assuming each j_β to be continuous, then $i_\alpha \otimes i_\beta$ is continuous and the final isomorphism likewise.

Summarizing we have

THEOREM 9.18. *Let $\mathcal{G}$ denote the category of ordinary abelian groups and their homomorphisms, and let $\mathcal{G}'$ denote the subcategory of groups having finite bases. Then the tensor product is defined in the following cases.*

(1)	$C \varepsilon \mathcal{G}_R$,	$G \varepsilon \mathcal{G}_R$,	*then*	$C \otimes G \varepsilon \mathcal{G}_R$,
(2)	$C \varepsilon \mathcal{G}$,	$G \varepsilon \mathcal{G}_R$,	*then*	$C \otimes G \varepsilon \mathcal{G}_R$,
(3)	$C \varepsilon \mathcal{G}'$,	$G \varepsilon \mathcal{G}_C$,	*then*	$C \otimes G \varepsilon \mathcal{G}_C$.

All the preceding results of this section are valid in cases 1 and 2. The same is true in case 3 except for the direct sum Theorem 9.5 which is valid when restricted to finite sums

10. GROUPS OF HOMOMORPHISMS

DEFINITION 10.1. Given two R-modules C and G, denote by $\mathrm{Hom}(C,G)$ the R-module of all homomorphisms

$$\phi.\ C \rightarrow G$$

with addition $\phi_1 + \phi_2$ defined by

$$(\phi_1 + \phi_2)(c) = \phi_1(c) + \phi_2(c),$$

and with the product $r\phi$ defined by

$$(r\phi)(c) = r(\phi(c)) = \phi(rc)$$

In a sense, $\mathrm{Hom}(C,G)$ is dual to $C \otimes G$ This duality becomes apparent if the results of this section are compared with the similarly numbered results of §9.

LEMMA 10 2. *If G is an R-module, the correspondence f. $\mathrm{Hom}(R,G) \rightarrow G$ defined by $f(\phi) = \phi(1)$ is an isomorphism*

PROOF. Since $(\phi_1 + \phi_2)(1) = \phi_1(1) + \phi_2(1)$, f is a homomorphism Since 1 generates R, $f(\phi) = 0$ implies $\phi = 0$. Since R is free, there is a ϕ having any prescribed $\phi(1)$.

DEFINITION 10 3. If $f.\ C' \rightarrow C$ and h: $G \rightarrow G'$ are homomorphisms, the correspondence $\phi \rightarrow h\phi f$ defines a homomorphism

$$\mathrm{Hom}(f,h).\ \mathrm{Hom}(C,G) \rightarrow \mathrm{Hom}(C',G').$$

The proof of the following proposition is immediate·

THEOREM 10 4 *If i: $C \rightarrow C$ and j. $G \rightarrow G$ are identity maps, then $\mathrm{Hom}(i,j)$ is the identity map of $\mathrm{Hom}(C,G)$ If f $C' \rightarrow C$, f': $C'' \rightarrow C'$, h': $G \rightarrow G'$, h': $G' \rightarrow G''$, then $\mathrm{Hom}(ff',h'h) = \mathrm{Hom}(f',h')\mathrm{Hom}(f,h)$*

Briefly, the theorem asserts that Hom is a functor contravariant in the first variable in $\mathcal{G}_R$, covariant in the second variable in $\mathcal{G}_R$ and with values in $\mathcal{G}_R$.

THEOREM 10 5 *Let C be represented as a direct sum $C \approx \sum_{\alpha \in M} C_\alpha$ by means of the injections i_α· $C_\alpha \rightarrow C$. Let G be represented as a direct product $G \approx \prod_{\beta \in N} G_\beta$ by means of the projections p_β $G \rightarrow G_\beta$ Then the projections*

$$\mathrm{Hom}(i_\alpha,p_\beta)\cdot\ \mathrm{Hom}(C,G) \rightarrow \mathrm{Hom}(C_\alpha,G_\beta)$$

represent Hom(C,G) *as a direct product*

$$\mathrm{Hom}(C,G) \approx \prod \mathrm{Hom}(C_\alpha,G_\beta), \qquad (\alpha,\beta) \in M \times N.$$

PROOF. The Hom (i_α,p_β) are the components of a homomorphism f: $\mathrm{Hom}(C,G) \to \prod \mathrm{Hom}(C_\alpha,G_\beta)$. Suppose $\phi \in \mathrm{Hom}(C,G)$ is not zero. Then for some α and $c_\alpha \in C_\alpha$, we have $\phi i_\alpha c_\alpha \neq 0$. Then there is a β such that $p_\beta\phi i_\alpha c_\alpha \neq 0$. Thus $\mathrm{Hom}(i_\alpha,p_\beta)$ applied to ϕ is not zero. Hence $f\phi \neq 0$ and the kernel of f is zero Suppose $\phi_{\alpha\beta} \in \mathrm{Hom}(C_\alpha,G_\beta)$ is given for each $(\alpha,\beta) \in M \times N$. For any $c \in C$, write c as a finite sum $c = \sum i_\alpha c_\alpha$ and define $\phi_\beta c = \sum \phi_{\alpha\beta}c_\alpha$. Then $\phi_\beta \in \mathrm{Hom}(C,G_\beta)$ for each $\beta \in M$. These are the components of a $\phi' \in \mathrm{Hom}(C, \prod G_\beta)$. The corresponding $\phi \in \mathrm{Hom}(C,G)$ has the property $\mathrm{Hom}(i_\alpha,p_\beta)\phi = \phi_{\alpha\beta}$. Thus f is an isomorphism.

LEMMA 10.6. *If C is a free R-module with base X, then* Hom(C,G) *is isomorphic with the module G^X of all functions $X \to G$, the isomorphism being obtained by assigning to each ϕ: $C \to G$ the function $X \to G$ defined by ϕ.*

The proof is obvious.

LEMMA 10 7. *If f is a homomorphism of B onto C, then the kernel of the induced homomorphism f':* $\mathrm{Hom}(C,G) \to \mathrm{Hom}(B,G)$ *is zero.*

PROOF. If $\phi \in \mathrm{Hom}(C,G)$ is not zero, choose $c \in C$ such that $\phi(c) \neq 0$. Then choose $b \in B$ such that $f(b) = c$. Then $(f'\phi)b = \phi f(b) \neq 0$ Hence $f'\phi$ is not zero.

LEMMA 10.8. *If*

$$0 \to A \xrightarrow{f} B \xrightarrow{h} C \to 0$$

is an exact sequence of R-modules and homomorphisms, then the induced sequence

$$0 \to \mathrm{Hom}(C,G) \xrightarrow{h'} \mathrm{Hom}(B,G) \xrightarrow{f'} \mathrm{Hom}(A,G)$$

is exact. If, further, the image of f is a direct summand of B, then the sequence

$$0 \to \mathrm{Hom}(C,G) \xrightarrow{h'} \mathrm{Hom}(B,G) \xrightarrow{f'} \mathrm{Hom}(A,G) \to 0$$

is exact and the image of h' is a direct summand of Hom(B,G).

PROOF. The second part of the lemma follows from the first and from 10 5. By 10.7, the kernel of h' is zero. It therefore remains to prove that image h' = kernel f' If $\phi \in \mathrm{Hom}(C,G)$, then $(f'h'\phi)(a) = \phi hf(a) = 0$ since $hf = 0$. Therefore image $h' \subset$ kernel f'. Suppose

$\phi \varepsilon$ kernel f'. Then $\phi f(a) = 0$ for $a \varepsilon A$, and $\phi(b) = 0$ for $b \varepsilon$ kernel h. Hence ϕ induces a homomorphism ϕ' of B/kernel h into G. Since h is onto, B/kernel h is naturally isomorphic to C, and ϕ' can be regarded as an element of Hom(C,G). Clearly $h'\phi' = \phi$, and we have image h' = kernel f'.

LEMMA 10.9. *If p is an integer, the correspondence f. Hom$(J/pJ,G) \to G$ defined by $f(\phi) = \phi(1)$ is an isomorphism of Hom$(J/pJ,G)$ onto the subgroup of G of elements of order p.*

The proof is immediate (compare with 10 2).

LEMMA 10 10. *If p and q are integers, and r is their greatest common divisor, then*

$$\mathrm{Hom}(J/pJ,J/qJ) \approx J/rJ.$$

This follows from 10.9 and the isomorphism

$$\frac{q}{r}(J/qJ) \approx J/rJ$$

So far we have regarded Hom as a functor on $\mathcal{G}_R$ and $\mathcal{G}_R$ with values in $\mathcal{G}_R$ As in the case of $\otimes$, we shall also consider Hom as a functor on $\mathcal{G}$ (= the category of ordinary abelian groups) and $\mathcal{G}_R$ with values in $\mathcal{G}_R$ Indeed, let C be an abelian group and G an R-module Define Hom(C,G) regarding G as an abelian group Then convert Hom(C,G) into an R-module by setting for $\phi\cdot\ C \to G$ and $r \varepsilon R$,

$$(r\phi)(c) = r(\phi c).$$

If $f.\ C' \to C$ and $h.\ G \to G'$, then Hom(f,g) is a homomorphism of R-modules Indeed,

$$[r\ \mathrm{Hom}(f,h)]\phi = r(h\phi f) = h(r\phi f) = \mathrm{Hom}(f,h)(r\phi)$$

We shall also consider Hom as a functor on $\mathcal{G}$ and $\mathcal{G}_C$ with values in $\mathcal{G}_C$ Indeed, let C be an ordinary abelian group and let G be a compact abelian group. Define Hom(C,G) regarding G as an abelian group (without topology). Then convert Hom(C,G) into a topological abelian group by treating it as a subgroup of the compact group G^C of all functions $\psi\ C \to G$. Since the projections of a cartesian product are continuous, $\psi(c)$ is a continuous function of ψ in G^C and therefore the condition $\psi(c_1 + c_2) - \psi(c_1) - \psi(c_2) = 0$, for fixed $c_1,c_2 \varepsilon C$, defines a closed subset of G^C. Thus Hom(C,G), as an intersection of closed subsets of G^C, is a closed subgroup and hence is compact

It must be checked that Hom(f,h) is continuous for $f\colon\ C' \to C$ and $h\ G \to G'$ with G,G' compact and h continuous. Let U be a rectangular open set in Hom(C',G') defined by the conditions $\phi c_i \varepsilon V_i$

where $c_i \in C'$, V_i is open in G' for $i = 1, \cdots, n$. The set $\mathrm{Hom}(f,h)^{-1}U$ is then a rectangular open set defined by $\psi f c_i \in h^{-1}(V_i)$, $(i = 1, \cdots, n)$.

Summarizing we have

THEOREM 10.11. *The group* $\mathrm{Hom}(C,G)$ *is defined in the following cases*:

(1)	$C \in \mathcal{G}_R$,	$G \in \mathcal{G}_R$,	*then*	$\mathrm{Hom}(C,G) \in \mathcal{G}_R$,
(2)	$C \in \mathcal{G}$,	$G \in \mathcal{G}_R$,	*then*	$\mathrm{Hom}(C,G) \in \mathcal{G}_R$,
(3)	$C \in \mathcal{G}$,	$G \in \mathcal{G}_C$,	*then*	$\mathrm{Hom}(C,G) \in \mathcal{G}_C$

The preceding results of this section hold in all these cases.

11. HOMOLOGY GROUPS OF A CHAIN COMPLEX OVER A COEFFICIENT GROUP

The tensor product operation extends to chain complexes as follows

DEFINITION 11.1. If $K = \{C_q(K),\partial_q\}$ is a chain complex and G is a group, define $K \otimes G$ to be the chain complex $\{C_q(K) \otimes G,\partial'_q\}$ where ∂'_q is induced by ∂_q, i.e. $\partial'_q = \partial_q \otimes i$ where i is the identity map of G. If f: $K \to K'$ is a map of chain complexes, let f': $K \otimes G \to K' \otimes G$ be the map induced by f, i.e. $f'_q = f_q \otimes i$. The resulting functor from chain complexes to chain complexes is denoted by $\otimes G$. By 9.18 there are three cases:

(1)	$G \in \mathcal{G}_R$,	*then*	$\otimes G$: $\partial\mathcal{G}_R \to \partial\mathcal{G}_R$,
(2)	$G \in \mathcal{G}_R$,	*then*	$\otimes G$: $\partial\mathcal{G} \to \partial\mathcal{G}_R$,
(3)	$G \in \mathcal{G}_C$,	*then*	$\otimes G$: $\partial\mathcal{G}' \to \partial\mathcal{G}_C$.

Naturally one must show that $K \otimes G$ is a chain complex. This is proved by tensoring $\partial_{q-1}\partial_q = 0$ with i, and applying 9.4 to obtain $\partial'_{q-1}\partial'_q = 0$. Similarly $f_{q-1}\partial_q = \partial_q f_q$ implies $f'_{q-1}\partial'_q = \partial'_q f'_q$; hence f' is a map.

THEOREM 11.2 *If* $\partial\mathcal{G}_R,\partial\mathcal{G},\partial\mathcal{G}'$ *are regarded as h-categories in the sense of direct couples* (*see* 3.1), *then, in all three cases,* $\otimes G$ *is a covariant h-functor.*

PROOF. If (ϕ,ψ): $L \to K \to M$ is a direct couple, then by definition

$$0 \to L \xrightarrow{\phi} K \xrightarrow{\psi} M \to 0$$

is exact and the image of ϕ is a direct summand of K. By 9.8 the same is true of the induced sequence

$$0 \to L \otimes G \xrightarrow{\phi'} K \otimes G \xrightarrow{\psi'} M \otimes G \to 0,$$

so that (ϕ',ψ') is again a direct couple. Thus $\otimes G$ is a c-functor.

Since the functor $\otimes G$ carries an isomorphism into an isomorphism, it follows immediately that $\otimes G$ carries excisions into excisions and points into points. It remains to show that it carries homotopies into homotopies. Let $f,g\colon K \to L$, and let $D\colon f \simeq g$. By definition D is a sequence of homomorphisms $D_q\colon C_q(K) \to C_{q+1}(L)$ such that

$$\partial_{q+1}D_q + D_{q-1}\partial_q = g_q - f_q \tag{4}$$

Let $D'_q = D_q \otimes i$ where i is the identity map of G. If we tensor (4) with i and apply the relations of 9.4, we obtain (4) with primes on each homomorphism. Therefore $D'\colon f' \simeq g'$, and the proof is complete.

DEFINITION 11.3. According to IV,9.5, the composition of the h-functor $\otimes G$ with the homology theory H on $\partial\mathcal{G}_R$ [or $\partial\mathcal{G}_C$] (see 4.9) is a new homology theory defined on the domain of $\otimes G$. It is called the *homology theory with coefficient group G*. For any chain complex K, the group $H_q(K \otimes G)$ is customarily written $H_q(K;G)$ and is called the *q-dimensional homology group of K with coefficients in G*, or *the qth homology group of K over G*. As in 11.1, there are three cases:

(1)	$G \in \mathcal{G}_R$,	$K \in \partial\mathcal{G}_R$,	*then*	$H_q(K;G) \in \mathcal{G}_R$,
(2)	$G \in \mathcal{G}_R$,	$K \in \partial\mathcal{G}$,	*then*	$H_q(K;G) \in \mathcal{G}_R$,
(3)	$G \in \mathcal{G}_C$,	$K \in \partial\mathcal{G}'$,	*then*	$H_q(K;G) \in \mathcal{G}_C$

CONVENTIONS AND INTERPRETATIONS 11.4. The group of chains, cycles, and boundaries of K with coefficients in G are written $C_q(K;G)$, $Z_q(K;G)$, and $B_q(K;G)$ rather than $C_q(K \otimes G)$, etc. In keeping with this notation the chain $c \otimes g$, where $c \in C_q(K)$ and $g \in G$, will be written gc. Then, by 9.1, any element of $C_q(K;G)$ is a linear combination $\sum g_ic_i$ of elements of $C_q(K)$ with coefficients in G, and any relation is a consequence of relations of the form

$$(g_1 + g_2)c = g_1c + g_2c, \qquad g(c_1 + c_2) = gc_1 + gc_2,$$

and $(rg)c = r(gc) = g(rc)$ in case 1, or $(rg)c = r(gc)$ in case 2. Likewise the boundary operator of $K \otimes G$ is given by

$$\partial'_q \sum g_ic_i = \sum g_i(\partial_q c_i),$$

and, if $f\colon K \to K'$, then $f'\colon K \otimes G \to K' \otimes G$ is given by

$$f'_q \sum g_ic_i = \sum g_i(f_qc_i).$$

REMARK. In the case when $G = J$ is the group of integers, we agreed in §9 to identify $C \otimes J$ with C for any group C. Consequently $K \otimes J = K$ and $H_q(K;J) = H_q(K)$. Thus we shall regard the ordinary homology groups as being those based on integer coefficients.

An analogous discussion for cochain complexes may be obtained by a simple application of the sign-changing trick If K is a cochain complex, then $H = \{H^q(K \otimes G), f_*, \delta^*\}$ is a mixed theory, i.e. a covariant δ-functor satisfying the Homotopy, Excision, and Dimension axioms.

12. COHOMOLOGY GROUPS OF A CHAIN COMPLEX OVER A COEFFICIENT GROUP

The operation Hom of §10 extends to chain complexes as follows

DEFINITION 12.1. If $K = \{C_q(K), \partial_q\}$ is a chain complex and G is a group, define $\mathrm{Hom}(K,G)$ to be the cochain complex $\{\mathrm{Hom}(C_q(K),G), \delta^q\}$ where

$$\delta^q\colon\ \mathrm{Hom}(C_q(K),G) \to \mathrm{Hom}(C_{q+1}(K),G)$$

is induced by $\partial_{q+1}\colon\ C_{q+1}(K) \to C_q(K)$, i e. $\delta^q = \mathrm{Hom}(\partial_{q+1}, i)$ where i is the identity map of G. If $f\colon\ K \to K'$ is map of chain complexes, let

$$f'\colon\ \mathrm{Hom}(K',G) \to \mathrm{Hom}(K,G)$$

be the map of cochain complexes induced by f, i e $f'^q = \mathrm{Hom}(f_q, i)$. The resulting contravariant functor from chain complexes to cochain complexes is denoted by $\mathrm{Hom}(\ ,G)$. By 10.11, there are three cases:

(1)	$G \in \mathcal{G}_R$,	*then*	$\mathrm{Hom}(\ ,G)\ \ \partial\mathcal{G}_R \to \delta\mathcal{G}_R$,
(2)	$G \in \mathcal{G}_R$,	*then*	$\mathrm{Hom}(\ ,G)\colon\ \partial\mathcal{G} \to \delta\mathcal{G}_R$,
(3)	$G \in \mathcal{G}_C$,	*then*	$\mathrm{Hom}(\ ,G)\colon\ \partial\mathcal{G} \to \delta\mathcal{G}_C$.

If we apply $\mathrm{Hom}(\ ,i)$ to the relation $\partial_{q+1}\partial_{q+2} = 0$ and use 10 4, we obtain $\delta^{q+1}\delta^q = 0$; hence $\mathrm{Hom}(K,G)$ is a cochain complex. Similarly $f_{q-1}\partial_q = \partial_q f_q$ implies $\delta^{q-1}f'^{q-1} = f'^q\delta^{q-1}$; hence f' is a map.

THEOREM 12 2 *If $\partial\mathcal{G}_R$, $\partial\mathcal{G}$ are regarded as h-categories in the sense of direct couples (see 3.1), then, in all three cases,* $\mathrm{Hom}(\ ,G)$ *is a contravariant h-functor.*

PROOF. If $(\phi,\psi)\cdot\ L \to K \to M$ is a direct couple, then by definition

$$0 \to L \xrightarrow{\phi} K \xrightarrow{\psi} M \to 0$$

is exact and the image of ϕ is a direct summand of K. By 10.8, the induced sequence

$$0 \to \mathrm{Hom}(M,G) \xrightarrow{\psi'} \mathrm{Hom}(K,G) \xrightarrow{\phi'} \mathrm{Hom}(L,G) \to 0$$

is likewise exact and the image of ψ' is a direct summand. Hence (ψ',ϕ') is a direct couple This proves that $\mathrm{Hom}(\ ,G)$ is a contravariant c-functor.

Since $\text{Hom}(f_q,i)$ is an isomorphism whenever f_q,i are isomorphisms, it follows that $\text{Hom}(\ ,G)$ carries excisions into excisions.

If the chain complex K is pointlike, i e ∂_q: $C_q(K) \approx C_{q-1}(K)$ for q even and >0, and for q odd and <0, then δ^{q-1} is likewise an isomorphism under the same conditions Therefore $\text{Hom}(K,G)$ is a pointlike cochain complex since, after application of the sign-changing trick, we obtain a pointlike chain complex

It remains to show that $\text{Hom}(\ ,G)$ carries homotopies into homotopies Let f,g: $K \to L$, and let D: $f \simeq g$ Then D is a sequence D_q: $C_q(K) \to C_{q+1}(L)$ such that

$$\partial_{q+1}D_q + D_{q-1}\partial_q = g_q - f_q \tag{4}$$

Let $D'^q = \text{Hom}(D_{q-1},i)$ where i is the identity map of G. Then

$$D'^q\text{: } \text{Hom}(C_q(L),G) \to \text{Hom}(C_{q-1}(K),G).$$

Applying $\text{Hom}(\ ,i)$ to (4) and using 10.4, we obtain

$$D'^{q+1}\delta^q + \delta^{q-1}D'^q = g'^q - f'^q.$$

Therefore D': $f' \simeq g'$, and the proof is complete.

Definition 12.3. According to 9 5c, the composition of the contravariant h-functor $\text{Hom}(\ ,G)$ with the covariant δ-functor H on $\delta\mathcal{G}_R$ [or $\delta\mathcal{G}_C$] (see 4 9c) is a cohomology theory defined on the domain of $\text{Hom}(\ ,G)$. It is called *the cohomology theory of chain complexes with coefficient group G* For any chain complex K, the group $H^q(\text{Hom}(K,G))$ is customarily written $H^q(K;G)$ and is called the *q-dimensional cohomology group of K with coefficient group G*, or *the qth cohomology group of K over G*. As in 12 1, there are three cases:

(1)	$G \in \mathcal{G}_R$,	$K \in \partial\mathcal{G}_R$,	*then*	$H^q(K;G) \in \mathcal{G}_R$,
(2)	$G \in \mathcal{G}_R$,	$K \in \partial\mathcal{G}$,	*then*	$H^q(K;G) \in \mathcal{G}_R$,
(3)	$G \in \mathcal{G}_C$,	$K \in \partial\mathcal{G}$,	*then*	$H^q(K;G) \in \mathcal{G}_C$.

Conventions and interpretations 12.4. The group of cochains, cocycles, and coboundaries of $\text{Hom}(K,G)$ are written $C^q(K;G)$, $Z^q(K;G)$, and $B^q(K,G)$ rather than $C^q(\text{Hom}(K;G))$, etc. By definition, a q-cochain of K is a homomorphism ϕ: $C_q(K) \to G$ Its coboundary $\delta^q\phi$ is the $(q+1)$-cochain $\phi\partial_{q+1}$: $C_{q+1}(K) \to G$. It follows that ϕ is a q-cocycle (i.e. $\delta^q\phi = 0$) if and only if ϕ maps $B_q(K)$ into zero. It also follows that a q-coboundary maps $Z_q(K)$ into zero; however, in general, this property is not sufficient to characterize a coboundary. If f. $K \to K'$ is a map of chain complexes, then the induced f': $\text{Hom}(K',G) \to \text{Hom}(K,G)$ is given by

$$(f'^q\phi')x = \phi' f_q x, \qquad \phi' \in C^q(K';G), \qquad x \in C_q(K).$$

An analogous discussion for cochain complexes may be obtained by a simple application of the sign-changing trick If K is a cochain complex, then $H = \{H_q(\text{Hom}(K,G)), f^*, \partial_*\}$ is a contravariant ∂-functor satisfying the Homotopy, Excision, and Dimension axioms.

13. COMPARISON OF VARIOUS COEFFICIENT GROUPS

In this section we shall establish various results bearing on the homology and cohomology groups of a chain complex for various coefficient groups The so-called "universal coefficient theorem" has not been included since its statement requires new group operations, besides $\otimes$ and Hom, which are not of sufficient importance in the sequel to be studied here (see Exercises v,G).

LEMMA 13 1 *If K is a chain complex composed of free abelian groups [or of vector spaces over a field F], then $Z_q(K)$ is a direct summand of $C_q(K)$.*

PROOF As a subgroup of a free group, the group $B_{q-1}(K)$ is free (see 6.8). Since ∂_q maps $C_q(K)$ onto $B_{q-1}(K)$ with $Z_q(K)$ as kernel, it follows from 6 3 that $Z_q(K)$ is a direct summand of $C_q(K)$.

LEMMA 13 2 *If K is a chain complex such that $Z_q(K)$ is a direct summand of $C_q(K)$ and such that $H_q(K) = 0$ for all q, then there exist homomorphisms D_q. $C_q(K) \to C_{q+1}(K)$ such that*

$$\partial_{q+1}D_qx + D_{q-1}\partial_qx = x$$

for all $x \in C_q(K)$.

PROOF. Let $C_q(K) = Z_q(K) + W_q$ be a direct sum decomposition. Then ∂_q maps W_q isomorphically onto $B_{q-1}(K)$ Since $H_{q-1}(K) = 0$, we have $B_{q-1}(K) = Z_{q-1}(K)$. Let $\bar{\partial}_q$: $W_q \to Z_{q-1}(K)$ denote the map defined by ∂_q. Define D_q: $C_q(K) \to C_{q+1}(K)$ as follows

$$\begin{aligned} D_qx &= 0 && \text{for } x \in W_q, \\ D_qx &= \bar{\partial}_{q+1}^{-1}x && \text{for } x \in Z_q(K). \end{aligned}$$

For $x \in W_q$ we have $\partial Dx + D\partial x = \bar{\partial}^{-1}\partial x = x$, while for $x \in Z_q$ we have $\partial Dx + D\partial x = \partial\bar{\partial}^{-1}x = x$. Thus the relation $\partial Dx + D\partial x = x$ holds for all $x \in C_q(K)$.

THEOREM 13 3 *Let K and L be chain complexes composed of free abelian groups [or of vector spaces over a field F] and let f $K \to L$ be a map In order that f be a homotopy equivalence* (IV,9.2), *it is necessary and sufficient that*

$$f_*\colon\ H_q(K) \approx H_q(L)$$

for all dimensions q.

PROOF. The necessity of the condition is an immediate consequence of 4.4. In order to prove the sufficiency we construct a new chain complex $\hat{f}$ as follows:

$$C_q(\hat{f}) = C_{q-1}(K) + C_q(L) \qquad \text{(direct sum)},$$
$$\partial_q(x,y) = (-\partial_{q-1}x, \partial_q y + f_{q-1}x), \qquad x \in C_{q-1}(K), \qquad y \in C_q(L).$$

Since $\partial\partial(x,y) = \partial(-\partial x, \partial y + fx) = (\partial\partial x, \partial\partial y + \partial fx - f\partial x) = 0$, this definition indeed yields a chain complex $\hat{f}$.

Next we show that the condition of the theorem implies $H_q(\hat{f}) = 0$ for all q. Indeed, let $(x,y) \in Z_q(\hat{f})$. Then $\partial x = 0$ and $-\partial y = fx$. Then $x \in Z_{q-1}(K)$ and $fx \in B_{q-1}(L)$. Since the kernel of f_* is zero, it follows that $x \in B_{q-1}(K)$. Thus there exists an $x' \in C_q(K)$ with $\partial x' = x$. Then $\partial(y + fx') = \partial y + f\partial x' = \partial y + fx = 0$, so that $y + fx' \in Z_q(L)$. Since f_* maps $H_q(K)$ onto $H_q(L)$, there exist $x'' \in Z_q(K)$ and $y' \in C_{q+1}(L)$ such that $fx'' + \partial y' = y + fx'$. Then

$$\partial(x'' - x', y') = (\partial x' - \partial x'', \partial y' + fx'' - fx') = (x,y),$$

and $H_q(\hat{f}) = 0$.

In view of 13.1, the groups $Z_q(\hat{f})$ are direct summands of $C_q(\hat{f})$, and therefore 13.2 implies the existence of homomorphisms D_q: $C_q(\hat{f}) \to C_{q+1}(\hat{f})$ such that

(i) $$\partial_{q+1}D_q(x,y) + D_{q-1}\partial_q(x,y) = (x,y).$$

Each of the homomorphisms D_q yields four homomorphisms

$$D'_{q-1}\colon\ C_{q-1}(K) \to C_q(K), \qquad D''_q\colon\ C_q(L) \to C_{q+1}(L),$$
$$h_q\colon\ C_q(L) \to C_q(K), \qquad E_{q-1}\colon\ C_{q-1}(K) \to C_{q+1}(L),$$

such that

$$D_q(x,y) = (D'_{q-1}x + h_q y, E_{q-1}x + D''_q y).$$

Upon computation, condition (i) yields

(ii) $x = -D'\partial x + h\partial y + hfx - \partial D'x - \partial hy$,
(iii) $y = -E\partial x + D''\partial y + D''fx + \partial Ex + \partial D''y + fD'x + fhy$.

Substituting $x = 0$ in (ii) yields $h\partial y = \partial hy$ so that h: $L \to K$ is a map. Substituting $y = 0$ in (ii) and $x = 0$ in (iii) yields

$$\partial D'x + D'\partial x = hfx - x, \qquad \partial D''y + D''\partial y = y - fhy.$$

Thus D': $hf \simeq i_1$ and $-D''$: $fh \simeq i_2$ where i_1 and i_2 are the identity maps of K and L respectively, so that f is a homotopy equivalence.

THEOREM 13.4. *Let K and L be chain complexes composed of free abelian groups [or of vector spaces over a field F], and let f: $K \to L$ be a map such that*

$$f_*\colon\ H_q(K) \approx H_q(L)$$

for all q. Then

$$f_*: \ H_q(K;G) \approx H_q(L;G), \qquad f^*: \ H^q(L;G) \approx H^q(K;G)$$

for any coefficient group G [or any vector space G over F].

PROOF. By 13.3, f is a homotopy equivalence. Since the functors $\otimes G$ and Hom(,G) carry homotopies into homotopies, it follows that they carry homotopy equivalences into homotopy equivalences Thus the maps

$$f': \ K \otimes G \to L \otimes G, \qquad f'': \ \mathrm{Hom}(L,G) \to \mathrm{Hom}(K,G)$$

are homotopy equivalences. Therefore the induced homomorphisms f_* and f^* are isomorphisms.

NOTE

Abstract cell complexes One often encounters chain complexes K in which each group $C_q(K)$ is a free abelian group with a given base $\{\sigma_i^q\}$. In defining such an "abstract cell complex," it suffices to indicate the cells in each dimension and the boundary

$$\partial\sigma_i^q = \sum_j [\sigma_i^q:\sigma_j^{q-1}]\sigma_j^{q-1}$$

of each "cell" σ_i^q. The coefficients $[\sigma_i^q:\sigma_j^{q-1}]$ are called "incidence numbers"; for a fixed i, only a finite number of them are different from zero The condition $\partial\partial = 0$ is equivalent to

(1) $$\sum_j [\sigma_i^q:\sigma_j^{q-1}][\sigma_j^{q-1}:\sigma_k^{q-2}] = 0$$

for any pair $\sigma_i^q, \sigma_k^{q-2}$.

Let A^q denote the matrix whose $(i,j)^{th}$ element is $[\sigma_i^q:\sigma_j^{q-1}]$. Then each row of A^q corresponds to a q-cell and each column of A^q corresponds to a $(q-1)$-cell. Condition (1) becomes

$$A^qA^{q-1} = 0$$

If K and L are two such complexes with cells σ and τ respectively, then a map f: $K \to L$ is determined by the coefficients c_{ij}^q in $f\sigma_i^q = \sum_j c_{ij}\tau_j^q$. If we denote by C^q the matrix of the integers c_{ij}^q, we find that the condition $\partial f = f\partial$ is equivalent to

$$C^qB^q = A^qC^{q-1}$$

where $\{B^q\}$ are the incidence matrices of the complex L.

EXERCISES

A. Homomorphisms of exact sequences.

Let $f.\ K \to K'$ be a map of the exact lower sequence K into another such sequence K' Regard both K and K' as chain complexes and denote their homomorphisms by ∂ and ∂'. The kernel of f is a subcomplex L of K, and the image of f is a subcomplex L' of K'. This gives couples

$$0 \to L \to K \to K/L \to 0, \qquad 0 \to L' \to K' \to K'/L' \to 0.$$

1. For each index q, the various subgroups of $C_q(K)$ form a lattice as follows:

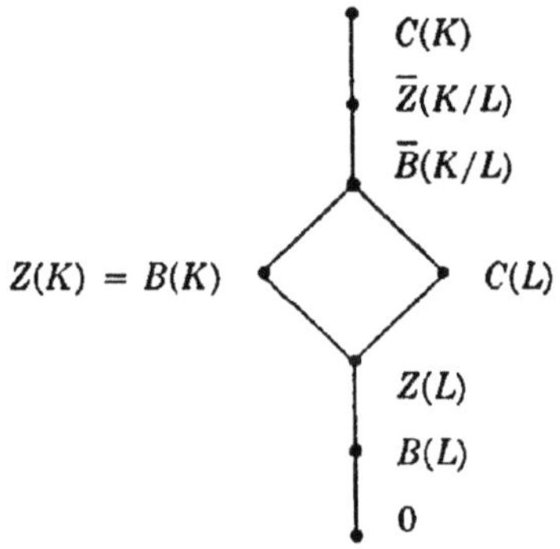

2. Show that Exercise 1 holds with K',L' in place of K,L.
3. Show that the above lattices of subgroups of $C_q(K)$, $C_q(K')$ form a closed family under the six operations ∂, ∂^{-1}, ∂', ∂'^{-1}, f, and f^{-1}.
4. For each index q, we have the isomorphisms

$$H_{q+1}(K'/L') \overset{\partial'}{\approx} H_q(L') \overset{(g_*)^{-1}}{\approx} H_q(K/L) \overset{\partial}{\approx} H_{q-1}(L)$$

where $g\ \ K/L \to L'$ is induced by f.

5. Derive from 4 the following proposition due to R. H Fox: If $\{f_q\}$ is a homomorphism of the exact sequence $\{C_q,\phi_q\}$ into another exact sequence $\{C'_q,\phi'_q\}$, then

$$\phi'^{-1}_{q+1}(\text{image } f_q)/(\text{image } f_{q+1}) \cup (\text{image } \phi'_{q+2})$$
$$\approx (\text{kernel } f_{q-1}) \cap (\text{kernel } \phi_{q-1})/\phi_q(\text{kernel } f_q).$$

B. CHAIN HOMOTOPIES.

1. Let f: $K \to K'$ be a map of chain complexes, and let $D = \{D_q\}$ be a sequence of homomorphisms D_q: $C_q(K) \to C_{q+1}(K')$. Show that

$$g_q = f_q + \partial_{q+1}D_q + D_{q-1}\partial_q$$

defines a map g: $K \to K'$ and that D: $f \simeq g$.

2. Let f,f': $K \to K'$, g,g'. $K' \to K$ be maps of chain complexes Suppose that g is the homotopy inverse of f and that $f \simeq f'$. Prove that g' is a homotopy inverse of f' if and only if $g' \simeq g$.

3. Let f: $K \to K'$, g: $K' \to K''$. If any two of the three maps f, g, and gf are homotopy equivalences, so is the third

C. FREE GROUPS AND COMPLEXES.

1. Show that, for abelian groups, a group H is free if and only if 6 2 holds

2. If F is free and $0 \to G' \to G \to G'' \to 0$ is exact, then the same is true in the induced sequences

$$0 \to F \otimes G' \to F \otimes G \to F \otimes G'' \to 0,$$
$$0 \to \mathrm{Hom}(F,G') \to \mathrm{Hom}(F,G) \to \mathrm{Hom}(F,G'') \to 0.$$

3. Let K be a chain complex composed of free abelian groups, and $0 \to G' \to G \to G'' \to 0$ an exact sequence of groups. Show that the induced sequences

$$0 \to K \otimes G' \to K \otimes G \to K \otimes G'' \to 0,$$
$$0 \to \mathrm{Hom}(K,G'') \to \mathrm{Hom}(K,G) \to \mathrm{Hom}(K,G') \to 0.$$

are exact and write down the exact sequences of the couples that they represent. Give an explicit definition of the homomorphisms

$$H_q(K,G'') \to H_{q-1}(K;G'), \qquad H^q(K,G') \to H^{q+1}(K,G'')$$

and examine their properties.

4. Let $0 \to L \to K \to M \to 0$ be an exact sequence of complexes composed of free abelian groups and $0 \to G' \to G \to G'' \to 0$ an exact sequence of coefficient groups There arise six homology sequences Arrange these sequences in a single diagram. Carry out a similar discussion of the cohomology sequences.

5. Let K and M be chain complexes composed of free abelian groups and let f: $K \to M$ be a map of K onto M. Establish the equivalence of the following conditions:

(a) f_*: $H_q(K) \approx H_q(M)$ for all q.

(b) If L is the kernel of f, then $H_q(L) = 0$ for all q

(c) There exists a map g: $M \to K$ such that $fh = i_K$, and there is a chain homotopy $D\colon hf \simeq i_M$ such that $fD = 0$ (i.e. the values of D are in N). Here i_K and i_M denote appropriate identity maps.

D. Homology groups of maps.

Definition. Given a map $f\colon K \to L$ of chain complexes, consider the chain complex $\hat{f}$ defined in the proof of 13.3. The groups $H_q(\hat{f})$ may be called the homology groups of the map f and denoted by $H_q(f)$.

1 Condider the maps $k\colon C_q(L) \to C_q(\hat{f})$ and $l\colon C_q(\hat{f}) \to C_{q-1}(K)$ defined by $ky = (0,y)$, $l(x,y) = -x$. Show that k is a map $k.\ L \to \hat{f}$, and that l lowers the dimension by 1 and commutes with ∂ Establish the exactness of the sequence

$$\cdots \to H_q(K) \xrightarrow{f_*} H_q(L) \xrightarrow{k_*} H_q(\hat{f}) \xrightarrow{l_*} H_{q-1}(K) \to$$

where l_* is induced by l in the obvious manner

E Properties of $\otimes$.

In all the subsequent exercises it is assumed that all groups are in the category $\mathcal{G}$. The restatements for other categories are left to the reader

1 Establish the natural isomorphisms

$$A \otimes B \approx B \otimes A, \qquad (A \otimes B) \otimes C \approx A \otimes (B \otimes C)$$

for ordinary abelian groups A,B,C

2 If $C' \subset C$ and $G' \subset G$, then the natural homomorphisms

$$\begin{array}{ccccc} & & C' \otimes G & & \\ & & \downarrow & & \\ C \otimes G' & \to & C \otimes G & \to & C \otimes (G/G') \\ & & \downarrow & & \downarrow \\ & & (C/C') \otimes G & \to & (C/C') \otimes (G/G') \end{array}$$

arise. Show that $C \otimes G \to (C/C') \otimes (G/G')$ is onto, and its kernel is the group

$$\text{image } (C' \otimes G) \cup \text{image } (C \otimes G')$$

3. If all elements of C or of G are of finite order, the same is true of $C \otimes G$.

Definition A group G is *infinitely divisible* if, for every $g \in G$ and every integer $n \neq 0$, there is a $g' \in G$ such that $ng' = g$.

4. If C or G is infinitely divisible, then so is $C \otimes G$.

5. If all elements of C are of finite order and G is infinitely divisible, then $C \otimes G = 0$.

6. Let $G = \prod G_\beta$ be a direct product and $p_\beta\colon G \to G_\beta$ the pro-

jection. The induced homomorphisms p'_β. $C \otimes G \to C \otimes G_\beta$ are the components of a homomorphism p': $C \otimes G \to \prod C \otimes G_\beta$ Show that p' has kernel zero, and show, by an example, that p' need not be onto If C is finitely generated, then p' is an isomorphism. If, further, the G_β are compact groups, then p' is continuous.

7. State and prove the analogs of 9.5 and 10 5 for direct products instead of direct sums assuming that M is finite, and each C_α is finitely generated. Include the case of compact groups G_β.

F. RELATIONS BETWEEN $\otimes$ AND Hom.

DEFINITION A *bihomomorphism* ϕ of the groups C_1 and C_2 with values in a group G is a function $\phi(c_1,c_2) \in G$, $c_1 \in C_1$, $c_2 \in C_2$ such that

$$\phi(c_1 + c'_1,c_2) = \phi(c_1,c_2) + \phi(c'_1,c_2), \ \phi(c_1,c_2 + c'_2) = \phi(c_1,c_2) + \phi(c_1,c'_2)$$

Let $\text{Hom}(C_1,C_2;G)$ denote the group of all such bihomomorphisms We assume that C_1 and C_2 are in the category $\mathcal{G}$ while both G and $\text{Hom}(C_1,C_2,G)$ are in one of the categories $\mathcal{G}_R$ or $\mathcal{G}_C$.

1. Establish the following natural isomorphisms:

$$\begin{aligned}\text{Hom}(C_1,C_2;G) &\approx \text{Hom}(C_1 \otimes C_2,G)\\ &\approx \text{Hom}(C_1, \text{Hom}(C_2,G))\\ &\approx \text{Hom}(C_2, \text{Hom}(C_1,G)).\end{aligned}$$

2. If K is a chain complex, then $H^q(K, \text{Hom}(C,G)) \approx H^q(K \otimes C,G)$

3. Let J be the group of integers, and define a homomorphism

$$\theta \quad C \otimes G \to \text{Hom}(\text{Hom}(C,J),G)$$

by setting for each $c \otimes g \in C \otimes G$ and each ϕ: $C \to J$

$$\theta_{c\otimes g}(\phi) = \phi(c)g.$$

Prove that θ is a natural transformation of the appropriate functors Prove that, if C is a free group on a finite base, then θ is an isomorphism

4. Let K be a chain complex such that each of the groups $C_q(K)$ is a free group on a finite base Define a cochain complex $\overline{K} = \text{Hom}(K,J)$ Prove that $K \otimes G \approx \text{Hom}(\overline{K},G)$ and $\text{Hom}(K,G) \approx \overline{K} \otimes G$, where G is a group in $\mathcal{G}$, $\mathcal{G}_C$, or $\mathcal{G}_F$. Consequently the homology [cohomology] groups of K over any coefficient group coincide with those homology [cohomology] groups of $\overline{K}$

G. UNIVERSAL COEFFICIENT THEOREMS.

1. Let F be a free group, R a subgroup of F, and G any group. Consider the homomorphisms

$$i': \ R \otimes G \to F \otimes G, \qquad i'': \ \text{Hom}(F,G) \to \text{Hom}(R,G)$$

induced by the inclusion i: $R \subset F$. Show that the groups

$$\text{kernel } i', \qquad \text{Hom}(R,G)/\text{image } i''$$

are essentially functions of the groups $H = F/R$ and G Denote these new groups by $\text{Tor}(H,G)$ and $\text{Ext}(H,G)$ and examine their properties

2 Let K be a chain complex composed of free groups. Establish exact sequences

$$\to B_q \otimes G \to Z_q \otimes G \to H_q(K,G) \to B_{q-1} \otimes G \to$$

$$\to \text{Hom}(Z^{q-1},G) \to \text{Hom}(B^{q-1},G) \to H^q(K,G) \to \text{Hom}(Z^q,G) \to$$

(Hint. Consider the exact sequence $0 \to Z \to K \to K/Z \to 0$ where $Z = \{Z_q(K),0\}$)

3 Using the results of 2 establish exact sequences

$$0 \to H_q(K) \otimes G \xrightarrow{\alpha} H_q(K,G) \xrightarrow{\beta} \text{Tor}(H_{q-1}(K),G) \to 0$$

$$0 \to \text{Ext}(H_{q-1}(K),G) \xrightarrow{\bar\beta} H^q(K,G) \xrightarrow{\bar\alpha} \text{Hom}(H_q(K),G) \to 0$$

and show that the images of α and $\bar\beta$ are direct summands Examine the behavior of these exact sequences under maps f $K \to K'$

The above results yield the following isomorphisms for the chain complex K:

$$H_q(K;G) \approx H_q(K) \otimes G + \text{Tor}(H_{q-1}(K),G),$$

$$H^q(K,G) \approx \text{Hom}(H_q(K),G) + \text{Ext}(H_{q-1}(K),G)$$

These are known as the "Universal Coefficient Theorems" since they express $H_q(K,G)$ and $H^q(K;G)$ over any coefficients in terms of $H_q(K)$ and $H_{q-1}(K)$ with integral coefficients For more details see S Eilenberg and S. MacLane, [*Group extensions and homology*, Annals of Math 43 (1942), 757-831], see also H Cartan and S Eilenberg, *Homological Algebra* (Princeton University Press) 1956

4 Using the above results show that, if f: $K \to K'$ is a map of a free chain complex into another such complex, and if f_* $H_q(K) \approx H_q(K')$ and f_*. $H_{q-1}(K) \approx H_{q-1}(K')$, then f_* $H_q(K,G) \approx H_q(K,G)$ and f^*: $H^q(K;G) \approx H^q(K',G)$ for any coefficient domain G

CHAPTER VI

Formal homology theory of simplicial complexes

1. INTRODUCTION

In this chapter we develop the formal homology and cohomology theories of simplicial complexes This is achieved by associating a chain complex with each simplicial complex and then using the definitions and results of Chapter v The formalism leading from simplicial to chain complexes is strongly influenced by the results of Chapter iv These definitions point the way to the existence proofs of Chapters vii and ix

The homology theory of simplicial complexes is constructed in two ways. The classical procedure attaches to each simplicial complex K a chain complex K_a which we call the *alternating* chain complex of K. The definition of K_a is completely motivated by the results of Chapter iii. The other procedure attaches to each K the *ordered* chain complex K_o There is a natural mapping $K_o \to K_a$ which induces isomorphisms of their homology groups The use of K_o is advantageous in proofs of general theorems It is formally simpler since one does not need to worry about relations in groups and possible degeneracy On the other hand, the groups $C_q(K_o)$ are unnecessarily large Consequently, the chain complex K_a is used whenever it is necessary to compute the groups of a complex

Since this chapter deals almost exclusively with the formal relations in a complex K, and the underlying space $|K|$ is not utilized, the assumption that the complex is finite may be dropped. To this end we reproduce a definition already made in the Exercises of Chapter ii

DEFINITION 1 1. Let W be an infinite set of objects called vertices A *complex K with vertices in W* is a collection of (finite dimensional) simplexes whose vertices are in W, subject to the condition that a face of simplex in the collection is also in the collection.

The concepts of "subcomplex" and "simplicial map" are introduced in the obvious way. The category of pairs of infinite complexes and simplicial maps will be denoted by $\mathcal{K}$. The term "infinite" is always used in the sense of "finite or infinite," so that the finite complexes form a subcategory $\mathcal{K}'_s$ of $\mathcal{K}$.

2. THE ORDERED CHAIN COMPLEX OF A SIMPLICIAL COMPLEX

DEFINITION 2.1 If K is a simplicial complex, an array $A^0 \cdots A^q$ ($q \geq 0$) of vertices of K, included among the vertices of some simplex of K, is called an *elementary q-chain of K* Precisely an elementary q-chain is a function which, to each integer $i = 0, \cdots, q$, assigns a vertex A^i of K such that $A^0, \cdots, A^q$ all lie in a simplex of K. The free group generated by this set of elementary q-chains of K (see v,6 5) will be denoted by $C_q(K_o)$ By definition, $C_q(K_o) = 0$ for $q < 0$.

For each elementary q-chain $A^0 \cdots A^q$ $(q > 0)$, define

$$\partial_q(A^0 \cdots A^q) = \sum_{i=0}^{q} (-1)^i A^0 \cdots \hat{A}^i \cdots A^q$$

where the circumflex over a vertex indicates that the vertex is omitted Having defined ∂_q for the generators of $C_q(K_o)$, a homomorphism

$$\partial_q\colon\ C_q(K_o) \to C_{q-1}(K_o)$$

is uniquely determined If $q \leqq 0$, then $\partial_q = 0$, by definition

LEMMA 2 2 $\partial_{q-1}\partial_q = 0$

PROOF If $q \leqq 1$, then $\partial_{q-1} = 0$, by definition Suppose therefore that $q \geqq 2$ It is sufficient to verify that $\partial_{q-1}\partial_q(A^0 \cdots A^q) = 0$ for $q \geqq 2$ Let $0 \leqq k < l \leqq q$ Since

$$\partial_{q-1}\partial_q(A^0 \cdots A^q) = \sum_{i=0}^{q} (-1)^i \partial_{q-1}(A^0 \cdots \hat{A}^i \cdots A^q),$$

the symbol $A^0 \cdots \hat{A}^k \cdots \hat{A}^l \cdots A^q$ will occur in the expression for $(-1)^k\partial_{q-1}(A^0 \cdots \hat{A}^k \cdots A^q)$ with the coefficient $(-1)^k(-1)^{l-1}$ and in the expression for $(-1)^l\partial_{q-1}(A^0 \cdots \hat{A}^l \cdots A^q)$ with the sign $(-1)^l(-1)^k$ Hence the two terms cancel, which proves the proposition

It follows from 2 2 that $\{C_q(K_o),\partial_q\}$ is a chain complex This chain complex will be denoted by K_o If L is a subcomplex of K, then $C_q(L_o)$ is generated by a subset of the set generating $C_q(K_o)$, and ∂_q on L_o agrees with ∂_q on K_o Therefore L_o is a subcomplex of K_o and is a direct summand of K_o Given $c \varepsilon C_q(K_o)$ we shall write $c \subset L$ to denote that $c \varepsilon C_q(L_o)$

DEFINITION 2 3 For each simplicial pair (K,L), the chain complex K_o/L_o is called the *ordered* chain complex of the pair (K,L). The groups $C_q(K_o/L_o)$ are free groups, and if K is finite, K_o/L_o is a finite chain complex in the sense of v,8 1.

LEMMA 2 4 *If f: $(K,L) \to (K',L')$ is simplicial, the homomorphisms f_q: $C_q(K_o) \to C_q(K'_o)$ and f_q $C_q(L_o) \to C_q(L'_o)$ defined by*

$$f_q(A^0 \cdots A^q) = f(A^0) \cdots f(A^q)$$

define a map

$$f_o\colon\ K_o/L_o \to K'_o/L'_o.$$

Moreover, if $f\colon (K,L) \to (K,L)$ is the identity, then f_o is the identity, and, if $f\colon (K,L) \to (K',L')$, $g\colon (K',L') \to (K'',L'')$, then $(gf)_o = g_o f_o$.

The proof only requires the verification of the commutativity relation $\partial_q f_q = f_{q-1}\partial_q$ which is an immediate consequence of the definitions.

The lemma states that K_o/L_o and f_o form a covariant functor O on the category $\mathcal{K}_s$ of simplicial pairs and simplicial maps with values in the category $\partial\mathcal{G}$ of chain complexes (recall that $\mathcal{G}$ is the category of ordinary abelian groups, i.e. $\mathcal{G} = \mathcal{G}_R$ where R = the integers). We now convert $\mathcal{K}_s$ into a c-category by defining couples (i,j) to consist of the inclusion maps $i\colon L \to K$, $j\colon K \to (K,L)$ for each pair (K,L) in $\mathcal{K}_s$. Since i_o is the inclusion map $L_o \to K_o$, it follows that the sequence

$$0 \to L_o \xrightarrow{i_o} K_o \xrightarrow{j_o} K_o/L_o \to 0$$

is exact, thus (i_o,j_o) is a couple on the category $\partial\mathcal{G}$. Since, as was observed earlier, L_o is a direct summand of K_o, it follows that the couple (i_o,j_o) is direct in the sense of v,3.1.

Summarizing we have

THEOREM 2.5. *The pair $K_o/L_o, f_o$ forms a covariant c-functor O on the category $\mathcal{K}_s$ of simplicial complexes and simplicial maps with values in the c-category $\partial\mathcal{G}$ of chain complexes (with only direct couples considered).*

3. HOMOTOPIES, EXCISIONS, POINTS

DEFINITION 3.1. Two simplicial maps $f,g\colon (K,L) \to (K',L')$ are called *contiguous* if, for every simplex s of K [of L], the simplexes $f(|s|)$ and $g(|s|)$ are faces of a single simplex of K' [of L']. This relation will play the role of homotopy in the category $\mathcal{K}_s$.

We use the term "contiguity" instead of homotopy to avoid confusion with the homotopy of the maps $f,g\colon (|K|,|L|) \to (|K'|,|L'|)$ of the associated topological spaces (when K and K' are finite complexes). Indeed if f and g are contiguous, then they are also homotopic, and $h(\alpha,t) = (1 - t)f(\alpha) + tg(\alpha)$ for $\alpha \in |K|$, $0 \leqq t \leqq 1$ is a homotopy. The converse is generally false: f and g may be homotopic without being contiguous.

THEOREM 3.2. *If the simplicial maps $f,g\colon (K,L) \to (K',L')$ are contiguous, then the induced chain maps $f_o,g_o\colon K_o/L_o \to K'_o/L'_o$ are chain homotopic.*

PROOF. Given an elementary chain $c = A^0 \cdots A^q$ in K, define

$$D_q c = \sum_{i=0}^{q} (-1)^i fA^0 \cdots fA^i gA^i \cdots gA^q,$$

This defines homomorphisms D_q: $C_q(K_o/L_o) \to C_{q+1}(K'_o/L'_o)$. It remains to verify that

$$\partial D_q c = g_o c - f_o c - D_{q-1}\partial c$$

For simplicity, let $B^i = fA^i$, $C^i = gA^i$. Then

$$\partial D_q c = \sum_{i=0}^{q} (-1)^i \Big[\sum_{j=0}^{i-1} (-1)^j B^0 \cdots \hat{B}^j \cdots B^i C^i \cdots C^q$$
$$+ (-1)^i B^0 \cdots B^{i-1} C^i \cdots C^q + (-1)^{i+1} B^0 \cdots B^i C^{i+1} \cdots C^q$$
$$+ \sum_{j=i+1}^{q} (-1)^{j+1} B^0 \cdots B^i C^i \cdots \hat{C}^j \cdots C^q \Big].$$

The terms on the middle line, when summed on i, cancel in pairs except for the initial and end terms.

$$C^0 \quad C^q = g_o c, \qquad -B^0 \quad B^q = -f_o c$$

If the order of summation of the terms on the first and third lines are interchanged, one obtains $-D_{q-1}\partial c$

A second and more conceptual proof of 3 2 will be given at the end of §5.

DEFINITION 3 3 If K' and K'' are subcomplexes of a chain complex K, we denote by $K' \cap K''$ and $K' \cup K''$ the subcomplexes of K defined by

$$C_q(K' \cap K'') = C_q(K') \cap C_q(K''),$$
$$C_q(K' \cup K'') = C_q(K') \cup C_q(K'')$$

where the last group is the least subgroup of $C_q(K)$ containing $C_q(K')$ and $C_q(K'')$ (sometimes written as $C_q(K') + C_q(K'')$)

The following lemma is an immediate consequence of the definitions:

LEMMA 3 4 *If K',K'' are subcomplexes of the simplicial complex K, then*

$$(K' \cap K'')_o = K'_o \cap K''_o, \qquad (K' \cup K'')_o = K'_o \cup K''_o$$

DEFINITION 3 5 Let K' and K'' be subcomplexes of a simplicial complex K. The inclusion map

$$i.\ \ (K',K' \cap K'') \to (K' \cup K'',K'')$$

is called an *excision*.

THEOREM 3.6. *If i $(K',K' \cap K'') \to (K' \cup K'',K'')$ is an excision, then*

$$i_o\colon\ K'_o/(K' \cap K'')_o \to (K' \cup K'')_o/K''_o$$

is an isomorphism.

PROOF. In view of 3 4, the map i_o consists of the homomorphisms

$$i_q.\quad C_q(K')/C_q(K') \cap C_q(K'') \to C_q(K') \cup C_q(K'')/C_q(K'')$$

induced by the inclusion homomorphism $C_q(K') \subset C_q(K') \cup C_q(K'')$ By the Noether isomorphism theorem i_q is an isomorphism This proves 3.6

THEOREM 3 7. *If P is a simplicial complex consisting of a single vertex, then the chain complex P_o is pointlike in the sense of* v,4 7

PROOF The only elementary q-chain of P_o is $c_q = A^0 \cdots A^q$ where $A^0 = \cdots = A^q = P$. For q even and >0, we have $\partial_q c_q = c_{q-1}$ so that $\partial_q\ C_q(P_o) \approx C_{q-1}(P_o)$ For $q < 0$ we have $C_q(P_o) = 0$ Thus the conditions of v,4 7 are fulfilled

Having defined the concepts of homotopy, excision, and point in the c-category $\mathcal{K}_s$, it becomes an h-category in the sense of IV,9 1 Then, by IV,9 4, Theorems 2.5, 3 2, 3 6, and 3 7 are summarized by

THEOREM 3 8. *If $\partial\mathcal{G}$ is regarded as an h-category in the sense of direct couples, then $O\quad \mathcal{K}_s \to \partial\mathcal{G}$ is a covariant h-functor*

DEFINITION 3 9 According to IV,9.5, the composition of the h-functor O with the homology theory of $\partial\mathcal{G}$ with coefficient group G (see v,11.3) yields an homology theory on $\mathcal{K}_s$ called *the homology theory of $\mathcal{K}_s$ with coefficient group G.* Likewise the composition of O with the cohomology theory of $\partial\mathcal{G}$ with coefficient group G (see v,12 3) yields a cohomology theory on $\mathcal{K}_s$ called *the cohomology theory of $\mathcal{K}_s$ with coefficient group G.* For any simplicial pair (K,L) the homology and cohomology groups $H_q(K_o/L_o;G)$, $H^q(K_o/L_o;G)$ (see v,11.4, 12 4) will be written $H_q(K,L;G)$ and $H^q(K,L;G)$, respectively, and are called the qth *homology and cohomology groups of (K,L) with coefficient group G.* According to v,11 3 and 12.3, we have the following cases

(1)	$G \in \mathcal{G}_R,$	$(K,L) \in \mathcal{K}_s,$	then	$H_q(K,L;G) \in \mathcal{G}_R,$
(2)	$G \in \mathcal{G}_R,$	$(K,L) \in \mathcal{K}_s,$	then	$H^q(K,L;G) \in \mathcal{G}_R,$
(3)	$G \in \mathcal{G}_C,$	$(K,L) \in \mathcal{K}_s,$	then	$H^q(K,L,G) \in \mathcal{G}_C.$

When K is a finite complex, i e $(K,L) \in \mathcal{K}'_s$, we have observed in 2.3 that K_o/L_o is a finite chain complex, hence $K_o/L_o \in \partial\mathcal{G}'$. Thus, by v,11 3, we have a fourth case·

(4)	$G \in \mathcal{G}_C,$	$(K,L) \in \mathcal{K}'_s,$	then	$H_q(K,L,G) \in \mathcal{G}_C.$

4. DIRECT DESCRIPTION OF THE BASIC CONCEPTS

The customary procedure for defining the groups $H_q(K,L,G)$ is to give a direct definition of the chain groups $C_q(K,L;G) = C_q(K_o/L_o \otimes G)$

and of the boundary operator, and then to set H_q equal to the q-cycles reduced by the q-boundaries The procedure we have followed presents H_q as the result of composing three functors O, $\otimes G$, and the homology group of a chain complex The advantages of our procedure are two-fold: it provides an analysis of the construction, and it enables us to handle a variety of cases with a minimum of repetition The disadvantage is that the direct definition has been obscured. The objective of this section is to rectify this situation by presenting the direct descriptions The statements to be made hardly require any proof since they are the results of assembling definitions.

The group $C_q(K,L,G)$ is by definition the tensor product $C_q(K_o/L_o) \otimes G$. By 2 1, a base for $C_q(K_o)$ is provided by the elementary chains $c = A^0 \cdots A^q$. By 2 3, a base for $C_q(K_o/L_o)$ is provided by those c's not on L By v,9 6, $C_q(K,L,G)$ is generated by the elements $c \otimes g$ which will be written gc in accordance with the convention v,11 4 Thus we have

THEOREM 4 1 *The group $C_q(K,L,G)$ is generated by the elements $gA^0 \cdots A^q$ where $g \in G$, and $A^0, \cdots, A^q$ are vertices of K, all contained in a simplex of K, with the relations*

$$(g_1 + g_2)A^0 \cdots A^q = g_1A^0 \cdots A^q + g_2A^0 \cdots A^q$$

and

$$gA^0 \cdots A^q = 0$$

whenever $A^0, \cdots, A^q$ are contained in a simplex of L The boundary is given by the formula

$$\partial_q(gA^0 \cdots A^q) = \sum_{i=0}^{q} (-1)^i gA^0 \cdots \hat{A}^i \cdots A^q$$

For a simplicial map f $(K,L) \to (K',L')$ the chain transformation f_q: $C_q(K,L;G) \to C_q(K',L',G)$ is given by

$$f_q(gA^0 \cdots A^q) = gf(A^0) \cdots f(A^q).$$

If this theorem were adopted as a definition, it would be necessary to verify that f_q and ∂_q are compatible with the relations and commute

The remainder of the description is a strict paraphrase of what was done in III, and v,2 The group $Z_q(K,L;G)$ of cycles mod L, consists of chains $c \in C_q$ with $\partial_q c = 0$, and the group $B_q(K,L;G)$ of boundaries mod L, consists of the boundaries of elements of C_{q+1}. Finally $H_q(K,L,G) = Z_q/B_q$. The homomorphisms f_q associated with f: $(K,L) \to (K',L')$ carry $Z_q(K,L,G)$ and $B_q(K,L,G)$ into $Z_q(K',L';G)$ and $B_q(K',L',G)$ respectively, and thereby induce the homomorphism

f_*: $H_q(K,L,G) \to H_q(K',L';G)$ The direct interpretation of ∂. $H_q(K,L,G) \to H_{q-1}(L;G)$ is as follows If $h \in H_q(K,L;G)$, choose $z \in Z_q(K,L,G)$ in the coset (= homology class) h Write z as a formal sum and interpret the resulting expression as an element x of $C_q(K;G)$. Then $\partial_q x$ is in $Z_{q-1}(L,G)$ and is in the coset ∂h.

If $G = J$ is the group of integers, then we write $C_q(K,L)$, etc , instead of $C_q(K,L,G)$, etc

Passing to cohomology we begin with the group $C^q(K,L;G) = C^q(K_o/L_o,G)$ of cochains of K mod L over G This group is by definition the group $\mathrm{Hom}(C_q(K_o/L_o),G)$, and therefore v,10.6 and v,12 imply

THEOREM 4 1c. *The group $C^q(K,L;G)$ is the group of functions ϕ defined for each array of vertices $A^0, \cdots, A^q$ all of which are on some simplex of K, the value $\phi(A^0, \cdots, A^q)$ is in G, and is zero if $A^0, \cdots, A^q$ are in a simplex of L The coboundary $\delta^q\phi$ is defined by*

$$(\delta^q\phi)(A^0, \cdots, A^{q+1}) = \sum_{i=0}^{q+1} (-1)^i\phi(A^0, \cdots, \hat{A}^i, \cdots, A^{q+1}).$$

For a simplicial map f: $(K,L) \to (K',L')$, the cochain transformation f^q. $C^q(K',L',G) \to C^q(K,L,G)$ is given for $\phi \in C^q(K',L',G)$ by the formula

$$(f^q\phi)(A^0, \cdots, A^q) = \phi(f(A^0), \cdots, f(A^q)).$$

Continuing in the classical vein, the group $Z^q(K,L,G)$ of cocycles mod L consists of cochains $\phi \in C^q(K,L,G)$ such that $\delta^q\phi = 0$; and the group $B^q(K,L,G)$ of coboundaries mod L consists of coboundaries of elements of $C^{q-1}(K,L,G)$ Then $H^q(K,L,G) = Z^q/B^q$ The direct interpretation of δ. $H^q(L,G) \to H^{q+1}(K,L;G)$ is the following If u is in $H^q(L,G)$, choose ϕ in $Z^q(L,G)$ belonging to the coset u Extend ϕ to a function $\phi' \in C^q(K,G)$ by defining $\phi'(A^0, \cdots, A^q)$ arbitrarily when $A^0, \cdots, A^q$ are not on a simplex of L Then $\delta^q\phi'$ is zero on L, and is in $Z^{q+1}(K,L;G)$ Its cohomology class is δu

The reason for adopting the linear form notation for chains and the functional notation for cochains appears here for the first time. the formulas for ∂, δ, f_q, f^q have closed forms. For example, if we were to use the functional notation for chains, then ∂ is given by

$$(\partial\phi)(A^0, \cdots, A^{q-1}) = \sum_A \sum_{i=0}^{q-1} (-1)^i\phi(A^0, \cdots, A^{i-1}, A, A^i, \cdots, A^{q-1})$$

where the sum on A extends over all vertices of all simplexes which include $A^0, \cdots, A^{q-1}$.

Following III,7 2 and v,3 2, we could also consider the groups $\bar{Z}_q(K,L,G)$ and $\bar{B}_q(K,L,G)$ which are the counterimages of Z_q and B_q under the natural mapping $C_q(K;G) \to C_q(K,L;G)$. Thus, by v,3 2, $\bar{Z}_q$ consists of all chains $c \in C_q(K,G)$ with $\partial c \in C_{q-1}(L;G)$, while $\bar{B}_q =$

$B_q(K,G) \cup C_q(L,G)$. The group $\overline{H}_q = \overline{Z}_q/\overline{B}_q$ is isomorphic with $H_q(K,L,G)$. This alternative description of the relative homology groups is very common, and has the advantage of allowing a much simpler definition of the operator ∂ $H_q(K,L,G) \to H_{q-1}(L;G)$

Most of the time we shall write ∂ and δ for ∂_q and δ^q, and f for f_q and f^q.

If P is a simplicial complex consisting of a single vertex, then by 4.1 and 4.1c

$$C_0(P;G) \approx G, \qquad C^0(P,G) \approx G,$$

and $\partial_1 = 0$, $\partial_0 = 0$. Thus H_0 and C_0 and $H^0 = C^0$. Consequently

$$H_0(P;G) \approx G, \qquad H^0(P;G) \approx G.$$

This shows that the homology [cohomology] theory on simplicial complexes constructed over a group G actually has the group G as a coefficient group in the sense of I,6 1.

5. REDUCED GROUPS, ACYCLICITY, ALGEBRAIC MAPPINGS

With the homology and cohomology theories for simplicial complexes fully established, we turn to the discussion of some special features of these groups First we discuss the question of reduced homology and cohomology groups These could be defined exactly as in I,7; however a more direct and explicit definition is useful for the applications.

Since the space consisting of a single point P is a simplicial complex and for every simplicial complex K the map $f.$ $K \to P$ is simplicial, it would be in accord with I,7 3 to define the reduced 0th homology group $\tilde{H}_0(K;G)$ as the kernel of the induced homomorphism f_*: $H_0(K,G) \to H_0(P,G)$ If $c = \sum g_i A^i$ is any 0-chain in $C_0(K,G)$, then $f(c) = (\sum g_i)P$. This suggests the following definitions.

Definition 5.1. For each element $c = \sum g_i A^i$ of $C_0(K,G)$, the *index* In(c) is

$$\text{In}(c) = \sum g_i \in G.$$

Clearly In: $C_0(K,G) \to G$ is a homomorphism, and In$\partial = 0$.

Definition 5.2. The *augmented* ordered chain complex $\tilde{K}_o$ of the simplicial complex K is defined as follows. If $K = 0$, then $\tilde{K}_o = K_o = 0$. If $K \neq 0$, then

$$\begin{aligned} C_q(\tilde{K}_o) &= C_q(K_o) && \text{for } q \neq -1, \\ C_{-1}(\tilde{K}_o) &= J && (= \text{additive group of integers}), \\ \tilde{\partial}_q &= \partial_q && \text{for } q \neq 0,-1, \\ \tilde{\partial}_0 &= \text{In}: \ C_0(\tilde{K}) \to J, \\ \tilde{\partial}_{-1} &= 0. \end{aligned}$$

The *reduced* 0th homology and cohomology groups of K are defined as

$$\tilde{H}_0(K;G) = H_0(\tilde{K}_o,G), \qquad \tilde{H}^0(K;G) = H^0(\tilde{K}_o;G)$$

Clearly

$$H_q(\tilde{K}_o,G) = H_q(K_o,G) = H_q(K;G) \qquad \text{for } q > 0$$

In the dimension 0, $\tilde{Z}_0(K;G) = Z_0(\tilde{K},G)$ is a subgroup of $Z_0(K_o;G) = C_0(K_o;G)$ and consists of the 0-chains c with $\text{In}(c) = 0$. Hence, by the argument preceding 5 1, we have

THEOREM 5 3 *The reduced* 0th *homology group* $\tilde{H}_0(K,G)$ *is the kernel of the homomorphism* f_*: $H_0(K;G) \to H_0(P,G)$ *where* f: $K \to P$ *and* P *consist of one point*

For cohomology we find that

$$H^q(\tilde{K}_o,G) = H^q(K_o;G) = H^q(K;G) \qquad \text{for } q \neq 0$$

while $\tilde{H}^0(K,G) = Z^0(K,G)/\tilde{B}^0(K;G)$, where $\tilde{B}^0(K,G)$ consists of those cochains $\phi \in C^0(K,G)$ which are constant on the vertices of K This implies

THEOREM 5 3c *The reduced* 0th *cohomology group* $\tilde{H}^0(K,G)$ *is the factor group of* $H^0(K,G)$ *by the image of the homomorphism* f^*: $H^0(P,G) \to H^0(K,G)$ *where* f $K \to P$ *and* P *consists of one point*

So far we have considered only those maps f $K_o \to K'_o$ which were induced by a simplicial f $K \to K'$ Such maps satisfy the condition $\text{In}(fc) = \text{In}(c)$ for every $c \in C_0(K)$. In the sequel we shall have occasion to consider maps which are not necessarily induced by simplicial maps (e g the subdivision operator for chains in §7 below).

DEFINITION 5 4. Let K,K' be simplicial complexes. An *algebraic map* f: $K \to K'$ is a chain map f: $K_o \to K'_o$ such that $\text{In}(fc) = \text{In}(c)$ for each $c \in C_0(K)$ A chain homotopy D between algebraic maps will be called an *algebraic* homotopy

Clearly an algebraic map f: $K \to K'$ can be extended to a map $\tilde{f}$ $\tilde{K}_o \to \tilde{K}'_o$ of the augmented ordered chain complexes by defining $\tilde{f}_{-1}$ to be the identity map of $J = C_{-1}(\tilde{K}_o) = C_{-1}(\tilde{K}'_o)$ Similarly an algebraic homotopy D can be extended by setting $D_{-1} = 0$

DEFINITION 5 5. A function C which to each simplex s of a simplicial complex K assigns a nonempty subcomplex $C(s)$ of a simplicial complex K' is called a *carrier function* if, for every face s' of s, $C(s')$ is a subcomplex of $C(s)$. If f $K \to K'$ is an algebraic map such that $c \in C_q(K)$ and $c \subset s$ imply $fc \subset C(s)$, then C is called a *carrier* of f. Similarly, if D: $f \simeq g$ is an algebraic homotopy such that $c \in C_q(K)$ and $c \subset s$ imply $Dc \subset C(s)$, then C is called a *carrier* of D.

DEFINITION 5 6. A simplicial complex K is called *acyclic* if

$H_q(K) = 0$ for $q \neq 0$ and $\tilde{H}_0(K) = 0$. A carrier function C is called *acyclic* if, for each simplex s, the complex $C(s)$ is acyclic.

THEOREM 5.7. *Let K and K' be simplicial complexes, let C be an acyclic carrier function defined on K with values in K', and let L be a subcomplex of K. Any algebraic map $L \to K'$ with carrier C can be extended to an algebraic map $K \to K'$ with carrier C. If f,g: $K \to K'$ are algebraic maps with carrier C, then any algebraic homotopy between $f|L$ and $g|L$ with carrier C can be extended to an algebraic homotopy between f and g with carrier C.*

PROOF. Suppose f: $L \to K'$ is an algebraic map. For each vertex A of K which is not in L we select a vertex $f(A)$ of $C(A)$. This extends f to an algebraic map f: $K^0 \cup L \to K'$. Let A^0A^1 be an elementary 1-chain of K and let s be the least simplex containing A^0A^1. If s is in L, then $f(A^0A^1)$ is already defined. If s is not in L, then

$$\text{In}(f\partial(A^0A^1)) = \text{In}(fA^1) - \text{In}(fA^0) = 0.$$

Thus $f\partial(A^0A^1) \in \tilde{Z}_0(C(s))$. Since $\tilde{H}_0(C(s)) = 0$, there is a chain $f(A^0A^1) \in C_1(C(s))$ such that $\partial f(A^0A^1) = f\partial(A^0A^1)$. This extends f to an algebraic map f: $K^1 \cup L \to K'$.

From here we proceed by induction and assume that an algebraic map f: $K^q \cup L \to K'$ with carrier C is already given ($q > 0$). Let c be an elementary $(q + 1)$-chain of K and let s be the least simplex of K containing c. If $s \subset L$, then fc is already defined. If s is not in L, then $\partial f\partial c = f\partial\partial c = 0$, so that $f\partial c \in Z_q(C(s))$. Since $H_q(C(s)) = 0$ we may choose $fc \in C_{q+1}(C(s))$ so that $\partial fc = f\partial c$. This extends f to an algebraic map f: $K^{q+1} \cup L \to K'$ with carrier C.

Let f,g: $K \to K'$ be algebraic maps with carrier C, and let D: $f|L \simeq g|L$ be an algebraic homotopy with carrier C. For each vertex A of K which is not in L, we have $\text{In}(gA - fA) = 0$. Thus $gA - fA \in \tilde{Z}_0(C(A))$, and since $\tilde{H}_0(C(A)) = 0$, we may choose $D_0(A) \in C_1(C(A))$ so that $\partial DA = gA - fA$. This extends D to an algebraic homotopy D: $f|K^0 \cup L \simeq g|K^0 \cup L$ with carrier C.

From here we proceed by induction and assume that an algebraic homotopy D: $f|K^{q-1} \cup L \simeq g|K^{q-1} \cup L$ with carrier C is already given ($q > 0$). Let c be an elementary q-chain of K, and let s be the least simplex of K containing c. If $s \subset L$, then Dc is already defined. If s is not in L, then define $z = gc - fc - D\partial c$. Clearly $z \subset C(s)$ and

$$\partial z = \partial gc - \partial fc - \partial D\partial c = g\partial c - f\partial c - (g\partial c - f\partial c + D\partial\partial c) = 0.$$

Thus $z \in Z_q(C(s))$. Since $H_q(C(s)) = 0$, there is a chain $Dc \in C_{q+1}(C(s))$ such that $\partial Dc = z$. Then $\partial Dc + D\partial c = gc - fc$ and we find that D is extended to an algebraic homotopy D: $f|K^q \cup L \simeq g|K^q \cup L$ with carrier C.

THEOREM 5.8. *Let f,g: $K \to K'$ be algebraic maps with an acyclic carrier C. Let L,L' be subcomplexes of K,K' respectively such that $s \subset L$ implies $C(s) \subset L'$. Then the maps f,g induce chain maps f_0,g_0: $K_*/L_* \to K'_*/L'_*$ which are chain homotopic. In particular, $f_{0*} = g_{0*}$ and $f^*_0 = g^*_0$ or homology and cohomology groups over any coefficient group.*

PROOF. By 5.7 there exists an algebraic homotopy D: $f \simeq g$ with carrier C. Each homomorphism D_q: $C_q(K) \to C_{q+1}(K')$ carries $C_q(L)$ into $C_{q+1}(L')$ thus inducing a homomorphism D_{0q}: $C_q(K)/C_q(L) \to C_{q+1}(K')/C_{q+1}(L')$. The homomorphisms D_{0q} yield the desired homotopy D_0.

DEFINITION 5.9. Given a vertex A of a simplicial complex K, the *star* St(A) of A in K is the subcomplex of K consisting of the simplexes which have A as a vertex, and all their faces. (This is to be distinguished from the open star st(A) defined in II,3.6 which is an open subset of $|K|$)

DEFINITION 5.10. Given a chain $c \in C_q(K)$ such that $c \subset \text{St}(A)$, define a $(q+1)$-chain $Ac \in C_{q+1}(K)$, called the *join* of A with c, as follows: If

$$c = \sum a_i A_i^0 \cdots A_i^q,$$

then

$$Ac = \sum a_i A A_i^0 \cdots A_i^q.$$

LEMMA 5.11. *If $c \in C_q(K)$ and $c \subset \text{St}(A)$, then*

$$\partial(Ac) = c - A(\partial c) \qquad \text{if } q > 0,$$
$$\partial(Ac) = c - \text{In}(c)A \qquad \text{if } q = 0.$$

PROOF. Because of linearity, we may assume that $c = A^0 \cdots A^q$ is an elementary q-chain. Then, for $q > 0$,

$$\partial Ac = \partial(AA^0 \cdots A^q) = A^0 \cdots A^q - \sum_{i=0}^{q} (-1)^i AA^0 \cdots \hat{A}^i \cdots A^q$$
$$= c - A\partial c,$$

and, for $q = 0$,

$$\partial Ac = \partial AA^0 = A^0 - A = c - \text{In}(c)A.$$

THEOREM 5.12. *For any vertex A of a simplicial complex K, the subcomplex St(A) is acyclic.*

PROOF Let $z \in Z_q(\text{St}(A))$ for $q > 0$. Then, by 5.11, $\partial(Az) = z$;

hence $z \in B_q(\mathrm{St}(A))$ and $H_q(\mathrm{St}(A)) = 0$. If $z \in \tilde{Z}_0(\mathrm{St}(A))$, then $\mathrm{In}(z) = 0$, so that $\partial Az = z$ and $z \in B_0(\mathrm{St}(A))$. Thus $\tilde{H}_0(\mathrm{St}(A)) = 0$.

THEOREM 5.13. *A simplex is acyclic.*

This follows from 5.12 and the fact that, if A is any vertex of the simplex s, then $s = \mathrm{St}(A)$.

Using 5.8 and 5.13 we can give a second and less computational proof of 3.2. Let f,g: $(K,L) \to (K',L')$ be contiguous simplicial maps. For each simplex s of K, let $C(s)$ be the least simplex of K' containing both fs and gs. By 5.13, $C(s)$ is acyclic. Further if s is in L then $C(s)$ is in L'. Then C is an acyclic carrier for the algebraic maps f_*,g_*. Thus, by 5.8, $f_* \simeq g_*$.

6. THE ALTERNATING CHAIN COMPLEX OF A SIMPLICIAL COMPLEX

DEFINITION 6.1 Let K be a simplicial complex. A chain $c \in C_q(K_o)$ will be called an *elementary degenerate chain* if it has either of the two following forms: $c = A^0A^1 \cdots A^q$ where $A^0 = A^1$, or $c = v + v'$ where $v = A^0 \cdots A^q$ is an elementary chain, and v' is the elementary chain obtained from v by interchanging two neighboring vertices A^i,A^{i+1} for some $i = 0,1, \cdots, q - 1$. A chain which is a linear combination of elementary degenerate chains will be called *degenerate.* The subgroup of degenerate chains is denoted by $D_q(K_o)$.

LEMMA 6.2. $\partial_q[D_q(K_o)] \subset D_{q-1}(K_o)$.

LEMMA 6 3 *If L is a subcomplex of K, then*

$$D_q(L_o) = D_q(K_o) \cap C_q(L_o).$$

LEMMA 6.4. *If f: $K \to K'$ is simplicial, then*

$$f_q[D_q(K_o)] \subset D_q(K'_o).$$

These lemmas are direct consequences of the definitions. The following three lemmas follow directly from the fact that every permutation is a product of transpositions (i.e. permutations interchanging two adjacent elements)

LEMMA 6 5. *If $i_0, \cdots, i_q$ is a permutation of the integers $0, \cdots, q$ and $\epsilon = \pm 1$ according as this permutation is even or odd, then the chain*

$$A^0 \cdots A^q - \epsilon A^{i_0} \cdots A^{i_q}$$

is degenerate

LEMMA 6.6. *If some vertex occurs at least twice in $c = A^0 \cdots A^q$, then c is degenerate*

LEMMA 6.7. *If the vertices of each q-simplex s_i of K have been ordered* *in a definite way* $A_i^0 < \cdots < A_i^q$, *then for each q-chain* $c \in C_q(K_o)$ *there is a unique q-chain* $c' \in C_q(K_o)$ *of the form*

$$c' = \sum_i a_i A_i^0 \cdots A_i^q$$

such that $c - c' \in D_q(K_o)$ *Moreover* $c' = 0$ *if and only if* $c \in D_q(K_o)$. *If* $c \subset L$, *where L is a subcomplex, then* $c' \subset L$

The analogy of the last three lemmas with III,5 5, 5 6, and 5 8 suggests the consideration of the groups

$$C_q(K_a) = C_q(K_o)/D_q(K_o).$$

In view of 6 2 the homomorphisms $\partial_q \; C_q(K_o) \to C_{q-1}(K_o)$ induces a homomorphism $\partial_q \; C_q(K_a) \to C_{q-1}(K_a)$ so that $K_a = \{C_q(K_a),\partial_q\}$ is a chain complex. If L is a subcomplex of K, then L_a may be regarded, in a natural fashion, as a subcomplex of K_a and is then a direct summand of K_a

DEFINITION 6 8. For each simplicial pair (K,L) the chain complex K_a/L_a is called the *alternating* chain complex of the pair (K,L).

All the statements made in §2 can now be repeated with the ordered complexes K_o/L_o replaced by the alternating complexes K_a/L_a throughout. The functor $K_a/L_a,f_a$ is denoted by A. Theorem 2.5 and the subsequent discussion apply with the functor O replaced by the functor A.

There are almost no changes to be made in §3 The proof of 3 2 as given in §3 remains valid provided the vertices of K are simply ordered and the elementary (alternating) chains of K are always written with the vertices in order The second proof given in §5 carries over without modification The proof of 3 7 is even easier, for, if P is the simplicial complex consisting of a single vertex, then $C_q(P_a) = 0$ for $q \neq 0$

The discussion of §4 carries over to the alternating language with the following two modifications

In 4 1 the set of relations that the symbols $gA^0 \cdots A^q$ satisfy should be augmented by the following two kinds of relations:

$$gA^0 \cdots A^q = 0$$

if $A^0, \cdots, A^q$ are not all distinct, and

$$gA^0 \cdots A^q = \epsilon g A^{i_0} \cdots A^{i_q}$$

if $i_0, \cdots, i_q$ is a permutation of $0, \cdots, q$ and ϵ is the sign of this permutation

The corresponding additions in 4 1c are: $\phi(A^0, \cdots, A^q) = 0$ if

$A^0, \cdots, A^q$ are not all distinct and $\phi(A^0, \cdots, A^q) = \epsilon\phi(A^{i_0}, \cdots, A^{i_q})$ if $i_0, \cdots, i_q$ is a permutation of $0, \cdots, q$ of sign ϵ.

The above remarks show that a second collection of homology and cohomology theories on the h-category $\mathcal{K}_s$ is obtained by replacing the functor O by the functor A. It will be shown below (see 6.9) that these two approaches yield isomorphic theories, so that the use of one or the other is entirely a matter of convenience. The alternating approach is closer to the geometry as is indicated by the alternating character of the chains and cochains obtained in III,5. The alternating approach is also more convenient for the actual computation of the homology groups, since the groups $C_q(K,L,G)$ are "smaller". The ordered approach is closer to the singular homology theory of Chapter VII and is also useful in many problems connected with abstract algebra. In the sequel, unless specific mention is made, either of the two approaches could be used.

To compare the ordered and the alternating theories, for each simplicial pair (K,L), define a map

$$(1) \qquad \alpha: K_o/L_o \rightarrow K_a/L_a$$

as follows. Let

$$\alpha_q: C_q(K_o) \rightarrow C_q(K_a) = C_q(K_o)/D_q(K_o)$$

be the natural homomorphism. By 6.2, α_q and ∂ commute, thus yielding a map $\alpha: K_o \rightarrow K_a$. This map carries L_o into L_a, and thus induces the map (1).

THEOREM 6.9. *The map α induces isomorphisms*

$$\alpha_*: H_q(K_o/L_o,G) \approx H_q(K_a/L_a,G),$$
$$\alpha^*: H^q(K_a/L_a,G) \approx H^q(K_o/L_o,G)$$

for any coefficient group G. These isomorphisms yield an isomorphism between the homology (or cohomology) theories derived from the ordered and the alternating approach.

PROOF. We first verify that α_* is a homomorphism of the ordered homology theory into the alternating one. This requires the verification of commutativity relations in the diagrams

$$\begin{array}{ccc} H_q(K_o/L_o,G) & \xrightarrow{\alpha_*} & H_q(K_a/L_a,G) \\ \downarrow f_* & & \downarrow f_* \\ H_q(K'_o/L'_o;G) & \xrightarrow{\alpha_*} & H_q(K'_a/L'_a;G) \end{array} \qquad \begin{array}{ccc} H_q(K_o/L_o,G) & \xrightarrow{\alpha_*} & H_q(K_a/L_a;G) \\ \downarrow \partial & & \downarrow \partial \\ H_{q-1}(L_o,G) & \xrightarrow{\alpha_*} & H_{q-1}(L_a,G) \end{array}$$

where f. $(K,L) \to (K',L')$ is simplicial. These commutativity relations are consequences of commutativity relations in the diagrams

$$
\begin{array}{ccc}
K_o/L_o & \xrightarrow{\alpha} & K_a/L_a \\
\downarrow{\scriptstyle f_o} & & \downarrow{\scriptstyle f_a} \\
K'_o/L'_o & \xrightarrow{\alpha} & K'_a/L'_a
\end{array}
\qquad
\begin{array}{ccccc}
L_o & \xrightarrow{i_o} & K_o & \xrightarrow{j_o} & K_o/L_o \\
\downarrow{\scriptstyle \alpha} & & \downarrow{\scriptstyle \alpha} & & \downarrow{\scriptstyle \alpha} \\
L_a & \xrightarrow{i_a} & K_a & \xrightarrow{j_a} & K_a/L_a
\end{array}
$$

where i: $L \subset K$ and j: $K \subset (K,L)$.

A similar argument shows that α^* yields a homomorphism of cohomology theories.

The fact that α_* and α^* are isomorphisms is a consequence of the following theorem:

THEOREM 6.10. *The map* α: $K_o/L_o \to K_a/L_a$ *is a homotopy equivalence* (see IV,9.2).

PROOF. Select a partial order for the vertices of K such that the vertices of any simplex of K are simply ordered. A q-chain of K_o will be called *normal* if it is a linear combination of chains $A^0 \cdots A^q$ with $A^0 < \cdots < A^q$. Clearly, if c is normal, so is $\partial_q c$.

Given a q-chain $c \in C_q(K_a)$, there exists in view of 6.7 a unique normal chain $\bar{\alpha}_q c \in C_q(K_o)$ such that $\alpha_q \bar{\alpha}_q c = c$. Since $\partial_q \bar{\alpha}_q c$ is normal, and $\alpha_{q-1}\partial_q\bar{\alpha}_q c = \partial_q \alpha_q \bar{\alpha}_q c = \partial_q c$, it follows that $\partial_q \bar{\alpha}_q c = \bar{\alpha}_{q-1}\partial_q c$. Further, if $c \subset L$ (i e. if $c \in C_q(L_a)$), then, by 6.7, $\bar{\alpha}_q c \subset L$. Hence the homomorphisms $\bar{\alpha}_q$: $C_q(K_a) \to C_q(K_o)$ define a map

$$\bar{\alpha}: \ K_a/L_a \to K_o/L_o$$

and $\alpha\bar{\alpha}$ = identity.

Now consider the map $\bar{\alpha}\alpha$: $K_o \to K_o$. If $c \in C_q(K)$ and $c \subset s$ where s is a simplex of K, then $\bar{\alpha}\alpha c \subset s$. Thus, setting $C(s) = s$, we find that C is a common carrier for $\bar{\alpha}\alpha$ and the identity map i: $K_o \to K_o$. Since the carrier C is acyclic by 5.13, and since both $\bar{\alpha}\alpha$ and i are algebraic maps, it follows from 5 8 that the map $\bar{\alpha}\alpha$: $K_o/L_o \to K_o/L_o$ is homotopic to the identity Thus $\bar{\alpha}$ is a homotopy inverse of α, proving that α: $K_o/L_o \to K_a/L_a$ is a homotopy equivalence.

Note that, in the above proof, the construction of $\bar{\alpha}$ depended on the ordering of the vertices of K. However the induced homomorphisms $\bar{\alpha}_*$ (or $\bar{\alpha}^*$) are independent of this ordering since they are the inverses of α_* (or α^*).

7. EFFECT OF BARYCENTRIC SUBDIVISION

In this section we shall assume that the complexes are finite. Let K be a finite simplicial complex and $(l_K, \text{Sd } K)$ its barycentric subdivision (II,6).

We begin with the construction of an algebraic map

$$\text{Sd}\colon\ K_c \to (\text{Sd } K)_c$$

called the subdivision operator for chains and satisfying the following properties for $c \in C_q(K)$:

(1) If $c \subset s$, where s is a simplex of K, then $\text{Sd } c \subset \text{Sd } s$.

(2) $$\partial \,\text{Sd } c = \text{Sd } \partial c$$

Define $\text{Sd } c = 0$ for $q < 0$. Since every vertex of K is also a vertex of $\text{Sd } K$, the group $C_0(K)$ is a subgroup of $C_0(\text{Sd } K)$. Define $\text{Sd}\colon C_0(K) \to C_0(\text{Sd } K)$ to be the inclusion homomorphism. From here proceed by induction and assume that Sd has been defined for dimensions $i < q$ $(q \geqq 1)$ so that (1) and (2) hold. Let c be an elementary q-chain of K, and let s be the least simplex of K such that $c \subset s$. Then $\partial c \subset s$ and therefore $\text{Sd } \partial c \subset \text{Sd } s$. Since b_s is a vertex of $\text{Sd } s$, and $\text{Sd } s \subset \text{St } b_s$, the join operation of 5.10 may be used to define

$$\text{Sd } c = b_s \text{Sd } (\partial c).$$

Clearly $\text{Sd } c \subset \text{Sd } s$. Further, by 5.11

$$\begin{aligned} \partial \,\text{Sd } c &= \text{Sd } (\partial c) - b_s\, \partial \,\text{Sd } (\partial c) \\ &= \text{Sd } (\partial c) - b_s \text{Sd } (\partial\partial c) = \text{Sd } (\partial c). \end{aligned}$$

Having defined Sd for the generators of $C_q(K)$ so that (1) and (2) hold, the homomorphism $\text{Sd}\colon C_q(K) \to C_q(\text{Sd } K)$ is well defined and satisfies (1) and (2). If L is a subcomplex of K, then Sd carries L_c into $(\text{Sd } L)_c$. There results a proper map

$$\text{Sd}\colon\ K_c/L_c \to (\text{Sd } K)_c/(\text{Sd } L)_c.$$

It should be noted that, although the subdivision operator for chains was defined by induction, the procedure is quite definite and could be replaced by a closed formula.

In addition to the linear map $l_K\colon \text{Sd } K \to K$, we shall define simplicial maps

$$\pi\colon\ \text{Sd } K \to K$$

called projections, as follows. For every vertex b_s of $\text{Sd } K$, let $\pi(b_s)$ be one of the vertices of the simplex s of K. If $b_{s_0}, \cdots, b_{s_q}$ are vertices of a simplex of Sd K, with s_i a face of s_{i+1}, $i = 0, \cdots, q-1$, then $\pi(b_{s_0}), \cdots, \pi(b_{s_q})$ are contained in the set of vertices of the simplex s_q. Thus, by II,4.4, the vertex map π of vertices of Sd K into the vertices of K extends in a unique fashion to a simplicial map π: Sd $K \to K$. If L is a subcomplex of K, then π carries Sd L into L so that

$$\pi\colon\ (\text{Sd } K, \text{Sd } L) \to (K,L).$$

THEOREM 7.1. *For every (finite) simplicial pair (K,L) the subdivision operator*

$$\text{Sd}\colon\ K_o/L_o \to (\text{Sd } K)_o/(\text{Sd } L)_o$$

is a homotopy equivalence. The map

$$\pi_o\colon\ (\text{Sd } K)_o/(\text{Sd } L)_o \to K_o/L_o$$

induced by any projection π: (Sd K, Sd L) $\to (K,L)$ is a homotopy inverse of Sd.

PROOF. We shall establish chain homotopies

$$\pi_o \,\text{Sd} \simeq i_o, \tag{3}$$
$$\text{Sd}\, \pi_o \simeq j_o \tag{4}$$

where i_o and j_o are the identity chain maps of K_o/L_o and $(\text{Sd } K)_o/(\text{Sd } L)_o$ respectively.

First observe that, if $c \in C_q(K)$ and $c \subset s$, then Sd $c \subset$ Sd s and π_q Sd $c \subset s$. Hence $C(s) = s$ defines a carrier C for both π_o Sd and i_o. Since the carrier is acyclic, (3) follows from 5.11.

For every simplex s of Sd K, let $\bar{s}$ denote the least simplex of K containing all the vertices of s, and let $C(s) = \text{Sd } \bar{s}$. If $c \in C_q(\text{Sd } K)$ and $c \subset s$, then $\pi_q c \subset \bar{s}$ and Sd $\pi_q c \subset C(s)$. Since $s \subset C(s)$, it follows that $C(s)$ is a common carrier for Sd π and j. Since $C(s) = \text{Sd } \bar{s} = \text{St }(\bar{s})$ (the star is taken in the complex Sd $\bar{s}$), it follows from 5.12 that $C(s)$ is acyclic. Thus (4) is a consequence of 5.8.

REMARK. It is easily seen that $c - \pi$ Sd c is a degenerate chain. Hence π Sd c induces the identity map on the alternating chain complex K_a. However this is not true of the operation which Sd π induces in $(\text{Sd } K)_a$.

As an immediate consequence of 7.1 we have

COROLLARY 7.2. *A projection*

$$\pi\colon\ (\text{Sd } K, \text{Sd } L) \to (K,L)$$

induces isomorphisms

$$\pi_*\colon\ H_q(\text{Sd } K, \text{Sd } L;G) \approx H_q(K,L;G),$$
$$\pi^*\colon\ H^q(K,L;G) \approx H^q(\text{Sd } K, \text{Sd } L;G).$$

These isomorphisms are independent of the choice of the projection.

8. UNIQUENESS OF SIMPLICIAL HOMOLOGY THEORIES

Let H be a homology theory defined on the h-category $\mathcal{K}'_f$ of *finite* simplicial complexes It will be shown how a uniqueness theorem analogous to III,10 1 may be established for such simplicial homology theories

First we observe that all the results of Chapter I with the exception of those of §11 carry over to simplicial homology theories with the obvious modifications in formulation Proposition I,11 5 that a contractible space is homologically trivial has the following analog:

If the complex K is the star (see 5 9) of one of its vertices, then K is homologically trivial.

Indeed, let $K = \text{St}\,A$. Consider the map f $K \to A$ and the identity map g $A \to K$ Then fg $A \to A$ is the identity while gf $K \to K$ is contiguous to the identity. Hence f_* is an isomorphism $H_q(K) \approx H_q(A)$, and since A is homologically trivial, so is K

With this done, all the propositions of III,2 through III,9 carry over to simplicial homology theories without any change except in notation The first part of the proof of the Uniqueness theorem III,10 1 (the construction of the homomorphism $\mathfrak{r}$) yields the following

THEOREM 8 1 (*Uniqueness theorem for simplicial homology theories*). *Given two simplicial homology theories H and $\bar{H}$ on the h-category $\mathcal{K}'_f$ of finite simplicial complexes, and given a homomorphism*

$$h_0\colon\ G \to \bar{G}$$

of their coefficient groups there exists a unique homomorphism

$$h\quad H \to \bar{H} \qquad \text{on} \qquad \mathcal{K}'_f$$

which is an extension of h_0 When h_0 is an isomorphism, so also is h.

Thus, as far as simplicial homology theories are concerned, the theory is complete both existence and uniqueness are established

This theorem casts light on the structure of the proof of the Uniqueness theorem of III,10 1 Each homology theory H (in the sense of Chapter I) defined on an admissible category $\mathcal{a}$ containing the category $\mathfrak{I}$ of triangulable pairs, leads to a simplicial homology theory H_s (defined on $\mathcal{K}'_f$) by setting $H_{s,q}(K,L) = H_q(|K|,|L|)$. The axioms for H_s follow directly from the corresponding axioms for H except for the Excision axiom which is the Theorem III,2.2

If $H,\bar{H}$ are two homology theories (defined on $\mathfrak{I}$) and h_0: $G \to \bar{G}$ is a homomorphism of their coefficient groups, then 8 1 yields a unique extension of h_0 to a homomorphism of the associated simplicial theories:

$$h_s\colon\ H_s \to \bar{H}_s \qquad \text{on} \qquad \mathcal{K}'_f,$$

The second part of the proof of III,10.1 then shows how one can pass from the homomorphism h_s to a homomorphism

$$h\quad H \to \overline{H} \qquad \text{on} \qquad \mathfrak{J}.$$

This is the only stage in the proof of the Uniqueness theorem which makes use of the Simplicial approximation theorem (II,7)

An analogous discussion can be given for cohomology; the details are left to the reader.

NOTES

The classical invariance theorem. We shall give here a modernized version of the classical invariance theorem

Let (X,A) be a triangulable pair If T,T' are triangulations of (X,A), we shall write $T' < T$ if, for each vertex B' of X (or of A) in the triangulation T', there is a vertex B of X (or of A) in the triangulation T such that

$$\operatorname{st}(B') \subset \operatorname{st}(B)$$

We note the following formal properties:

(1) $T' < T$ and $T'' < T'$ imply $T'' < T$

(2) ${}^nT < T$ where ${}^nT$ is the nth barycentric subdivision of T $(n \geqq 0)$

(3) For any T and T', there exists an integer n such that ${}^nT < T'$ (follows from II,6 5)

(4) For any T_1,T_2, there is a T with $T < T_1$, $T < T_2$ (follows from (2) and (3)).

Suppose $T' < T$. If, for each vertex B' of T', we select a vertex $p(B') = B$ as above, we obtain a simplicial map $p\quad T' \to T$ called a *projection*. Any two projections are contiguous, and therefore, by 3 2 and V,4.4, they induce the same homomorphism

$$\alpha(T,T'):\ H_q(T') \to H_q(T)$$

Here $H_q(T)$, where $T = \{t,(K,L)\}$ is a triangulation of (X,A), denotes the homology group $H_q(K,L)$ in either the ordered or the alternating approach.

If $T'' < T' < T$ and p: $T' \to T$, p'. $T'' \to T'$ are projections, then pp'. $T'' \to T$ is also a projection This yields·

(5) If $T'' < T' < T$, then $\alpha(T,T')\alpha(T',T'') = \alpha(T,T'')$.

The crucial fact is the following:

(6) $\alpha(T,T')$ is an isomorphism for any $T' < T$

In the special case when $T' = {}^1T$ is the barycentric subdivision of T, (6) follows from 7 2. Thus (5) implies that (6) is valid for $T' = {}^nT$.

For any $T' < T$, choose n so that ${}^nT < T' < T$. Then $\alpha(T,{}^nT) = \alpha(T,T')\alpha(T',{}^nT)$ is an isomorphism. It follows that $\alpha(T,T')$ is onto. Therefore $\alpha(T',{}^nT)$ is also onto. This implies that $\alpha(T,T')$ is an isomorphism as desired.

Let now T_1,T_2 be any two triangulations of (X,A). Choose T with $T < T_1$, $T < T_2$ and define

$$\beta_T(T_2,T_1) = \alpha(T_2,T)\alpha^{-1}(T_1,T) \quad H_q(T_1) \approx H_q(T_2).$$

If $T' < T$, then

$$\begin{aligned}\beta_{T'}(T_2,T_1) &= \alpha(T_2,T')\alpha^{-1}(T_1,T') \\ &= \alpha(T_2,T)\alpha(T,T')\alpha^{-1}(T,T')\alpha^{-1}(T,T_1) = \beta_T(T_2,T_1)\end{aligned}$$

This fact, together with (4), implies that $\beta_T(T_2,T_1)$ is independent of the choice of T. We thus obtain unique isomorphisms

(7) $\beta(T_2,T_1) \quad H_q(T_1) \approx H_q(T_2)$

In particular $\beta(T_2,T_1) = \alpha(T_2,T_1)$ if $T_1 < T_2$. If T_1,T_2,T_3 are three triangulations, then, choosing T with $T < T_i$ $(i = 1,2,3)$, we find

(8) $\beta(T_3,T_1) = \beta(T_3,T_2)\beta(T_2,T_1)$

It follows that the groups $H_q(T)$ together with the isomorphisms (7) form a transitive system of groups as defined in I,6. In this sense, the groups $H_q(T)$ are independent of the triangulation T, and may be denoted by $H_q(X,A)$.

The classical invariance proof limited itself to showing that $H_q(T_1)$ and $H_q(T_2)$ were isomorphic. It did not exhibit a "unique" isomorphism (7) satisfying (8).

Having constructed a group $H_q(X,A)$, we could continue in the same vein and construct ∂ and f_*, and then verify the axioms. This would provide a proof of the existence of a homology theory on the category of triangulable spaces. It is not worthwhile to do this in view of subsequent existence proofs for much larger categories.

The development of the invariance theorem. Much of the development of homology theory during the period 1895-1925 centered around the question of the topological invariance of the homology groups. The first fully satisfactory solution was achieved by Alexander near the end of this period through the development of simplicial complexes and their techniques [Trans. Amer. Math. Soc. 28 (1926), 301-329].

Prior to this a broader concept of *cell* complex was used (see Chapter XIII of the second volume). Homology groups were defined by choosing a cellular decomposition of the space, orienting the cells (i.e. selecting incidence numbers—the coefficients in the boundary relations), and taking the homology groups of the resulting chain complex. This procedure gave rise to a series of problems: (1) a clear-cut definition of cell complex was needed; (2) the orientability must be proved (i e show that $\partial\partial = 0$); (3) the independence of the homology groups on the choice of orientation must be shown, and (4) the homology groups must be proved independent of the choice of the cellular decomposition The last is the invariance problem

By using too broad a definition of complex, the second problem became too difficult Problems (1), (2), and (3) were solved by using polyhedra for complexes A polyhedron is a collection of cells in a euclidean space, each cell is defined by a system of linear equalities and inequalities. A step in solving (4) was the proof of invariance under subdivision. This led to the Hauptvermutung. *If two polyhedra are homeomorphic then they have isomorphic subdivisions.* This plus invariance under subdivision would imply topological invariance But it remains unproved to this day

The wording of the invariance problem is such that one feels constrained to use only topological maps in its solution However the homology groups are invariant under homotopy equivalences It was perhaps the realization of this fact which led Alexander to abandon the refined techniques of topological equivalence, and to replace them by the rough techniques of simplicial approximation and deformation, and thereby achieve a satisfactory solution.

Although simplicial complexes appear to eliminate most theoretical difficulties, they are highly inefficient for the computation of homology groups of simple spaces. A triangulation of a space usually demands a large number of simplexes As noted in III, Exer. D6, a triangulation of a torus requires at least 42 simplexes The n-simplex has $2^n - 1$ faces. Polyhedral and cellular decompositions are far more efficient (see Chapters XIII and XIV of the second volume).

The Mayer homology groups. Let K be a simplicial complex Consider the free abelian group $M_n(K)$ whose generators are sequences $(A^0, \cdots, A^n)$ of vertices of K, such that $A^0, \cdots, A^n$ are in a simplex of K Two such sequences are regarded as equal if they differ only by a permutation of coordinates Define the homomorphism F: $M_n(K) \rightarrow M_{n-1}(K)$ by setting

$$F(A^0, \cdots, A^n) = \sum_{i=0}^{n} (A^0, \cdots, \hat{A}^i, \cdots, A^n).$$

Let p be a prime and G an abelian group with $pG = 0$ (i e. $pg = 0$ for all $g \in G$). Define $M_n(K;G) = M_n(K) \otimes G$ and consider the homomorphism F $M_n(K;G) \rightarrow M_{n-1}(K;G)$ induced by the operator F above.

Then the iteration F^p of F is zero Thus for each integer $0 < q < p$ the composition

$$M_{n+p-q} \xrightarrow{F^{p-q}} M_n \xrightarrow{F^q} M_{n-q}$$

is zero The Mayer homology group $H_{n,q}(K,G)$ is defined to be (kernel F^q)/(image F^{p-q}). These groups were defined by W Mayer [*A new homology theory* I, II, Annals of Math 43 (1942) 370-380, 594-605] He established a large number of their properties, but was unable to determine their relation with the ordinary homology groups This was done by E Spanier [*The Mayer homology theory*, Bulletin A M S 55 (1949) 102-112], making essential use of our Uniqueness theorem VI,8 1 In this order, Spanier defines (1°) relative groups $H_{n,q}(K,L,G)$ where L is a subcomplex of K, (2°) a homomorphism f_* $H_{n,q}(K,L,G) \rightarrow H_{n,q}(K',L',G)$ for each simplicial map f $(K,L) \rightarrow (K',L')$, and (3°) a boundary operator F^q $H_{n,q}(K,L,G) \rightarrow H_{n-q,p-q}(L,G)$ He then shows that the groups $H_{n,q}$, suitably reindexed, provide several homology theories on the h-category of simplicial complexes The coefficient group of each theory is shown to be either 0 or G depending on arithmetic properties of the indices involved Thus the Uniqueness theorem implies that $H_{n,q}(K,L,G)$ is either 0 or isomorphic with the ordinary homology group $H_r(K,L,G)$ for a suitable r The precise result is that $H_{n,q}(K,L,G) \approx H_r(K,L;G)$ in the following two cases

$$n + 1 \equiv q \pmod p, \quad \text{and} \quad r = 2(n + 1 - q)/p,$$
$$n + 1 \equiv 0 \pmod p, \quad \text{and} \quad r = 2(n + 1)/p - 1.$$

In all other cases $H_{n,q}$ is zero

EXERCISES

A SIMPLICIAL APPROXIMATIONS

1 If g_1,g_2 $K \rightarrow K_1$ are both simplicial approximations to a map f $|K| \rightarrow |K_1|$, then g_1 and g_2 are contiguous

2 Show that a projection π. Sd $K \rightarrow K$ (in the sense of §7) is a simplicial approximation to the linear map l_K: Sd $K \rightarrow K$ involved in the definition of subdivision

B. LOCALLY FINITE COMPLEXES.

DEFINITION Let K be a (possibly infinite) simplicial complex Given an elementary chain A^0 A^q in $C_q(K_o)$, define an integral cochain $\phi \in C^q(K_o;J)$ so that $\phi(A^0, \cdots, A^q) = 1$ and $\phi(c) = 0$ for all other elementary chains in $C_q(K_o)$ The cochains ϕ thus obtained generate a free subgroup of $C^q(K_o,J)$ denoted by $C^q(K)$ and called the group of *finite integral* cochains of K

1. Show that K is locally finite if and only if the coboundary of every finite cochain is again a finite cochain. Denote the resulting cochain complex by K°.

2 A simplicial map $f.\ K \to K_1$ of a locally finite complex K into another such complex K_1 is called locally finite if, for every finite subcomplex L of K_1, $f^{-1}(L)$ is a finite subcomplex of K Show that a locally finite simplicial map induces a map $f^\circ\ \ K_1^\circ \to K^\circ$.

3 Let L be a subcomplex of the locally finite complex K, and let $i\cdot\ L \to K$, j: $K \to (K,L)$ be inclusion maps Denote by $K^\circ - L^\circ$ the kernel of the map $i^\circ\cdot\ K^\circ \to L^\circ$. Show that a locally finite simplicial map f: $(K,L) \to (K_1,L_1)$ induces a map f°: $K_1^\circ - L_1^\circ \to K^\circ - L^\circ$. Show that with suitable definitions the category of pairs of locally finite complexes and locally finite maps constitutes an h-category $\mathcal{L}$ and that the pair $K^\circ - L^\circ, f^\circ$ then yields a contravariant h-functor O' on $\mathcal{L}$ with values in the category of $\delta\mathcal{G}$ of cochain complexes (direct couples)

4. Use the h-functor O' to define homology and cohomology theories $\mathcal{H}_q(K,L;G)$, $\mathcal{H}^q(K,L;G)$. Give a detailed description of these theories analogous to that of §3 What are the limitations on the coefficient group G?

5. Adopt the alternating approach and define a functor A' analogous to O'. Show that the homology and cohomology theories obtained using O' and A' are isomorphic

6. Compare the H and $\mathcal{H}$ theories on finite complexes Correlate the result with v, Exer F4.

7. Show that, in the $\mathcal{H}$ theories, the reduced groups are defined only for finite complexes.

Remarks concerning terminology. The homology groups H_q of an infinite complex based on the chain complexes K_o (or K_a) will be called the *direct* homology groups, while the corresponding cohomology groups H^q will be called the *inverse* cohomology groups In a locally finite complex the cohomology groups $\mathcal{H}^q$ based on the chain complexes K° (or K^a) will be called the *direct cohomology groups* while the corresponding homology groups $\mathcal{H}_q$ will be called the *inverse homology groups.* The reasons for this terminology will become clear in Exercises VIII,F The direct homology and cohomology groups are not defined for compact coefficients. If the linear form notation were used for all chains and cochains, then the direct groups would be based on finite linear forms while the inverse groups would utilize infinite linear forms. For reasons explained in the remark following 4.1c, it is convenient to use the linear form notation for chains and the functional notation for cochains, regardless of whether the chain or cochain is finite or infinite

CHAPTER VII

The singular homology theory

1. INTRODUCTION

The objective of this chapter is to establish the existence of homology and cohomology theories on the largest admissible category, namely: the category $\mathcal{A}_1$ of all pairs (X,A) and all maps of such pairs. For homology [cohomology] theory, the existence is established for any prescribed coefficient group in any of the categories $\mathcal{G}_R$ [$\mathcal{G}_R$ or $\mathcal{G}_C$], and the theory has values in the same category. A singular homology theory with values in $\mathcal{G}_C$ is not constructed.

The method is the following. Using mappings of ordered simplexes into the space X, a chain complex $S(X)$ is obtained, called the *singular complex* of the space X. Then the homology and cohomology groups of a pair (X,A) over any coefficient group are defined as the appropriate groups of the complex $S(X)/S(A)$, using the methods of Chapter v.

The verifications of all the axioms, save for Homotopy and Excision, are trivial consequences of the corresponding theorems about chain complexes. To prove the Homotopy axiom, it is necessary to construct a chain homotopy from the given homotopy. The proof of the Excision axiom requires some preliminary construction.

2. THE SINGULAR COMPLEX OF A SPACE

In the euclidean R^{q+1} space with coordinates $(x_0, \cdots, x_q)$ consider the unit simplex Δ_q (see II,2.2). The vertices $d^0, \cdots, d^q$ are the unit points on the coordinate axes of R^{q+1}. If we regard R^q as the subset of R^{q+1} given by $x_q = 0$, then Δ_{q-1} is a face of Δ_q and has vertices $d^0, \cdots, d^{q-1}$. In Δ_q the barycentric coordinates of a point coincide with its cartesian coordinates.

Consider the simplicial maps

$$e_q^i: \Delta_{q-1} \to \Delta_q, \qquad i = 0, \cdots, q,$$

given by the following vertex assignments:

$$e_q^i(d^j) = d^j \qquad \text{if } j < i,$$
$$e_q^i(d^j) = d^{j+1} \qquad \text{if } i \leqq j.$$

Thus e_q^i maps Δ_{q-1} simplicially onto the $(q-1)$-face of Δ_q not containing d^i and preserves the order of the vertices

LEMMA 2.1. $e_q^i e_{q-1}^j = e_q^j e_{q-1}^{i-1}$ for $0 \leqq j < i \leqq q$.

The proof is obvious

DEFINITION 2.2. A continuous mapping

$$T.\ \Delta_q \to X$$

of the unit q-simplex Δ_q into a topological space X is called a *singular q-simplex* in X The singular $(q-1)$-simplex

$$T^{(i)} = Te_q^i\colon\ \Delta_{q-1} \to X$$

is called the ith *face* of T, for $i = 0, \cdots, q$

By 2.1, we have

LEMMA 2.3 $(T^{(i)})^{(j)} = (T^{(j)})^{(i-1)}$ for $0 \leqq j < i \leqq q$

DEFINITION 2.4 The free abelian group generated by the singular q-simplexes in X is denoted by $C_q(X)$ and is called the group of *singular (integral) q-chains* of X. If $q < 0$, there are no q-simplexes and $C_q(X) = 0$.

The *boundary* homomorphism

$$\partial_q.\ C_q(X) \to C_{q-1}(X)$$

is defined as follows: If $q \leqq 0$, then $\partial_q = 0$ If $q > 0$ and T is a singular q-simplex in X, then

$$\partial_q T = \sum_{i=0}^{q} (-1)^i T^{(i)}.$$

LEMMA 2.5 $\partial_{q-1}\partial_q = 0$

PROOF If $q \leq 1$, then $\partial_{q-1} = 0$ by definition Assume therefore that $q \geqq 2$ It is sufficient to verify 2.5 for each generator T of $C_q(X)$ By definition

$$\begin{aligned}\partial_{q-1}\partial_q T &= \sum_{i=0}^{q} (-1)^i \partial_{q-1} T^{(i)} = \sum_{i=0}^{q}\sum_{j=0}^{q-1} (-1)^i(-1)^j (T^{(i)})^{(j)} \\ &= \sum_{0 \leqq j < i \leqq q} (-1)^{i+j}(T^{(i)})^{(j)} + \sum_{0 \leqq i \leqq j < q} (-1)^{i+j}(T^{(i)})^{(j)}.\end{aligned}$$

In view of 2.3, the first sum equals

$$\sum_{0 \leqq j < i \leqq q} (-1)^{i+j}(T^{(j)})^{(i-1)}.$$

Replacing $i - 1$ by j and j by i gives

$$-\sum_{0 \leqq i \leqq j < q} (-1)^{i+j}(T^{(i)})^{(j)}.$$

Thus the two sums cancel, and the lemma is proved.

It follows from 2.5 that $\{C_q(X),\partial_q\}$ is a chain complex. This chain complex will be denoted by $S(X)$. If A is a subspace of X, then $C_q(A)$ is generated by a subset of the set generating $C_q(X)$ and ∂_q on $C_q(A)$ agrees with ∂_q on $C_q(X)$. Therefore $S(A)$ is a subcomplex of $S(X)$ and it is a direct summand of $S(X)$. Given $c \varepsilon C_q(X)$ we shall write $c \subset A$ to indicate $c \varepsilon C_q(A)$.

DEFINITION 2.6. For each pair (X,A) consisting of a topological space X and a subset A, the chain complex $S(X)/S(A)$ is called the *singular complex of the pair* (X,A). The groups $C_q(S(X)/S(A)) = C_q(X)/C_q(A)$ are free groups.

REMARK. Strictly speaking $S(A)$ is not a subcomplex of $S(X)$, because a generator of $S(A)$ is a map $T: \Delta_q \to A$ and not a map $\Delta_q \to X$. The inclusion map $i: A \subset X$ induces an isomorphism of $S(A)$ with a direct summand of $S(X)$, and $S(X)/S(A)$ should be written $S(X)/iS(A)$.

LEMMA 2.7. *If $f: (X,A) \to (Y,B)$, the homomorphisms f_q: $C_q(X) \to C_q(Y)$ and $f_q: C_q(A) \to C_q(B)$ defined by*

$$f_q(T) = fT$$

for $T: \Delta_q \to X$ (or $T: \Delta_q \to A$), yields a map

$$f_S: S(X)/S(A) \to S(Y)/S(B)$$

Moreover, if $f: (X,A) \to (X,A)$ is the identity, then f_S is the identity, and, if $f: (X,A) \to (Y,B)$, $g: (Y,B) \to (Z,C)$, then $(gf)_S = g_S f_S$.

The proof requires only the verification of the commutativity relation $\partial_q f_q = f_{q-1}\partial_q$ which is an immediate consequence of the relation $(fT)^{(i)} = fTe_q^i = f(T^{(i)})$.

The lemma states that $S(X)/S(A)$ and f_S form a covariant functor S on the category $\mathcal{A}_1$ of pairs of topological spaces with values in the category $\partial\mathcal{G}$ of chain complexes (recall that $\mathcal{G}$ is the category of ordinary abelian groups, i.e. $\mathcal{G} = \mathcal{G}_R$ where $R =$ the integers). We now convert $\mathcal{A}_1$ into a c-category by defining couples (i,j) to consist of the inclusion maps $i: A \to X$, $j: X \to (X,A)$ for each pair (X,A) in $\mathcal{A}_1$. Since i_S is the inclusion map $S(A) \to S(X)$, it follows that the sequence

$$0 \to S(A) \xrightarrow{i_S} S(X) \xrightarrow{j_S} S(X)/S(A) \to 0$$

is exact, thus (i_S,j_S) is a couple in the category $\partial\mathcal{G}$. Since, as was observed earlier, $S(A)$ is a direct summand of $S(X)$, it follows that the couple (i_S,j_S) is *direct* in the sense of v,3.1. Thus S is a c-functor.

We now go a step further. The c-category $\partial\mathcal{G}$ has been converted in v,4 into an h-category. We shall do the same with the category $\mathcal{A}_1$. To this end we define homotopies and points in $\mathcal{A}_1$ to be actual homotopies and actual points, while excisions in $\mathfrak{A}_1$ are defined as follows A map f $(X',A') \to (X,A)$ is called an *excision* if (1) f is an inclusion map, (2) $X - A = X' - A'$, and (3) the closure of $X - X'$ lies in the interior of A. The last two conditions may be replaced by the more symmetric conditions

$$A' = X' \cap A, \qquad X = \text{Int } (X') \cup \text{Int } (A).$$

Note that this concept of excision is broader than that of I,3 in that $U = X - X'$ is not required to be open

With these definitions $\mathfrak{A}_1$ becomes an h-category, and we can state the main result of this chapter

THEOREM 2 8 *If $\partial\mathcal{G}$ is treated as an h-category in the sense of direct couples, then the functor S $\mathfrak{A}_1 \to \partial\mathcal{G}$ is a covariant h-functor* (*see* IV,9 4).

It has already been shown that S is a c-functor To prove the theorem we must show that S preserves homotopies, generalized excisions, and points. This will be done in Theorems 4 7, 7 1, and 9 1. Granting that this has been done, we apply IV,9 5 to obtain our chief objective.

DEFINITION 2 9 The composition of the h-functor S with the homology theory of $\partial\mathcal{G}$ with coefficient group G (see V,11 3) yields a homology theory on $\mathfrak{A}_1$ called *the singular homology theory of $\mathfrak{A}_1$ with coefficient group G* The composition of S with the cohomology theory of $\partial\mathcal{G}$ with coefficient group G (see V,12 3) yields a cohomology theory on $\mathfrak{A}_1$ called *the singular cohomology theory of $\mathfrak{A}_1$ with coefficient group G* For any given pair $(X,A) \in \mathfrak{A}_1$, the homology and cohomology groups $H_q(S(X)/S(A),G)$, $H^q(S(X)/S(A),G)$ (see V,11 4, 12 4) will be written $H_q(X,A,G)$, $H^q(X,A,G)$, respectively, and are called *the qth singular homology and cohomology groups of (X,A) with coefficient group G* According to V,11 3 and 12 3 we have the following cases

(1)	$G \in \mathcal{G}_R$,	then	$H_q(X,A;G) \in \mathcal{G}_R$,
(2)	$G \in \mathcal{G}_R$,	then	$H^q(X,A,G) \in \mathcal{G}_R$,
(3)	$G \in \mathcal{G}_C$,	then	$H^q(X,A,G) \in \mathcal{G}_C$

Observe that singular homology groups $H_q(X,A,G)$ are not defined when G is compact The reason for this is that $C_q(X)$ usually does not have a finite base (see V,11 3)

We note here the following alternative definition of the singular simplexes. Let s be an ordered q-simplex with vertices $A^0 < \cdots < A^q$. A map T. $|s| \to X$ will be called a singular q-simplex in X (in the new sense) If s' has vertices $B^0 < \cdots < B^q$ and T' $|s'| \to X$, then we say that T and T' are equivalent if $T = T'B$ where $B\cdot$ $s \to s'$ is the simplicial map with $B(A^i) = B^i$, $i = 0, \cdots, q$ In this definition, the group $C_q(X)$ is generated not by the individual singular q-simplexes, but by the equivalence classes of such q-simplexes Clearly every such equivalence class contains precisely one singular q-simplex in the sense of 2 2.

3. DIRECT DESCRIPTION OF THE BASIC CONCEPTS

This section serves the same purpose as Section 4 of Chapter VI. A direct description of the singular homology and cohomology theories will be given The opening remarks of VI,4 apply equally well here

The group $C_q(X,A,G)$ of singular q-chains of X mod A is by definition the tensor product $[C_q(X)/C_q(A)] \otimes G$ By 2 4, the singular simplexes T $\Delta_q \to X$ form a base for $C_q(X)$ By V,9 6, $C_q(X,A,G)$ is generated by the elements $T \otimes g$ which will be written gT in accordance with the convention V,11 4 Thus we have

THEOREM 3 1 *The group $C_q(X,A,G)$ is generated by the elements gT, where $g \in G$ and T $\Delta_q \to X$ is a singular q-simplex in X. These generators are subject to the relations*

$$(g_1 + g_2)T = g_1T + g_2T$$

and

$$gT = 0 \qquad \text{for} \qquad T(\Delta_q) \subset A.$$

The boundary is given by the formula

$$\partial_q(gT) = \sum_{i=0}^{q} (-1)^i gT^{(i)}.$$

For any map f $(X,A) \to (Y,B)$, the chain transformation f_q: $C_q(X,A,G) \to C_q(Y,B;G)$ is given by

$$f_q(gT) = g(fT).$$

If this theorem were adopted as a definition, it would be necessary to verify that f_q and ∂_q are compatible with the relations and commute.

The descriptions of $Z_q(X,A,G)$, $B_q(X,A,G)$, $H_q(X,A,G)$, f_*, and ∂_* are as in VI,4 with (K,L) replaced by (X,A)

Passing to cohomology we begin with the group $C^q(X,A;G) = C^q(S(X)/S(A);G)$ of cochains of X mod A over G. This group is by definition the group $\text{Hom}(C_q(X)/C_q(A),G)$, and therefore V,10 and V,12 imply

THEOREM 3.1c *The group $C^q(X,A,G)$ is the group of functions ϕ defined on singular q-simplexes T in X, the value $\phi(T)$ is in G, and is zero if T is in A. The coboundary $\delta^q\phi$ is defined by*

$$(\delta^q\phi)(T) = \sum_{i=0}^{q+1} (-1)^i\phi(T^{(i)}).$$

For any map $f\colon (X,A) \to (Y,B)$, *the cochain transformation* $f^q\colon C^q(Y,B;G) \to C^q(X,A;G)$ *is given for* $\phi \in C^q(Y,B;G)$ *by the formula*

$$(f^q\phi)(T) = \phi(fT).$$

The descriptions of $Z^q(X,A;G)$, $B^q(X,A;G)$, $H^q(X,A;G)$, f^*, and δ^* are as in VI,4 with (K,L) replaced by (X,A).

4. PRESERVATION OF POINTS, REDUCED GROUPS

THEOREM 4.1. *If P is a space consisting of a single point, then the chain complex $S(P)$ is pointlike in the sense of* V,4.7.

PROOF. For each $q \geqq 0$ there is only one singular q-simplex $T_q\colon \Delta_q \to P$. For q even and > 0, we have $\partial T_q = T_{q-1}$ so that $\partial_q\colon C_q(X) \approx C_{q-1}(X)$. For $q < 0$, we have $C_q(X) = 0$. Thus conditions of V,4.7 are fulfilled.

Observe further that by 3.1 and 3.1c we have

$$C_0(P;G) \approx G, \qquad C^0(P;G) \approx G,$$

and $\partial_1 = 0$, $\partial_0 = 0$. Thus $H_0 = C_0$ and $H^0 = C^0$. Consequently

$$H_0(P;G) \approx G, \qquad H^0(P;G) \approx G.$$

This shows that the singular homology and cohomology theories constructed over a group G actually have the group G as coefficient group in the sense of I,6.1.

Observe that, for any space X, a singular 0-simplex $T\colon \Delta_0 \to X$ is completely determined by the image point $x = T(\Delta_0)$. If we agree to write x instead of T, then every 0-chain $c \in C_0(X,G)$ will have the form of a finite formal sum

$$c = \sum g_i x_i$$

where $g_i \in G$, $x_i \in X$. Every cochain $\phi \in C^0(X,G)$ will be a function $\phi\colon X \to G$ (with no continuity assumption).

In analogy with VI,5.1 and VI,5.2, we introduce the following definitions:

DEFINITION 4.2. For each element $c = \sum g_i x_i$ of $C_0(X,G)$, the *index* $\text{In}(c)$ is

$$\text{In}(c) = \sum g_i \in G.$$

Clearly $\text{In}\colon C_0(X,G) \to G$.

DEFINITION 4.3. The *augmented* singular complex $\tilde{S}(X)$ of the space

X is defined as follows If $X = 0$, then $\tilde{S}(X) = S(X) = 0$ If $X \neq 0$, then

$$
\begin{aligned}
&C_q(\tilde{S}(X)) = C_q(X) && \text{for } q \neq -1, \\
&C_{-1}(\tilde{S}(X)) = J && (= \text{additive group of integers}), \\
&\tilde{\partial}_q = \partial_q && \text{for } q \neq 0, -1, \\
&\tilde{\partial}_0 = \text{In}.\ \ C_0(\tilde{S}(X)) \to J, && \\
&\tilde{\partial}_{-1} = 0. &&
\end{aligned}
$$

The *reduced* 0th homology and cohomology groups of X over G are defined as

$$\tilde{H}_0(X,G) = H_0(\tilde{S}(X);G), \qquad \tilde{H}^0(X,G) = H^0(\tilde{S}(X),G)$$

Clearly $H_q(\tilde{S}(X),G) = H_q(X,G)$ for $q \neq 0$ In the dimension 0, $\tilde{Z}_0(X,G) = Z_0(\tilde{S}(X),G)$ is a subgroup of $Z_0(X,G) = C_0(X,G)$ and consists of 0-chains c with $\text{In}(c) = 0$ This implies

THEOREM 4 4 *The reduced* 0th *homology group* $\tilde{H}_0(X,G)$ *is the kernel of the homomorphism* f_* $H_0(X,G) \to H_0(P,G)$ *where* f $X \to P$ *and* P *is a space consisting of one point*

Therefore the concept of reduced group coincides with that defined in I,7 for any homology theory

For cohomology, we find that $H^q(\tilde{S}(X),G) = H^q(X;G)$ for $q \neq 0$, and that $\tilde{H}^0(X,G) = Z^0(X,G)/\tilde{B}^0(X,G)$, where $\tilde{B}^0(X,G) = B^0(\tilde{S}(X),G)$ consists of those cochains $\phi \in C^0(X,G)$ which are constant on the singular 0-simplexes (i e the points) of X This implies

THEOREM 4 4c *The reduced* 0th *cohomology group* $\tilde{H}^0(X;G)$ *is the factor group of* $H^0(X,G)$ *by the image of the homomorphism* f^* $H^0(P;G) \to H^0(X;G)$ *where* f $X \to P$ *and* P *is a space consisting of one point*

The definition of "acyclic space" and "algebraic map f. $S(X) \to S(Y)$" are as in VI,5 6 and VI,5 4 A carrier of f is a function C which to each singular simplex T in X assigns a subset $C(T)$ in Y such that $f(T) \subset C(T)$ and $C(T^{(i)}) \subset C(T)$ The statement and proof of VI,5 7 carry over with only formal changes

5. THE LINEAR COMPLEX OF A CONVEX SET

We proceed to develop the tools to be used in the proofs that the functor S preserves homotopies and generalized excisions (given in §§7 and 9)

DEFINITION 5 1 Let V be a convex set in a euclidean space A singular simplex $T: \Delta_q \to V$ is said to be *linear*, if $\alpha_1,\alpha_2 \in \Delta_q$, $\lambda_1 \geqq 0$, $\lambda_2 \geqq 0$ and $\lambda_1 + \lambda_2 = 1$ imply

$$T(\lambda_1\alpha_1 + \lambda_2\alpha_2) = \lambda_1 T(\alpha_1) + \lambda_2 T(\alpha_2)$$

The linear simplexes form a subcomplex $Q(V)$ of $S(V)$.

The linear simplex T is completely determined by the points $v^i = T(d^i)$, $i = 0, \cdots, q$, and we shall therefore use the symbol $v^0 \cdots v^q$ to designate the linear simplex T. With this notation we have

$$\partial_q(v^0 \cdots v^q) = \sum_{i=0}^{q} (-1)^i v^0 \cdots \hat{v}^i \cdots v^q$$

DEFINITION 5.2. If $v \in V$ and $T = v^0 \cdots v^q$ is a linear q-simplex, the *join* vT is the linear $(q + 1)$-simplex $vv^0 \cdots v^q$. With v fixed, this operation on generators extends uniquely to a homomorphism $c \to vc$ of $C_q(Q(V))$ into $C_{q+1}(Q(V))$.

LEMMA 5.3. *If $v \in V$ and $c \in C_q(Q(V))$, then*

$$\partial(vc) = c - v(\partial c) \qquad \textit{if } q > 0,$$

$$\partial(vc) = c - \mathrm{In}(c)v \qquad \textit{if } q = 0$$

The proof is the same as that of VI,5.11.

6. PRISMS

The *unit q-prism* ($q > 0$) is the cartesian product

$$\pi_q = I \times \Delta_{q-1}$$

where Δ_{q-1} is the unit $(q - 1)$-simplex in R^q, and I is the closed interval $0 \leqq t \leqq 1$ in R^1. The prism π_q is a convex subset of $R^1 \times R^q = R^{q+1}$.

Consider the maps

$$r_q^i: \pi_{q-1} \to \pi_q, \qquad i = 0, \cdots, q - 1$$

defined for $(t,v) \in \pi_{q-1}$ by

$$r_q^i(t,v) = (t, e_{q-1}^i(v))$$

and the maps

$$l_q: \Delta_{q-1} \to \pi_q, \qquad u_q: \Delta_{q-1} \to \pi_q$$

defined for $v \in \Delta_{q-1}$ by

$$l_q(v) = (0,v), \qquad u_q(v) = (1,v).$$

As an immediate consequence of these definitions we have

LEMMA 6.1 $r_q^i r_{q-1}^j = r_q^j r_{q-1}^{i-1}$ for $0 \leqq j < i \leqq q - 1$,
$r_q^i l_{q-1} = l_q e_{q-1}^i$, $\quad r_q^i u_{q-1} = u_q e_{q-1}^i$, for $i = 0, \cdots, q - 1$.

DEFINITION 6 2 A continuous mapping

$$P: \pi_q \to X$$

of the unit q-prism π_q into the topological space X is called a *singular q-prism* in X ($q > 0$) The singular ($q - 1$)-prism

$$P^{(i)} = Pr_q^i: \pi_{q-1} \to X$$

is called the ith *face* of P for $i = 0, \cdots, q - 1$. A singular 1-prism has no faces The singular ($q - 1$)-simplexes

$$P_l = Pl_q: \Delta_{q-1} \to X, \qquad P_u = Pu_q: \Delta_{q-1} \to X$$

are called the *lower* and the *upper base* of P, respectively.

By 6 1, we have

LEMMA 6 3. $(P^{(i)})^{(j)} = (P^{(j)})^{(i-1)}$ for $0 \leqq j < i \leqq q - 1$,
$(P^{(i)})_l = (P_l)^{(i)}, \quad (P^{(i)})_u = (P_u)^{(i)},$ for $i = 0, \cdots, q - 1$,

and we may write $P_u^{(i)}, P_l^{(i)}$ *without ambiguity*

DEFINITION 6 4. The free group generated by the singular q-simplexes and the singular q-prisms in X is denoted by $C_q(SP(X))$ and is called the group of *singular (integral) prismatic q-chains* in X. If $q < 0$, then $C_q(SP(X)) = 0$ If $q = 0$, there are no singular 0-prisms so $C_0(SP(X) = C_0(X)$.

The boundary homomorphism

$$\partial_q: C_q(SP(X)) \to C_{q-1}(SP(X))$$

is defined as follows. If $q \leqq 0$, then $\partial_q = 0$ If $q > 0$ and P is a singular q-prism, then

$$\partial_q P = P_u - P_l - \sum_{i=0}^{q-1} (-1)^i P^{(i)}.$$

For a singular q-simplex T

$$\partial_q T = \sum_{i=0}^{q} (-1)^i T^{(i)}.$$

LEMMA 6 5 $\partial_{q-1}\partial_q = 0$

PROOF. Clearly it is sufficient to verify that $\partial_{q-1}\partial_q P = 0$ for every singular q-prism P Since $\partial_0 = 0$, we may assume that $q > 1$ By definition

$$\begin{aligned}\partial_{q-1}\partial_q P &= \partial_{q-1}\left(P_u - P_l - \sum_{i=0}^{q-1}(-1)^i P^{(i)}\right)\\ &= \sum_{i=0}^{q-1}(-1)^i P_u^{(i)} - \sum_{i=0}^{q-1}(-1)^i P_l^{(i)}\\ &\quad - \sum_{i=0}^{q-1}(-1)^i\left\{P_u^{(i)} - P_l^{(i)} - \sum_{j=0}^{q-2}(-1)^j (P^{(i)})^{(j)}\right\}.\end{aligned}$$

The single sums cancel because of the second part of 6 3 The double sum is zero by the same calculation as in the proof of 2 5, using the first part of 6 3.

DEFINITION 6 6 The chain complex

$$SP(X) = \{C_q(PS(X)),\partial_q\}$$

is called the *singular prismatic complex* of the space X.

Clearly $S(X)$ is a subcomplex of $SP(X)$. If A is a subset of X, then $SP(A)$ is a subcomplex of $SP(X)$ and $S(A) = S(X) \cap SP(A)$. Hence the inclusions $S(A) \subset S(X)$ and $SP(A) \subset SP(X)$ induce a homomorphism

$$\eta \quad S(X)/S(A) \to SP(X)/SP(A)$$

whose kernel is zero by virtue of the Noether isomorphism theorem. It will be convenient to regard $S(X)/S(A)$ as a subcomplex of $SP(X)/SP(A)$ so that η is an inclusion map.

THEOREM 6 7. *There exists a map*

$$r. \quad (SP(X),SP(A)) \to (S(X),S(A))$$

which is the identity on $S(X)$. In other words the pair of chain complexes $(S(X),S(A))$ is a retract of the pair $(SP(X),SP(A))$

PROOF For any linear n-simplex $v^0 \cdots v^n$ of Δ_{q-1}, define a linear $(n+1)$-chain in π_q by

$$(1) \qquad D_n^q(v^0 \cdots v^n) = \sum_{i=0}^{n} (-1)^i (l_q v^0) \cdots (l_q v^i)(u_q v^i) \cdots (u_q v^n).$$

Then D_n^q extends uniquely to a homomorphism

$$D_n^q. \quad C_n(Q(\Delta_{q-1})) \to C_{n+1}(Q(\pi_q)).$$

The same calculation as in the proof VI,3 2 shows that

$$(2) \qquad \partial D_n^q c = u_q c - l_q c - D_{n-1}^q \partial c, \qquad \text{for } c \in C_n(Q(\Delta_{n-1})).$$

If we use the second set of relations of 6 1 with (1), we obtain immediately

$$(3) \qquad r_q^i D_n^{q-1} c = D_n^q e_{q-1}^i c, \qquad \text{for } c \in C_n(Q(\Delta_{n-2}))$$

Define the homomorphism

$$r_q\colon \quad C_q(SP(X)) \to C_q(X)$$

as follows. If T is a singular q-simplex, then $r_q T = T$. If P is a singular q-prism, then

$$(4) \qquad r_q P = P_S D_{q-1}^q (d^0 \cdots d^{q-1}),$$

where P_S is defined by 2 7 Clearly r_q is the identity on $C_q(X)$, and it maps $C_q(SP(A))$ into $C_q(A)$ It remains to prove that r and ∂ commute. This is obvious when applied to a singular simplex. Hence it suffices to prove that

$$\partial r_q P = r_{q-1}\partial P$$

Apply ∂ to both sides of (4), use $\partial P_S = P_S\partial$ and then use (1) to obtain

$$(5)\qquad \partial r_q P = (P_S u_q - P_S l_q - P_S D^q_{q-2}\partial)(d^0 \cdots d^{q-1}).$$

By definition

$$(6)\qquad P_S u_q(d^0 \cdots d^{q-1}) = P_u, \qquad P_S l_q(d^0 \cdots d^{q-1}) = P_l.$$

Since P_u, P_l are singular simplexes,

$$(7)\qquad P_u = r_{q-1}P_u, \qquad P_l = r_{q-1}P_l.$$

Using (3) we obtain

$$\begin{aligned} D^q_{q-2}\partial(d^0 \cdots d^{q-1}) &= D^q_{q-2} \sum_{i=0}^{q-1} (-1)^i(d^0 \cdots \hat{d}^i \cdots d^{q-1}) \\ &= D^q_{q-1} \sum_{i=0}^{q-1} (-1)^i e^i_{q-1}(d^0 \cdots d^{q-2}) \\ &= \sum_{i=0}^{q-1} (-1)^i r^i_q D^{q-1}_{q-2}(d^0 \cdots d^{q-2}). \end{aligned}$$

Since, by 6 2, $P_S r^i_q = P^{(i)}_S$, we have

$$(8)\qquad \begin{aligned} P_S D^q_{q-2}\partial(d^0 \cdots d^{q-1}) &= \sum_{i=0}^{q-1} (-1)^i P^{(i)}_S D^{q-1}_{q-2}(d^0 \cdots d^{q-2}) \\ &= \sum_{i=0}^{q-1} (-1)^i r_{q-1}P^{(i)}. \end{aligned}$$

Combining (5), (6), (7), and (8) gives

$$\partial r_q P = r_{q-1}P_u - r_{q-1}P_l - \sum_{i=0}^{q-1} (-1)^i r_{q-1}P^{(i)} = r_{q-1}\partial P.$$

This completes the proof.

7. PRESERVATION OF HOMOTOPIES

THEOREM 7 1 *Two homotopic maps*

$$f,g \quad (X,A) \to (Y,B)$$

induce chain homotopic maps

$$f_S, g_S\colon\ S(X)/S(A) \to S(Y)/S(B).$$

PROOF. Let

$$h: \quad (X \times I, A \times I) \to (Y,B)$$

be a map such that for each $x \in X$

$$h(x,0) = f(x), \qquad h(x,1) = g(x).$$

For every singular q-simplex T: $\Delta_q \to X$, consider the singular $(q+1)$-prism

$$DT: \quad \pi_{q+1} \to Y$$

defined by

$$(DT)(x,t) = h(T(x),t) \qquad \text{for} \qquad x \in \Delta_q, \qquad 0 \leqq t \leqq 1.$$

This defines homomorphisms

$$D. \quad C_q(X) \to C_{q+1}(SP(Y))$$

which carries $C_q(A)$ into $C_{q+1}(SP(B))$ Thus D $C_q(X)/C_q(A) \to C_{q+1}(SP(Y))/C_{q+1}(SP(B))$ Clearly

$$(DT)_l = fT, \qquad (DT)_u = gT,$$
$$(DT)^{(i)} = D(T^{(i)}) \qquad \text{for } i = 0, \cdot\ , q.$$

Hence

$$\begin{aligned} \partial DT &= (DT)_u - (DT)_l - \sum_{i=0}^{q} (-1)^i (DT)^{(i)} \\ &= gT - fT - \sum (-1)^i D(T^{(i)}) \\ &= gT - fT - D\partial T. \end{aligned}$$

Let now r $SP(Y)/SP(B) \to S(Y)/S(B)$ be a retraction as asserted by 6.7. Then

$$rD \quad C_q(X)/C_q(A) \to C_{q+1}(Y)/C_{q+1}(B)$$

with

$$\begin{aligned} \partial rDT &= r\partial DT = rgT - rfT - rD\partial T \\ &= gT - fT - rD\partial T. \end{aligned}$$

Thus the operator rD establishes the desired chain homotopy.

8. A COVERING THEOREM

DEFINITION 8 1 Let F be a family of subsets of the space X. A singular simplex T. $\Delta_q \to X$ will be said to *belong* to F if the set

$T(\Delta_q)$ is contained in at least one of the sets of the family F. Clearly then the faces of T also belong to F. The singular simplexes of X that belong to F make up a subcomplex $S(X,F)$ of the singular complex $S(X)$. If A is a subset of X, then $S(A,F)$ is the subcomplex of $S(X,F)$ made up of the singular simplexes of A that belong to F, i e

$$S(A,F) = S(A) \cap S(X,F).$$

The inclusions $S(X,F) \subset S(X)$, $S(A,F) \subset S(A)$ define a homomorphism

$$\eta: \quad S(X,F)/S(A,F) \to S(X)/S(A)$$

which, in view of the Noether isomorphism theorem, has kernel zero. It will be convenient to regard $S(X,F)/S(A,F)$ as a subgroup of $S(X)/S(A)$ so that η is an inclusion map

THEOREM 8 2. *If F is a family of subsets of the space X such that every point of X is in the interior of at least one set of F, then, for every $A \subset X$, the inclusion map*

$$\eta \quad S(X,F)/S(A,F) \to S(X)/S(A)$$

is a homotopy equivalence (see IV,9 2)

The proof requires several preliminary constructions.

Given a linear r-simplex $\sigma = v^0 \cdots v^r$ in Δ_q the *barycenter* b_σ of σ is the point

$$b_\sigma = \frac{1}{r+1} v^0 + \cdots + \frac{1}{r+1} v^r .$$

We define two sequences of homomorphisms

$$\begin{aligned} \mathrm{Sd} \quad & C_r(Q(\Delta_q)) \to C_r(Q(\Delta_q)) \\ R: \quad & C_r(Q(\Delta_q)) \to C_{r+1}(Q(\Delta_q)) \end{aligned}$$

by recurrence as follows. If $r = 0$, then Sd is the identity map and $R = 0$. For every linear r-simplex σ in Δ_q with $r > 0$, we define using the join operation of 5.2

$$\mathrm{Sd}\, \sigma = b_\sigma \,\mathrm{Sd}\, \partial\sigma, \tag{1}$$
$$R\sigma = b_\sigma(\sigma - \mathrm{Sd}\, \sigma - R\partial\sigma). \tag{2}$$

We shall prove by induction with respect to r that

$$\partial\, \mathrm{Sd}\, c = \mathrm{Sd}\, \partial c, \tag{3}$$
$$\partial Rc = c - \mathrm{Sd}\, c - R\partial c. \tag{4}$$

for every linear r-chain c in Δ_q. If $q \leqq 0$, the propositions are obvious

Assuming that (3) and (4) hold for linear i-chains with $i < r$ $(r > 0)$, consider a linear r-simplex σ in Δ_q. From 5.4 we deduce

$$\begin{aligned}
\partial \operatorname{Sd} \sigma &= \partial b_\sigma \operatorname{Sd} \partial\sigma = \operatorname{Sd} \partial\sigma - b_\sigma \partial \operatorname{Sd} \partial\sigma \\
&= \operatorname{Sd} \partial\sigma - b_\sigma \operatorname{Sd} \partial\partial\sigma = \operatorname{Sd} \partial\sigma, \\
\partial R\sigma &= \partial b_\sigma(\sigma - \operatorname{Sd} \sigma - R\partial\sigma) \\
&= \sigma - \operatorname{Sd} \sigma - R\partial\sigma - b_\sigma(\partial\sigma - \partial \operatorname{Sd} \sigma - \partial R\partial\sigma) \\
&= \sigma - \operatorname{Sd} \sigma - R\partial\sigma - b_\sigma(\partial\sigma - \operatorname{Sd} \partial\sigma - \partial\sigma + \operatorname{Sd} \partial\sigma + R\partial\partial\sigma) \\
&= \sigma - \operatorname{Sd} \sigma - R\partial\sigma.
\end{aligned}$$

Formula (3) asserts that Sd is a map $Q(\Delta_q) \to Q(\Delta_q)$, and (4) asserts that R is a chain homotopy of Sd into the identity map.

Now let T: $\Delta_q \to X$ be a singular q-simplex in X. Define singular chains Sd T and RT by

$$\operatorname{Sd} T = T_S \operatorname{Sd} (d^0 \cdots d^q), \qquad RT = T_S R(d^0 \cdots d^q),$$

where T_S is defined by 2.7. The operations Sd and R defined on the base elements extend uniquely to homomorphisms

$$\operatorname{Sd}\colon\ C_q(X) \to C_q(X), \qquad R\colon\ C_q(X) \to C_{q+1}(X).$$

It is easy to see that formulas (3) and (4) are still valid for every singular q-chain c in X. As before we define the iterates Sd^n of Sd by setting $\operatorname{Sd}^0 c = c$, $\operatorname{Sd}^n c = \operatorname{Sd} \operatorname{Sd}^{n-1} c$ for $n > 0$. Formulas (3) and (4) then yield

$$\partial \operatorname{Sd}^n c = \operatorname{Sd}^n \partial c, \tag{5}$$

$$\partial \sum_{i=0}^{n-1} R \operatorname{Sd}^i c = c - \operatorname{Sd}^n c - \sum_{i=0}^{n-1} R \operatorname{Sd}^i \partial c. \tag{6}$$

If F is any family of subsets of X and c is a singular chain in the complex $S(X,F)$, then Sd c and Rc are both in the complex $S(X,F)$.

Let now F be a family of subsets of X having the property stated in 8.2. We shall prove that for every singular simplex T in X there is an integer n such that $\operatorname{Sd}^n T$ is in the complex $S(X,F)$.

Let T: $\Delta_q \to X$ and let F_T be the family of subsets of Δ_q of the form $T^{-1}(B)$ for $B \in F$. Since every point $x \in X$ is in the interior of one of the sets in F, it follows that every point $v \in \Delta_q$ in the interior of one of the sets of F_T. Since the set Δ_q is compact, it follows from II,7.5 that there exists a number $\epsilon > 0$ such that every subset of Δ_q of diameter $< \epsilon$ is in one of the sets of F_T.

The argument just concluded shows that it suffices to construct an integer n such that $\operatorname{Sd}^n (d^0, \cdots, d^q)$ is a linear combination of linear simplexes of diameter $< \epsilon$. To verify this observe that, if $\sigma = v^0 \cdots v^q$, then Sd σ has the form $\sum \pm \tau$ where $\tau = b_{\sigma_0} \cdots b_{\sigma_q}$, with $\sigma_0 = \sigma$, and σ_{i+1} is a face of σ_i. By II,6 8,

$$\operatorname{diam} \tau \leqq \frac{q}{q+1} \operatorname{diam} \sigma.$$

Hence $\mathrm{Sd}^n (d^0 \cdots d^q)$ is a linear combination of linear simplexes of diameter $\leqq (q/(q+1))^n \operatorname{diam} (d^0 \cdots d^q)$, and this is $< \epsilon$ for n sufficiently large

We are now ready to prove 8 2 Since the complexes $S(X)/S(A)$ and $S(X,F)/S(A,F)$ are composed of free groups, it suffices in view of v,13 3 to prove that

$$\eta_*: \ H_q(S(X,F)/S(A,F)) \approx H_q(S(X)/S(A))$$

for integer coefficients and for all dimensions q Since η_* is one of the maps in the mapping of the homology sequence of the couple

$$0 \to S(A,F) \to S(X,F) \to S(X,F)/S(A,F) \to 0$$

into the homology sequence of (X,A), it follows from the five lemma (I,4 3) that it suffices to prove that

$$\eta_*. \ H_q(S(X,F)) \approx H_q(S(X))$$

for all q By exactness this is equivalent to showing that

$$H_q(S(X)/S(X,F)) = 0 \tag{7}$$

for all dimensions q Let $c \in C_q(S(X))$ be a singular chain such that $\partial c \equiv 0 \mod S(X,F)$ By the result previously established there is an integer n such that $\mathrm{Sd}^n c$ is in $S(X,F)$ Since ∂c is in $S(X,F)$, the chains $R\ \mathrm{Sd}^i\ \partial c$ are also in $S(X,F)$ Thus formula (6) yields

$$\partial \sum_{i=0}^{n-1} R\ \mathrm{Sd}^i c \equiv c \qquad \mod S(X,F)$$

so that c is a bounding cycle mod $S(X,F)$. This proves (7) and concludes the proof of 8.2.

9 INVARIANCE UNDER EXCISION

THEOREM 9 1 *Let the inclusion map*

$$i: \ (X - U, A - U) \subset (X,A)$$

be given where $\overline{U} \subset \operatorname{Int} A$ Then the induced map

$$i_S \quad S(X - U)/S(A - U) \to S(X)/S(A)$$

is a homotopy equivalence.

Proof. Consider the family F consisting of the two sets A and $X - U$. Condition $\overline{U} \subset \text{Int}(A)$ implies $\text{Int}(X - U) \cup \text{Int}(A) = X$ Thus 8 2 applies. Observe that

$$S(A - U) = S(X - U) \cap S(A),$$
$$S(X,F) = S(X - U) \cup S(A), \; S(A,F) = S(A)$$

Thus the map i_S may be factored into two maps

$$j\colon \; S(X - U)/S(A - U) \to S(X,F)/S(A,F)$$
$$\eta\colon \; S(X,F)/S(A,F) \to S(X)/S(A)$$

both of which are induced by inclusions The map η is a homotopy equivalence by 8 2. The map j may be written as

$$j\colon \; S(X - U)/S(X - U) \cap S(A) \to S(X - U) \cup S(A)/S(A)$$

which, by the Noether isomorphism theorem, is an isomorphism Consequently j also is a homotopy equivalence. Since, in the category $\partial\mathcal{G}$ of chain complexes, the composition of homotopy equivalences is again a homotopy equivalence, it follows that $i_S = \eta j$ is a homotopy equivalence.

Theorem 9 1 together with 4.1 and 7 1 conclude the proof of the statement of 2.8 that S. $(X,A) \to S(X)/S(A)$ is an h-functor This concludes the construction of the singular homology and cohomology theories

It should be pointed out that the excision axiom that has been obtained for the singular theories is stronger than that required in I,3, namely, the assumption that U be an *open* subset of X was not made

10. SINGULAR THEORY OF TRIANGULATED SPACES

Now that the axioms for the singular homology and cohomology theories have been verified, it follows from the results of Chapter III that, in a triangulated space, the singular groups are isomorphic with the simplicial groups computed from the triangulation using one of the two procedures of Chapter VI. Nevertheless, it is useful to have a direct definition of an isomorphism between the singular and the simplicial groups. In this section such an isomorphism is exhibited.

Consider a triangulated pair (X,A) with a triangulation $T = \{t,(K,L)\}$. We define a map

$$\beta. \; K_o/L_o \to S(X)/S(A)$$

as follows Given an elementary q-chain $c = A^0 \cdots A^q$ in K_o, consider the linear map

$$L_c\colon \; \Delta_q \to K$$

defined by $L_c(d^i) = A^i$, $i = 0, \cdots, q$. Then $tL_c\colon \Delta_q \to X$ is a singular q-simplex. We define

$$\beta c = tL_c$$

This defines homomorphisms

$$\beta\colon C_q(K) \to C_q(X)$$

carrying $C_q(L)$ into $C_q(A)$ To prove that β and ∂ commute, consider $c^{(i)} = A^0 \cdots \hat{A}^i \cdots A^q$ Then $\beta c^{(i)} = \beta c e_q^i = (\beta c)^{(i)}$. Thus

$$\partial\beta c = \sum_{i=0}^{q} (-1)^i(\beta c)^{(i)} = \sum (-1)^i\beta c^{(i)} = \beta\partial c$$

The main result of this section is the following

THEOREM 10 1 *The map*

$$\beta\colon K_o/L_o \to S(X)/S(A)$$

induces isomorphisms

$$\beta_*\colon H_q(K,L;G) \approx H_q(X,A;G), \qquad \beta^*\colon H^q(K,L;G) \approx H^q(X,A;G)$$

over any coefficient group G. The groups of (K,L) are the groups of the chain complex K_o/L_o of Chapter VI *while the groups of (X,A) are the singular homology and cohomology groups*

In view of the "five lemma" I,4 3, it suffices to prove the case $A = 0$

The proof leans heavily on the concepts introduced in II,9 For every simplex s of K we shall consider the first regular neighborhood $N^1(s)$ In view of II,9 6 a point α of K is in $N^1(s)$ if and only if there is a vertex A of s such that $\alpha(A) > \alpha(B)$ for each vertex B of K not in s. This implies the following properties of $N^1(s)$

(1) $N^1(s)$ is an open set containing $|s|$

(2) Every vertex of K that is in $N^1(s)$ is a vertex of s

(3) $N^1(s_1) \cap N^1(s_2) = N^1(s_1 \cap s_2)$

Let F be the family of open sets $tN^1(s) \subset X$ for all simplexes s of K This family satisfies the condition stated in 8 2, and therefore the inclusion map

$$\eta\colon S(X,F) \subset S(X)$$

induces homomorphisms η_* and η^* which are isomorphisms Observe further that in virtue of (1), for each chain c of K_o, the singular chain βc is in $S(X,F)$ Thus β may be factored $\beta = \eta\gamma$ where

$$K_o \xrightarrow{\gamma} S(X,F) \xrightarrow{\eta} S(X)$$

with $\gamma c = \beta c$. It suffices therefore to show that γ_* and γ^* are isomorphisms.

For each point α of K select a vertex $n(\alpha)$ of K which is nearest in the sense that $\alpha(n(\alpha)) \geqq \alpha(B)$ for any vertex B of K. We note the following properties of $n(\alpha)$:

(4) If α is a vertex, then $n(\alpha) = \alpha$.

(5) If $\alpha \in N^1(s)$, then $n(\alpha)$ is a vertex of s.

Proposition (4) is obvious while (5) follows from the characterization of $N^1(s)$ quoted above.

Given a singular simplex $T: \Delta_q \to X$, consider the vertices

$$A^i = n(t^{-1}Td^i), \qquad i = 0, \cdots, q$$

of K. If $T(\Delta_q) \subset tN^1(s)$, then $t^{-1}Td^i \in N^1(s)$ and A^i is a vertex of s by (5). Thus, if T is a singular simplex in $S(X,F)$, then

$$\bar{\gamma}T = A^0 \cdots A^q$$

is an elementary q-chain in K_o. This yields homomorphisms $\bar{\gamma}$: $C_q(S(X,F)) \to C_q(K_o)$ which clearly commute with ∂. Thus we have a map

$$\bar{\gamma}: \quad S(X,F) \to K_o.$$

We examine the composed map $\bar{\gamma}\gamma$. Let $B^0 \cdots B^q$ be an elementary q-chain of K_o, and let $\bar{\gamma}\gamma T = A^0 \cdots A^q$. Then by definition of β we have $Td^i = tB^i$ and $A^i = n(t^{-1}Td^i) = n(B^i)$. Thus, by (4), $A^i = B^i$ so that $\bar{\gamma}\gamma$ is the identity map of K_o. Consequently

(6) $\bar{\gamma}_*\gamma_* =$ identity, $\gamma^*\bar{\gamma}^* =$ identity.

We now turn our attention to the map $\gamma\bar{\gamma}$. Let $T: \Delta_q \to X$ be a singular simplex of $S(X,F)$, and let s be the least simplex of K such that $T(\Delta_q) \subset tN^1(s)$. Such a simplex s exists and is unique by virtue of (3). Denote

$$C(T) = tN^1(s).$$

We shall establish the following properties of the sets $C(T)$ for T in $S(X,F)$:

(7) $C(T^{(i)}) \subset C(T)$

(8) T and $\gamma\bar{\gamma}T$ are singular simplexes in $C(T)$

(9) $C(T)$ is acyclic, i.e. $\tilde{H}_0(C(T)) = 0$ and $H_q(C(T)) = 0$ for $q > 0$.

Proposition (7) is obvious and so is the first half of (8). Let $C(T) = tN^1(s)$. It follows that $t^{-1}Td^i \in N^1(s)$, and, by (5), $n(t^{-1}Td^i)$ is a vertex of s. Thus $\bar{\gamma}T$ is in s and $\gamma\bar{\gamma}T$ is in $t(|s|)$. Since $|s| \subset N^1(s)$, it follows that $t(|s|) \subset C(T)$. Thus $\gamma\bar{\gamma}T$ is in $C(T)$.

To prove (9) observe first that, in view of II,9 8, $|s|$ is a deformation retract of $N^1(s)$, and, since $|s|$ is contractible to point, $N^1(s)$ also is contractible to a point Thus $C(T)$ is contractible to a point and therefore homologically trivial by I,11 5

With (7)-(9) established, an exact replica of the proof of VI,5 7 yields a chain homotopy of the map $\gamma\bar{\gamma}$ with the identity map of $S(X,F)$ Consequently

$$\gamma_*\bar{\gamma}_* = \text{identity}, \qquad \bar{\gamma}^*\gamma^* = \text{identity},$$

which, combined with (6), shows that γ_* and γ^* are isomorphisms with $\bar{\gamma}_*$ and $\bar{\gamma}^*$ as inverses

11. TRIADS

Let (X,X_1,X_2) be a triad, i e X is a topological space and X_1,X_2 are subspaces of X The homology sequence of a triad was defined in I,14 for *proper* triads only Within the setting of the singular homology, we shall define new homology groups $H_q(X,X_1,X_2)$ called the homology groups of the triad We shall imbed these groups in an exact sequence called the *triadic* homology sequence Whenever the triad is proper, the sequence reduces to the homology sequence of the triad

DEFINITION 11 1. The *homology* and *cohomology groups* $H_q(X,X_1,X_2;G)$ and $H^q(X,X_1,X_2,G)$ of the triad (X,X_1,X_2) are defined as the appropriate homology and cohomology groups of the free chain complex $S(X)/S(X_1) \cup S(X_2)$ A map f $(X,X_1,X_2) \to (Y,Y_1,Y_2)$ of triads induces a map f $S(X)/S(X_1) \cup S(X_2) \to S(Y)/S(Y_1) \cup S(Y_2)$ which in turn induces homomorphisms f_*: $H_q(X,X_1,X_2,G) \to H_q(Y,Y_1,Y_2,G)$ and f^* $H^q(Y,Y_1,Y_2,G) \to H^q(X,X_1,X_2,G)$

Note that the analogs of Axioms 1 and 2 are satisfied (identity)$_*$ = identity, $(gf)_* = g_*f_*$.

Now consider the homomorphisms

$$0 \to S(X_1)/S(X_1) \cap S(X_2) \xrightarrow{i'} S(X)/S(X_2) \xrightarrow{j'} S(X)/S(X_1) \cup S(X_2) \to 0$$

induced by inclusions Since i' may be factored into

$$S(X_1)/S(X_1) \cap S(X_2) \xrightarrow{i_1'} S(X_1) \cup S(X_2)/S(X_2) \xrightarrow{i_2} S(X)/S(X_2)$$

where i_1' is the Noether isomorphism, and since the maps i_2,j' form a direct couple, it follows that (i',j') form a direct couple.

DEFINITION 11.2. The *triadic* homology and cohomology sequences of $(X;X_1,X_2)$ over G

$$\cdots \leftarrow H_q(X,X_2;G) \xleftarrow{i'_*} H_q(X_1,X_1 \cap X_2;G) \xleftarrow{\partial} H_{q+1}(X;X_1,X_2;G) \xleftarrow{j'_*} H_{q+1}(X,X_2;G) \leftarrow \cdots$$

$$\cdots \rightarrow H^q(X;X_2;G) \xrightarrow{i'^*} H^q(X_1,X_1 \cap X_2;G) \xrightarrow{\delta} H^{q+1}(X;X_1,X_2,G) \xrightarrow{j'^*} H^{q+1}(X,X_2;G) \rightarrow \cdots$$

are defined as the appropriate sequences of the couple (i',j').

A map f: $(X,X_1,X_2) \rightarrow (Y,Y_1,Y_2)$ induces a map f $(i',j')_X \rightarrow (i',j')_Y$ of the appropriate couples Therefore the homomorphisms f_* $[f^*]$ yield a map f_{**} $[f^{**}]$ of the triadic homology [cohomology] sequences

It should be noted that, although the homology groups of the triads $(X;X_1,X_2)$ and $(X;X_2,X_1)$ are the same, their triadic homology sequences are distinct.

If, in the triad (X,X_1,X_2), we have $X_1 \supset X_2$, then the triadic homology sequence of (X,X_1,X_2) is easily seen to coincide with the homology sequence of the triple (X,X_1,X_2)

THEOREM 11 3. *For each triad (X,X_1,X_2) and each coefficient group G the following conditions are equivalent:*

(i) *The triad $(X;X_1,X_2)$ is proper in the singular homology theory with G as coefficient group.*

(ii) *The inclusion map k_2 $(X_1,X_1 \cap X_2) \rightarrow (X_1 \cup X_2,X_2)$ induces isomorphisms k_{2*}. $H_q(X_1,X_1 \cap X_2;G) \approx H_q(X_1 \cup X_2,X_2,G)$ for all q.*

(iii) *The homomorphism $S(X)/S(X_1) \cup S(X_2) \rightarrow S(X)/S(X_1 \cup X_2)$ induced by inclusion induces isomorphisms $H_q(X;X_1,X_2,G) \approx H_q(X,X_1 \cup X_2,G)$ for all q*

(iv) *$H_q(X_1 \cup X_2;X_1,X_2;G) = 0$ for all q*

(v) *The inclusion map $S(X_1) \cup S(X_2) \rightarrow S(X_2 \cup X_2)$ induces isomorphisms of all the homology groups over G.*

A similar statement holds for the cohomology groups over G.

PROOF. Throughout this proof we shall omit the symbol G in all formulas.

(ii) $\leftrightarrow$ (iii). Consider the inclusion map θ: $(X,X_1,X_2) \rightarrow (X;X_1 \cup X_2,X_2)$. The triadic homology sequence of the second triad coincides with the homology sequence of the triple $(X,X_1 \cup X_2,X_2)$ This yields the diagram

$$
\begin{array}{ccccccc}
 & & H_q(X_1,X_1 \cap X_2) & \xleftarrow{\partial} & H_{q+1}(X;X_1,X_2) & & \\
\nwarrow & \swarrow & & & & \nwarrow & \swarrow \\
H_q(X,X_2) & & \Big\downarrow k_{2*} & & \Big\downarrow \theta_* & & H_{q+1}(X,X_2) \\
\swarrow & \nwarrow & & & & \swarrow & \nwarrow \\
 & & H_q(X_1 \cup X_2,X_2) & \xleftarrow{\partial} & H_{q+1}(X,X_1 \cup X_2) & &
\end{array}
$$

where the upper edge is the triadic homology sequence of (X,X_1,X_2) and the lower one is the H S of the triple $(X,X_1 \cup X_2,X_2)$ An application of the "five lemma" (I,4.3) yields the equivalence of (ii) and (iii).

(i) $\leftrightarrow$ (ii) The condition that the triad $(X;X_1,X_2)$ be proper consists of (ii) and the condition (ii)′ obtained from (ii) by interchanging the indices 1 and 2 Therefore (i) $\to$ (ii) On the other hand (ii) is equivalent to (iii) which is symmetric in the indices 1 and 2 Thus (ii) $\to$ (iii) $\to$ (ii)′ Hence (ii) $\to$ (i).

(iii) $\leftrightarrow$ (iv). Consider the direct couple

$$
0 \to S(X_1 \cup X_2)/S(X_1) \cup S(X_2) \to S(X)/S(X_1) \cup S(X_2) \to S(X)/S(X_1 \cup X_2) \to 0
$$

induced by inclusions The H S of this couple is

$$
\cdots \leftarrow H_{q-1}(X_1 \cup X_2,X_1,X_2) \leftarrow H_q(X,X_1 \cup X_2) \leftarrow H_q(X;X_1,X_2) \leftarrow H_q(X_1 \cup X_2;X_1,X_2) \leftarrow \cdots
$$

Thus (iii) $\leftrightarrow$ (iv) by exactness

(iv) $\leftrightarrow$ (v). The proof is the same as the preceding using the direct couple

$$
0 \to S(X_1) \cup S(X_2) \to S(X_1 \cup X_2) \to S(X_1 \cup X_2)/S(X_1) \cup S(X_2) \to 0.
$$

We return for a moment to the diagram used in the proof of (ii) $\leftrightarrow$ (iii), and assume that (X,X_1,X_2) is a proper triad Then both k_{2*} and θ_* are isomorphisms Using either k_{2*}^{-1} or θ_*^{-1} we may insert the homomorphism $H_{q+1}(X,X_1 \cup X_2) \to H_q(X_1,X_1 \cap X_2)$ in the diagram. In this way the above diagram contains also the homology sequence of the proper triad $(X;X_1,X_2)$ as defined in I,14 3

The fact that a given triad $(X;X_1,X_2)$ is proper in the singular theory may depend on the coefficient group and whether homology or cohomology is employed The following theorem shows that, in a sense, homology groups with integer coefficients are universal.

THEOREM 11.4. *If the triad $(X;X_1,X_2)$ is proper in the singular homology theory with integer coefficients, then it is also proper in the singular homology and cohomology theory for arbitrary coefficients.*

PROOF. By 11.3 (i) and (iv), the triad $(X;X_1,X_2)$ is proper for homology with integer coefficients if and only if the chain complex $S(X_1 \cup X_2)/S(X_1) \cup S(X_2)$ has vanishing integral homology groups. Since the complex in question is a free chain complex, it follows from v,13 5 that all the homology and cohomology groups $H_q(X_1 \cup X_2; X_1,X_2;G)$ and $H^q(X_1 \cup X_2,X_1,X_2,G)$ vanish. Thus, again by 11 3, the triad $(X;X_1,X_2)$ is proper for singular homology and cohomology with any coefficients.

For the use of triads in homotopy theory, see note below.

NOTES

The development of the singular theory Perhaps the basic idea of the singular theory is that a singular chain on X is a collection consisting of a complex K, a map $K \to X$, and an ordinary chain on K This can be found in the book of Veblen [*Analysis Situs*, Colloq. Pub Amer Math Soc v (1921)] Elaborations of this approach have been given (Hurewicz, Dowker, and Dugundji, Annals of Math 49 (1948), 391-406); but they have the defect that the homology groups are not obtained from chain complexes, for the chains do not form groups

It was Lefschetz [Bull Amer Math. Soc 39 (1933), 124-129] who first defined a *group* of singular chains He considered only the maps of oriented simplexes into X Two maps f $s_1 \to X$, g $s_2 \to X$ were called *equivalent* if there exists a barycentric equivalence h $s_1 \to s_2$ preserving orientation such that $gh = f$. An equivalence class is called a *singular oriented simplex* Each such has a *negative* obtained by reversing the orientation of the original simplex Generators for the chain groups are found by choosing one of the two orientations of each singular simplex

It was Čech who pointed out a difficulty in this approach: *the two orientations of a singular simplex could coincide* An example is obtained by folding a 1-simplex about its midpoint and then mapping it into X. This meant that the groups of chains contained elements of order 2, and so were not free abelian As we have seen, the free abelian character of the chain groups is very useful in proving that the functor S is an h-functor. We defined homomorphisms of $S(X)$ by prescribing their values on the generators, namely, the singular simplexes If $S(X)$ were not free abelian, one would have to verify that each such assignment could be extended to a homomorphism

The final step in the development was the substitution of the *ordered* simplex for the oriented simplex [S Eilenberg, Annals of Math 45 (1944), 407-447]. The intuitive justification is contained in the theorem

(VI,6 10) that the ordered chain complex and the alternating chain complex assigned to a simplicial complex are chain equivalent

Homotopy groups of triads The groups and sequences of §11 have a natural counterpart in homotopy theory Given a triad $(X;X_1,X_2)$ with base point x_0 Blakers and Massey [*The homotopy groups of a triad I*, Annals of Math 53 (1951), 161-205] introduce homotopy groups $\pi_q(X;A,B)$ based upon the consideration of maps $(E^n;E^{n-1}_+,E^{n-1}_-) \to (X,X_1,X_2)$ The groups are defined for $n > 2$ and are abelian for $n > 3$. In analogy with the triadic sequences of 11 2, they construct an exact sequence

$$\leftarrow \pi_q(X,X_2) \leftarrow \pi_q(X_1,X_1 \cap X_2) \leftarrow \pi_{q+1}(X,X_1,X_2) \leftarrow \pi_{q+1}(X,X_2) \leftarrow$$

In particular, if $X = X_1 \cup X_2$, it follows that the "excision" homomorphism

$$\pi_q(X_1,X_1 \cap X_2) \to \pi_q(X_1 \cup X_2,X_2)$$

belongs to an exact sequence in which it is preceded by $\pi_{q+1}(X_1 \cup X_2,X_1,X_2)$ and followed by $\pi_q(X_1 \cup X_2,X_1,X_2)$ Thus the latter groups measure the extent by which the Excision axiom fails for homotopy groups Using the homotopy groups of a triad, Blakers and Massey have proved the suspension theorems of Freudenthal (see Note to Chapter I on homotopy groups) and various other results of the same general nature

Proof of theorem 8.2 The proof given in the text uses V,13 3 to deduce the fact that η is a chain equivalence from the fact that η_* is an isomorphism (for integral homology groups) We shall give here a somewhat longer but more direct proof, not using V,13 3. The advantage of this proof is that it applies in some situations (e g for local coefficients) in which the proof of the text breaks down

For a given singular simplex T in X, let $\{T_\alpha\}$ be the family of singular simplexes consisting of T, its faces, their faces, and so on Let $n(T)$ be the least integer such that $\mathrm{Sd}^{n(T)}\,(T_\alpha)$ is in $S(X,F)$ for every T_α in $\{T_\alpha\}$. This integer exists and has the following properties

$n(T^{(i)}) \leq n(T)$.
$\mathrm{Sd}^{n(T)}\, T$ *is in* $S(X,F)$.
If T *is in* $S(X,F)$, *then* $n(T) = 0$,
If T *is in* A, *then* $\mathrm{Sd}^{n(T)}\, T$ *is in* $S(A,F)$.

Now define for every singular q-simplex T in X

$$(1)\qquad \tau T = \mathrm{Sd}^{n(T)}\, T + \sum_{i=0}^{q} (-1)^i \sum_{j=n(T^{(i)})}^{n(T)-1} R\, \mathrm{Sd}^j\, (T^{(i)}),$$

$$(2)\qquad DT = \sum_{j=0}^{n(T)-1} R\, \mathrm{Sd}^j\, T.$$

We observe that τT is a q-chain in $S(X,F)$ and DT is a $(q+1)$-chain in $S(X)$. Thus homomorphisms

$$\tau\colon\ C_q(X) \to C_q(S(X,F)), \qquad D\colon\ C_q(X) \to C_{q+1}(S(X,F))$$

are obtained. Further

$$\tau[C_q(A)] \subset C_q(S(A,F)), \qquad D[C_q(A)] \subset C_{q+1}(S(A,F)).$$

From (2) we deduce

$$\partial DT = T - \mathrm{Sd}^{n(T)}\, T - \sum_{i=0}^{q} (-1)^i \sum_{j=0}^{n(T)-1} R\, \mathrm{Sd}^j\, T^{(i)},$$

$$D\partial T = \sum_{i=0}^{q} (-1)^i \sum_{j=0}^{n(T^{(i)})} R\, \mathrm{Sd}^j\, T^{(i)}$$

Adding these two formulas and comparing with (1), we find

$$\partial DT + D\partial T = T - \tau T.$$

Hence for every singular q-chain c in X

$$(3)\qquad \partial Dc = c - \tau c - D\partial c$$

This implies

$$\begin{aligned}\partial\tau c &= \partial(c - \partial Dc - D\partial c) = \partial c - \partial\partial Dc - \partial D\partial c\\ &= \partial c - \partial c + \tau\partial c + D\partial\partial c = \tau\partial c,\end{aligned}$$

which proves that τ is a map $S(X) \to S(X,F)$. Since τ carries $S(A)$ into $S(A,F)$, it defines a map

$$\tau.\ \ S(X)/S(A) \to S(X,F)/S(A,F),$$

and the composition $\tau\eta$ is the identity map of $S(X,F)/S(A,F)$. On the other hand $\eta\tau c = \tau c$, so that formula (3) may be rewritten as $\partial Dc + D\partial c = c - \eta\tau c$. Thus $\eta\tau$ is homotopic to the identity map of $S(X)/S(A)$. Hence τ is a homotopy inverse of η.

Spaces with operators. Let X be a topological space operated on by a group W (see II, Exer. C for a definition). Since each $w \in W$ yields a homeomorphism $w\colon X \to X$, it induces an isomorphism $w_S\colon\ S(X) \to$

$S(X)$ so that the chain complex has W as a group of (left) operators. If A is a subspace of X such that the operators of W carry A into A, then we say that W operates on the pair (X,A). Then W operates also on $S(X)/S(A)$.

Let G be an abelian group with W as a group of right operators. We define the tensor product $G \otimes_W S(X)/S(A)$ as the factor group of $G \otimes S(X)/S(A)$ obtained by identifying $gw \otimes c$ with $g \otimes wc$. The homology groups of this tensor product over W are called the *equivariant* homology groups $H_q^W(X,A,G)$.

If G is an abelian group with W as left operators we consider the cochain complex $\mathrm{Hom}_W(S(X)/S(A),G)$ which is the subcomplex of $\mathrm{Hom}(S(X)/S(A),G)$ consisting of equivariant homomorphisms (i.e. operator homomorphisms). The cohomology groups of this subcomplex are called the equivariant cohomology groups $H_W^q(X,A,G)$.

If $f: (X,A) \to (Y,B)$ is an equivariant map of pairs each having W as operator group, then the map $f_S: S(X)/S(A) \to S(Y)/S(B)$ is equivariant and induces homomorphisms f_* and f^* of the equivariant homology and cohomology groups. The operators

$$\partial_*: H_q^W(X,A;G) \to H_{q-1}^W(A,G), \qquad \delta^*: H_W^q(A;G) \to H_W^{q+1}(X,A;G)$$

can also be defined in the usual fashion.

With the basic concepts defined, it is possible to prove that one obtains in this way homology and cohomology theories on the category $\mathfrak{A}_W$ defined as follows: the objects are pairs (X,A) with W as operators, maps are equivariant mappings, couples are $i,j: A \to X \to (X,A)$, homotopies are provided by maps $F: (X \times I, A \times I) \to (Y,B)$ which are equivariant, i.e. satisfy $F(wx,t) = wF(x,t)$, excisions are inclusions $(X - U, A - U) \to (X,A)$ where $\bar{U} \subset \mathrm{Int}\, A$ and $w(U) = U$ for all $w \in W$, a *point* is a discrete space X on which W operates transitively and without fixed points (in particular, W itself is a *point*). Of course, the resulting category $\mathfrak{A}_W$ is not an admissible category in the sense of I,1.

If W operates simplicially on the simplicial pair (K,L) (see II,Exer. C), then equivariant homology and cohomology groups may be defined using either the complexes $G \otimes_W K_o/L_o$, $\mathrm{Hom}_W(K_o/L_o,G)$ or the complexes $G \otimes_W K_a/L_a$, $\mathrm{Hom}_W(K_a/L_a,G)$. The resulting groups are isomorphic with the equivariant homology and cohomology groups of the pair $(|K|,|L|)$ defined using the singular theory.

EXERCISES

A. Pathwise connectedness.

Definition. Let $x_0,x_1 \in X$. A mapping $f: I \to X$ of the closed unit interval I into X such that $f(0) = x_0$ and $f(1) = x_1$, is called a *path* in X joining x_0 and x_1

1 Show that each space X decomposes into disjoint pathwise connected sets $\{X_\alpha\}$, called the pathwise connected *components* of X, such that two points belonging to distinct components cannot be joined by a path in X.

2 Establish the isomorphisms

$$H_q(X;G) \approx \sum_\alpha H_q(X_\alpha,G), \qquad H^q(X;G) \approx \prod_\alpha H^q(X_\alpha,G)$$

where $\{X_\alpha\}$ are the pathwise connected components of X

3. Compute the groups $H_0(X;G)$, $\tilde{H}_0(X,G)$, $H^0(X,G)$, and $\tilde{H}^0(X;G)$

4. Carry out an analogous discussion for the groups of a pair (X,A).

B. INFINITE COMPLEXES.

1. Let K be an infinite complex In analogy with the transformation β of §10, define chain transformations

$$\beta_w: \ K_o \to S(|K|_w), \qquad \beta_m \ \ K_o \to S(|K|_m)$$

where $|K|_w$ and $|K|_m$ have the weak and the metric topology respectively (see II, Exer F) Show that both β_w and β_m induce isomorphisms of the homology and of the cohomology groups

2 Consider the map i. $|K|_w \to |K|_m$ induced by the identity map $K \to K$. Show that this map defines $S(|K|_w)$ as a subcomplex of $S(|K|_m)$, and that the inclusion map $S(|K|_w) \subset S(|K|_m)$ induces isomorphisms of the homology and the cohomology groups. Show by examples that $S(|K|_w)$ may be a proper subcomplex of $S(|K|_m)$.

C. THE MAPPING CYLINDER.

DEFINITION. Let f· $X \to Y$ be a continuous map. In the join $X \circ Y$ (see II, Exer D6) consider the set C consisting of X, Y, and all intervals joining $x \in X$ with $f(x) \in Y$. The set C is called the *mapping cylinder* of f

1. Establish the exactness of the sequence

$$\cdots \to H_q(X) \xrightarrow{f_*} H_q(Y) \xrightarrow{j_*} H_q(C,X) \xrightarrow{\partial} H_{q-1}(X) \to \cdots$$

where j. $Y \to (C,X)$ is the inclusion map.

2. Show the existence of homomorphisms D_q· $C_q(X) \to C_{q+1}(C)$ such that $\partial Dx + D\partial x = fx - x$ for $x \in C_q(X)$

3. Consider the map f_S. $S(X) \to S(Y)$ induced by f. $X \to Y$, and consider the complex $\hat{f}_S$ as introduced in V,13.3. The groups $H_q(\hat{f}_S)$ will be called the homology groups of f and will be denoted by $H_q(f)$.

Using a homotopy operator $D\colon\ S(X) \to S(C)$ as in 2, define a map $\phi\colon\ \hat{f}_S \to S(C)/S(X)$ by setting

$$\phi(x,y) = Dx + y \bmod S(X) \qquad \text{for } x \in C_{q-1}(X),\ y \in C_q(Y)$$

Show that $\phi_*\colon\ H_q(f) \approx H_q(C,X)$ is an isomorphism. Hint: use the diagram

$$\begin{array}{ccccccccc}
\cdots \longrightarrow & H_q(X) & \xrightarrow{f_*} & H_q(Y) & \xrightarrow{k_*} & H_q(f) & \xrightarrow{l_*} & H_{q-1}(X) & \longrightarrow \cdots \\
 & \downarrow{\scriptstyle m_*} & & \downarrow{\scriptstyle n_*} & & \downarrow{\scriptstyle \phi_*} & & \downarrow{\scriptstyle m_*} & \\
\longrightarrow & H_q(X) & \xrightarrow{i_*} & H_q(C) & \xrightarrow{j_*} & H_q(C,X) & \xrightarrow{\partial} & H_{q-1}(X) & \longrightarrow \cdots
\end{array}$$

where the upper row is the exact sequence of v, Exer D1, the lower row is the homology sequence of (C,X), the map m is the identity, and $n\colon\ Y \subset C$ is the inclusion map.

4. Carry out a similar discussion for a map $f\colon\ (X,A) \to (Y,B)$.

CHAPTER VIII

Systems of groups and their limits

1. INTRODUCTION

The material of this chapter is preparatory for the second existence proof of Chapter IX It is concerned almost entirely with the algebraic aspects of the limiting process used in that proof The notions of direct and inverse systems of groups are due to Pontrjagin In Čech's original construction of his homology theory, these ideas appeared only in implicit form There are several advantages to their explicit use First, the algebraic and geometric difficulties are separated Secondly, the reasons for the various limitations on the Čech theory become apparent. Finally, direct and inverse systems are of use in other connections, so that an independent treatment is not amiss

2. DIRECT AND INVERSE SYSTEMS

DEFINITION 2.1. A relation $\alpha < \beta$ in a set M is called a *quasi-order* if it is reflexive and transitive In general $\alpha < \beta$ and $\beta < \alpha$ does not imply $\alpha = \beta$. A *directed set* M is a quasi-ordered set such that for each pair $\alpha,\beta \in M$, there exists a $\gamma \in M$ for which $\alpha < \gamma$ and $\beta < \gamma$ A directed set M' is a *subset* of M ($M' \subset M$) if $\alpha \in M'$ implies $\alpha \in M$ and $\alpha < \beta$ in M' implies $\alpha < \beta$ in M A subset M' is *cofinal* in M if, for each $\alpha \in M$, there exists a $\beta \in M'$ such that $\alpha < \beta$. If M and N are directed sets, a *map* ϕ. $M \to N$ is an order-preserving function from M to N (i e $\alpha < \beta$ in M implies $\phi\alpha < \phi\beta$ in N)

DEFINITION 2.2 *A direct system of sets* $\{X,\pi\}$ *over a directed set* M is a function which attaches to each $\alpha \in M$ a set X^α, and, to each pair α,β such that $\alpha < \beta$ in M, a map

$$\pi_\alpha^\beta\colon\ X^\alpha \to X^\beta$$

such that, for each $\alpha \in M$,

$$\pi_\alpha^\alpha = \text{identity},$$

and for $\alpha < \beta < \gamma$ in M,

$$\pi_\beta^\gamma \pi_\alpha^\beta = \pi_\alpha^\gamma.$$

An *inverse system of sets* $\{X,\pi\}$ *over a directed set* M is a function which attaches to each $\alpha \in M$ a set X_α, and to each pair α,β such that $\alpha < \beta$ in M, a map

$$\pi_\alpha^\beta\colon\ X_\beta \to X_\alpha$$

such that

$$\begin{aligned} \pi_\alpha^\alpha &= \text{identity}, & \alpha &\in M, \\ \pi_\alpha^\beta \pi_\beta^\gamma &= \pi_\alpha^\gamma, & \alpha &< \beta < \gamma \text{ in } M. \end{aligned}$$

For both inverse and direct systems, the maps π_α^β are called *projections* of the system. If each X^α $[X_\alpha]$ is a topological space, or an R-module, or a topological group, and each projection is, respectively, continuous, or an R-homomorphism, or a continuous homomorphism, then $\{X,\pi\}$ is called a direct [inverse] system of, respectively, spaces, R-modules, or topological groups.

A directed set M becomes a category if each relation $\alpha < \beta$ is regarded as a map $\alpha \to \beta$. Then a direct [inverse] system over M is simply a covariant [contravariant] functor from M to the category of sets and maps, or to the category of R-modules and homomorphisms, etc.

DEFINITION 2.3. Let $\{X,\pi\},\{X',\pi'\}$ be direct systems over M and M' respectively. Then a *map*

$$\Phi\colon\ \{X,\pi\} \to \{X',\pi'\}$$

consists of a map $\phi\colon\ M \to M'$, and, for each $\alpha \in M$, a map

$$\phi^\alpha\colon\ X^\alpha \to X'^{\phi\alpha}$$

such that, if $\alpha < \beta$ in M, then commutativity holds in the diagram

$$\begin{array}{ccc} X^\alpha & \xrightarrow{\ \pi\ } & X^\beta \\ \Big\downarrow{\scriptstyle\phi} & & \Big\downarrow{\scriptstyle\phi} \\ X'^{\phi\alpha} & \xrightarrow{\ \pi'\ } & X'^{\phi\beta} \end{array}$$

Now let $\{X,\pi\},\{X',\pi'\}$ be inverse systems over M,M' respectively. Then a *map*

$$\Phi\colon\ \{X,\pi\} \to \{X',\pi'\}$$

consists of a map $\phi\colon\ M' \to M$, and, for each $\alpha' \in M'$, a map

$$\phi_{\alpha'}\colon\ X_{\phi\alpha'} \to X'_{\alpha'}$$

such that, if $\alpha' < \beta'$ in M', then commutativity holds in the diagram

$$\begin{array}{ccc} X_{\phi\alpha'} & \xleftarrow{\ \pi\ } & X_{\phi\beta'} \\ \downarrow{\scriptstyle\phi} & & \downarrow{\scriptstyle\phi} \\ X'_{\alpha'} & \xleftarrow{\ \pi'\ } & X'_{\beta'} \end{array}$$

Whenever both of the direct [inverse] systems are systems of topological spaces, or groups, etc., the components ϕ^α $[\phi_{\alpha'}]$ of the map Φ are required to be continuous, or homomorphic, etc. In case

$$\Phi: \{X,\pi\} \to \{X',\pi'\}, \qquad \Phi': \{X',\pi'\} \to \{X'',\pi''\}$$

are two maps of direct [inverse] systems, their *composition*

$$\Phi'\Phi: \{X,\pi\} \to \{X'',\pi''\}$$

is defined to consist of the compositions $\phi'\phi$ and $\phi'^{\phi\alpha}\phi^\alpha$, $\alpha \in M$ [$\phi\phi'$ and $\phi'_{\alpha''}\phi_{\phi'\alpha''}$, $\alpha'' \in M''$].

It is a simple exercise to verify that direct or inverse systems of any specified type and their maps form a category. The category of direct [inverse] systems of R-modules is denoted by $\mathrm{Dir}\mathcal{G}_R$ [$\mathrm{Inv}\mathcal{G}_R$]. Similar notations are used for the other categories (e.g. $\mathrm{Inv}\mathcal{G}_c$ = category of inverse systems of compact groups).

It is to be noted that the identity map Φ of $\{X,\pi\}$ is composed of the identities $\phi: M \to M$ and $\phi^\alpha: X^\alpha \to X^\alpha$.

DEFINITION 2.4. If M,M' are directed sets, $M' \subset M$ (see 2.1), and $\{X,\pi\}$ is a direct [inverse] system over M, then the sets and maps of $\{X,\pi\}$ which correspond to elements and relations in M' form a direct [inverse] system $\{X',\pi'\}$ over M', and is called the *subsystem* of $\{X,\pi\}$ over M'. If M' is cofinal in M, the subsystem is called *cofinal*. The identity map $\phi: M' \to M$ and identity maps

$$\phi: X'^\alpha \to X^\alpha, \qquad [\phi_\alpha: X_\alpha \to X'_\alpha], \qquad \alpha \in M'$$

form a map

$$\Phi: \{X',\pi'\} \to \{X,\pi\}, \qquad [\Phi: \{X,\pi\} \to \{X',\pi'\}]$$

called the *injection* of the subsystem into the system [system into the subsystem].

It is clear that the identity map of $\{X,\pi\}$ is an injection. Also the composition of two injections is itself an injection.

It is to be observed that, if the indexing set M reduces to a single element, the inverse or direct system indexed by M reduces to a single set or group Conversely, every set or group may be regarded as an inverse or direct system indexed by a set M consisting of a single element In particular, if in 2 3 we select M' to reduce to one element, we see that a mapping

$$\Phi \quad \{X,\pi\} \to X'$$

of a direct system $\{X,\pi\}$ indexed by set M into the set X' is a family of maps ϕ^α. $X^\alpha \to X'$ such that

$$\phi^\beta \pi^\beta_\alpha = \phi^\alpha.$$

Similarly, if in 2 3 M reduces to a point, we obtain a mapping

$$\Phi \quad X \to \{X',\pi'\}$$

If $\{X',\pi'\}$ is an inverse system, then Φ is a family of maps ϕ_α: $X \to X'_\alpha$ such that

$$\phi_\alpha \pi^\beta_\alpha = \phi_\beta$$

We could also consider maps $\{X,\pi\} \to X'$ with $\{X,\pi\}$ an inverse system, and maps $X \to \{X',\pi'\}$ where $\{X',\pi'\}$ is a direct system However, these two types of maps are not of much interest to us.

3. INVERSE LIMITS

DEFINITION 3 1 Let $\{X,\pi\}$ be an inverse system of sets over the directed set M The *inverse limit* X_∞ (briefly limit) of $\{X,\pi\}$ is the subset of the product $\prod X_\alpha$ (see v,5 1) consisting of those functions $x = \{x_\alpha\}$ such that, for each relation $\alpha < \beta$ in M,

(1) $$\pi^\beta_\alpha(x_\beta) = x_\alpha$$

Define the projection

(2) $$\pi_\beta \quad X_\infty \to X_\beta \qquad \text{by} \qquad \pi_\beta(x) = x_\beta,$$

so that π_β is the same as the function p_β of v,5 1 restricted to the subset X_∞ If $\{X,\pi\}$ is an inverse system of spaces, then X_∞ is assigned the topology it has as a subspace of $\prod X_\alpha$ If $\{X \pi\}$ is an inverse system of R-modules, it is easily seen that X_∞ is a subgroup of $\prod X_\alpha$, and X_∞ is assigned this structure of an R-module Similarly, an inverse limit of topological groups is a topological group.

LEMMA 3.2 *If $\alpha < \beta$, then $\pi_\alpha = \pi^\beta_\alpha \pi_\beta$*

PROOF By (2), $\pi_\beta(x) = x_\beta$ By (1), $\pi^\beta_\alpha(x_\beta) = x_\alpha$ Hence $\pi^\beta_\alpha\pi_\beta(x) = x_\alpha = \pi_\alpha(x)$ by (2)

LEMMA 3 3 *If $\{X,\pi\}$ is an inverse system of spaces [R-modules or* *topological groups*], *then each* π_α *is continuous* [*homomorphic or continuously homomorphic*]

These follow from the corresponding properties of the projection p_α (see v,5 2,5 4).

REMARK. If the directed set M is enlarged to a directed set M' by adjoining an element ∞ greater than each $\alpha \in M$, then the inverse system $\{X,\pi\}$ together with X_∞ and the projections π_α form an inverse system of M' The inverse limit of this enlarged system is in a natural 1-1 correspondence with X_∞

EXAMPLE As it stands the notion of inverse limit is derived from that of product Conversely, the product of an infinite number of sets can be represented as an inverse limit of finite products. Let $\{X_\iota\}$ be a collection of sets, indexed by an infinite set J Let M be the collection of finite subsets of J ordered by inclusion For each $\alpha \in M$ define $Y_\alpha = \prod_{\iota \in \alpha} X_\iota$ If $y \in Y_\beta$, then y is a function defined over β If $\alpha < \beta$, the function y defines a function on α, denoted by $\pi^\beta_\alpha(y)$, and the latter is in Y_α Thus π^β_α: $Y_\beta \to Y_\alpha$. It is easily seen that the inverse limit Y_∞ of $\{Y,\pi\}$ coincides with $\prod X_\iota$.

THEOREM 3.4. *If* $\{X,\pi\}$ *is an inverse system over* M *and, for each relation* $\alpha < \beta$ *in* M, π^β_α *is a 1-1 map of* X_β *into (onto)* X_α, *then, for each* $\alpha \in M$, π_α *is a 1-1 map of* X_∞ *into (onto)* X_α

PROOF. Suppose $x,y \in X_\infty$, and, for some α, $x_\alpha = y_\alpha$ Since π^β_α is a 1-1 into, it follows, for every $\beta > \alpha$, that $x_\beta = y_\beta$ Since M is a directed set, for any $\gamma \in M$, there is a $\beta > \gamma,\alpha$ Then $x_\beta = y_\beta$ implies $x_\gamma = y_\gamma$ Thus $x = y$, and π_α is 1-1 into Suppose now that each π^β_α is 1-1 onto, and for a fixed α, $x_\alpha \in X_\alpha$ For $\beta > \alpha$, define $x_\beta = (\pi^\beta_\alpha)^{-1}(x_\alpha)$. For any $\gamma \in M$, choose a $\beta > \alpha,\gamma$ and define $x_\gamma = \pi^\beta_\gamma(x_\beta)$ It remains to prove that $x = \{x_\gamma\}$ lies in X_∞. Suppose $\gamma_1 < \gamma_2$ Let $\beta_1 > \alpha,\gamma_1$ and $\beta_2 > \alpha,\gamma_2$ be the elements used in defining $x_{\gamma_1}, x_{\gamma_2}$. Choose $\beta > \beta_1,\beta_2$ Then $x_\alpha = \pi^\beta_\alpha(x_\beta) = \pi^{\beta_i}_\alpha \pi^\beta_{\beta_i}(x_\beta)$, and $x_\alpha = \pi^{\beta_i}_\alpha(x_{\beta_i})$ ($i = 1,2$) Since $\pi^{\beta_i}_\alpha$ is 1-1, it follows that $\pi^\beta_{\beta_i}(x_\beta) = x_{\beta_i}$. Therefore $\pi^\beta_{\gamma_i}(x_\beta) = \pi^{\beta_i}_{\gamma_i}\pi^\beta_{\beta_i}(x_\beta) = \gamma_i$ Finally, $\pi^{\gamma_2}_{\gamma_1}(x_{\gamma_2}) = \pi^{\gamma_2}_{\gamma_1}\pi^\beta_{\gamma_2}(x_\beta) = \pi^\beta_{\gamma_1}(x_\beta) = x_{\gamma_1}$, and the proof is complete.

REMARK. If the hypothesis of 3 4 is weakened by requiring that each projection be only a mapping onto, one might hope to prove that each π_α is a mapping onto If M is a countable set, or if M has a countable cofinal subset, this can be proved It is also proved for inverse systems of compact Hausdorff spaces in 3 9 That the result does not hold in the general case is shown by an example found by L. Henkin [Proc. Amer. Math. Soc 1 (1950), 224-225].

LEMMA 3 5 *If* $\{X,\pi\}$ *is an inverse system of Hausdorff spaces, then the limit space* X_∞ *is a closed subspace of* $\prod X_\alpha$.

Proof. Suppose $x \in \prod X_\alpha$ is not in X_∞. Then for some relation $\alpha < \beta$ in M, $\pi_\alpha^\beta(x_\beta) \neq x_\alpha$. Since X_α is a Hausdorff space, there are disjoint open sets U_α, V_α containing x_α and $\pi_\alpha^\beta(x_\beta)$ respectively. Let $U_\beta = (\pi_\alpha^\beta)^{-1}V_\alpha$. Replace X_α, X_β by U_α, U_β in the collection $\{X\}$. The product of the resulting collection is a rectangular neighborhood in $\prod X_\alpha$ (see v,5.2). This open set contains x, but no point of X_∞. Thus X_∞ is closed since its complement is open.

Theorem 3.6. *The limit space of an inverse system of compact spaces is a compact space. If each space of the inverse system is nonvacuous, then the limit space is nonvacuous.*

Proof. Since X_∞ is closed in $\prod X_\alpha$ (3.5), and $\prod X_\alpha$ is a compact space (v,5.4), it follows immediately that X_∞ is a compact space. (Recall that "compact" includes "Hausdorff".) For the second part, define, for each $\beta \in M$, the subset Y_β of $\prod X_\alpha$ to consist of those elements x such that $\pi_\gamma^\beta(x_\beta) = x_\gamma$ for all $\gamma < \beta$. Now Y_β is nonvacuous; for x_β can be specified arbitrarily in X_β, then set $x_\gamma = \pi_\gamma^\beta(x_\beta)$ for $\gamma < \beta$, and finally use the choice axiom to extend x over all other sets of the system. Next, Y_β is closed. The proof of this is similar to that of 3.5, and is omitted. If $\alpha < \beta$, then $Y_\alpha \supset Y_\beta$. This follows immediately from $\pi_\gamma^\beta = \pi_\gamma^\alpha \pi_\alpha^\beta$ where $\gamma < \alpha < \beta$. Since M is a directed set, it follows now that $\{Y_\alpha\}$ is a collection of closed sets in a compact space and any finite number of them have a nonvacuous intersection. It follows that their intersection is nonvacuous. Clearly, any point of this intersection belongs to X_∞, and the proof is complete.

Theorem 3.7. *If $\{X,\pi\}$ is an inverse system of compact spaces over M, $\alpha \in M$, and U is an open set of X_α containing $\pi_\alpha(X_\infty)$, then there exists a $\beta > \alpha$ such that $\pi_\alpha^\beta(X_\beta) \subset U$.*

Proof. For each $\beta > \alpha$, let $Y_\beta = X_\beta - (\pi_\alpha^\beta)^{-1}(U)$. Otherwise, $Y_\beta = X_\beta$. If $\beta < \gamma$ let ρ_β^γ: $Y_\gamma \to Y_\beta$ be the map defined by π_β^γ. Then $\{Y_\beta, \rho_\beta^\gamma\}$ is an inverse system of compact spaces. If the conclusion of 3.7 were false, each Y_β would be nonvacuous. Hence $Y_\infty \subset X_\infty$ would be nonvacuous (3.6). If $y \in Y_\infty$, then $y_\alpha = \rho_\alpha(y) = \pi_\alpha(y)$ would lie in both Y_α and U. As this is a contradiction, 3.7 is proved.

Corollary 3.8. *Under the hypotheses of 3.7,*

$$\pi_\alpha(X_\infty) = \bigcap_{\alpha<\beta} \pi_\alpha^\beta(X_\beta).$$

Proof. If $x_\alpha \in X_\alpha$ is not in $\pi_\alpha(X_\infty)$, then its complement U contains $\pi_\alpha^\beta(X_\beta)$ for some β. Therefore x_α is not in the intersection. This shows that the left side contains the right side. By 3.2, $\pi_\alpha(X_\infty) \subset \pi_\alpha^\beta(X_\beta)$ for each $\beta > \alpha$. This proves that the left side is included in the right.

COROLLARY 3.9 *If, to the hypothesis of 3.7, we adjoin the condition that each π_α^β is a mapping onto, then each π_α is a mapping onto.*

This follows immediately from 3.8.

DEFINITION 3.10. Let $\Phi: \{X,\pi\} \to \{X',\pi'\}$ be a map of one inverse system into another (see 2.3). The *inverse limit* ϕ_∞ of Φ is a map

$$\phi_\infty: X_\infty \to X'_\infty$$

defined as follows. If $x \in X_\infty$ and $\alpha \in M'$, set $x'_\alpha = \phi_\alpha(x_{\phi(\alpha)})$. If $\alpha < \beta$ in M', it follows from the commutativity condition of 2.3 that $\pi'^{\beta}_{\alpha}(x'_\beta) = x'_\alpha$. Therefore $x' = \{x'_\alpha\}$ is an element of X'_∞. Define $\phi_\infty(x) = x'$.

LEMMA 3.11. *If $\Phi: \{X,\pi\} \to \{X',\pi'\}$ and $\alpha \in M'$, then commutativity holds in the diagram*

$$\begin{array}{ccc} X_{\phi(\alpha)} & \xleftarrow{\ \pi\ } & X_\infty \\ \big\downarrow{\scriptstyle \phi_\alpha} & & \big\downarrow{\scriptstyle \phi_\infty} \\ X'_\alpha & \xleftarrow{\ \pi'\ } & X'_\infty \end{array}$$

This follows immediately from the definition.

LEMMA 3.12 *If $\{X,\pi\}$ is an inverse system of spaces, then the sets $\{\pi_\alpha^{-1}(U)\}$, where U runs over all open sets of X_α and α over M, form a base for the open sets of X_∞.*

PROOF. Let W be open in X_∞ and $x \in W$. By v,5.2, there exists a rectangular neighborhood V of x in $\prod X_\alpha$ such that $V \cap X_\infty \subset W$. Now V is determined by specifying open sets $V_i \subset X_{\alpha_i}$ $(i = 1, \cdots, n)$, and consists of all $y \in \prod X_\alpha$ such that $y_{\alpha_i} \in V_i$. Since M is a directed set, there exists a $\beta > \alpha_i$ $(i = 1, \cdots, n)$. Since $\pi^\beta_{\alpha_i}$ is continuous, and $\pi^\beta_{\alpha_i}(x_\beta) \in V_i$, it follows that

$$U = \bigcap_{i=1}^{n} (\pi^\beta_{\alpha_i})^{-1}(V_i)$$

is an open set of X_β containing x_β. Hence $x \in \pi_\beta^{-1}(U)$. Suppose $y \in \pi_\beta^{-1}(U)$. Then $y_\beta \in U$. This implies $y_{\alpha_i} \in V_i$ $(i = 1, \cdots, n)$, and this in turn implies $y \in V \cap X_\infty \subset W$. Thus $x \in \pi_\beta^{-1}(U) \subset W$, and the lemma is proved.

THEOREM 3.13 *If $\Phi: \{X,\pi\} \to \{X',\pi'\}$ and $\{X,\pi\},\{X',\pi'\}$ are both inverse systems of spaces [R-modules or topological groups], then the limit ϕ_∞ of Φ is continuous [homomorphic or continuously homomorphic].*

PROOF. To prove continuity it suffices to show that ϕ_∞^{-1} maps each open set of a base for the open sets of X'_∞ into an open set of X_∞. By

3.12, the sets $\pi_\alpha'^{-1}(U)$, where $\alpha \in M'$, and U is open in X_α', form such a base. By 3.11,

$$\phi_\infty^{-1}\pi_\alpha'^{-1}(U) = \pi_{\phi(\alpha)}^{-1}\phi_\alpha^{-1}(U).$$

Since ϕ_α and $\pi_{\phi(\alpha)}$ are continuous, this is an open set. The proof that, in the case of R-modules, ϕ_α is homomorphic is left to the reader.

THEOREM 3.14. *Let $\mathfrak{a}$ be any one of the following categories: sets and maps, spaces and continuous maps, compact spaces and continuous maps, R-modules and homomorphisms, or compact groups and continuous homomorphisms. Let* Inv($\mathfrak{a}$) *denote the category of inverse systems and maps of such all of whose elements belong to $\mathfrak{a}$. Then the operations of assigning an inverse limit X_∞ to each inverse system $\{X,\pi\}$ and an inverse limit ϕ_∞ to each map Φ: $\{X,\pi\} \to \{X',\pi'\}$ form a covariant functor from* Inv($\mathfrak{a}$) *to* $\mathfrak{a}$.

PROOF. The fact that $X_\infty \in \mathfrak{a}$ is trivial in all cases except those involving compactness, and this was proved in 3.8. That $\phi_\infty \in \mathfrak{a}$ is 3.13. Condition 1° of IV,4 for a covariant functor requires that ϕ_∞: $X_\infty \to X_\infty'$, and this follows from its definition. Condition 2° reads: If Φ: $\{X,\pi\} \to \{X,\pi\}$ is the identity, so also is ϕ_∞. But Φ is composed of identity maps, so this is immediate. The final condition 3° requires that the limit of a composition of maps Φ,Φ' shall be the composition of their limits. This follows readily from the fact that $\phi_\infty(x)$ is defined by mapping the coordinates of x by means of the coordinate functions ϕ_α, and the fact that the composition $\Phi'\Phi$ was defined by composing coordinate functions.

Propositions 3.10, 3.11, and 3.13 can be applied in the special case when the system $\{X,\pi\} = X$ is a system indexed by a set consisting of a single point. The limit of a system consisting of a single set being the set itself, we see that a mapping Φ: $X \to \{X',\pi'\}$ induces a mapping ϕ_∞: $X \to X_\infty'$ such that $\pi_\alpha'\phi_\infty = \phi_\alpha$.

THEOREM 3.15. *Let $\{X,\pi\},\{X',\pi'\}$ be inverse systems over directed sets M,M', and let Φ: $\{X,\pi\} \to \{X',\pi'\}$ be a map. If there exists a directed subset N of M' such that* (1°) *N is cofinal in M' (see* 2.1), (2°) *$\phi(N)$ is cofinal in M, and* (3°) *ϕ_β is a 1-1 map of $X_{\phi(\beta)}$ onto X_β' for each $\beta \in N$, then ϕ_∞ is a 1-1 map of X_∞ onto X_∞'.*

PROOF. If $\beta \in M'$, abbreviate $\phi(\beta)$ by β'. Suppose $x,y \in X_\infty$ are distinct. Then, for some $\alpha \in M$, $x_\alpha \neq y_\alpha$. By 2°, there is a $\beta \in N$ such that $\alpha < \beta'$. Then $x_{\beta'} \neq y_{\beta'}$; and by 3°, $\phi_\beta(x_{\beta'}) \neq \phi_\beta(y_{\beta'})$. But these are the β-coordinates of $\phi_\infty(x)$ and $\phi_\infty(y)$. Thus ϕ_∞ is 1-1 into. It remains to prove that ϕ_∞ is onto. Suppose $x' \in X_\infty'$. For each $\alpha \in M$, choose a $\beta \in N$ such that $\alpha < \beta'$, and define

$$x_\alpha = \pi^{\beta'}_\alpha\phi_\beta^{-1}(x_\beta') \tag{3}$$

If $\beta < \gamma$ in N, commutativity in the diagram

$$\begin{array}{ccc} X_{\beta'} & \xleftarrow{\ \pi\ } & X_{\gamma'} \\ \downarrow{\scriptstyle \phi_\beta} & & \downarrow{\scriptstyle \phi_\gamma} \\ X'_\beta & \xleftarrow{\ \pi'\ } & X'_\gamma \end{array}$$

and the 1-1 properties of ϕ_β, ϕ_γ yield $\pi^{\gamma'}_{\beta'}\phi_\gamma^{-1}(x'_\gamma) = \phi_\beta^{-1}(x'_\beta)$ It follows that (3) with γ in place of β yields the same value of x_α. Since N is directed, it follows that (3) is independent of the choice of β We must show that $x = \{x_\gamma\}$ lies in X_∞ Suppose $\alpha < \gamma$ in M. Choose $\beta \in N$ such that $\gamma < \beta'$. As just shown

$$x_\alpha = \pi^{\beta'}_{\alpha}\phi_\beta^{-1}(x'_\beta), \qquad x_\gamma = \pi^{\beta'}_{\gamma}\phi_\beta^{-1}(x'_\beta).$$

Since $\pi^{\beta'}_\alpha = \pi^\gamma_\alpha \pi^{\beta'}_\gamma$, it follows that $\pi^\gamma_\alpha(x_\gamma) = x_\alpha$ Finally, $\phi_\infty(x) = x'$ To prove this it suffices to show that they have the same β-coordinate for each $\beta \in M'$ By 1°, it suffices to prove this for each $\beta \in N$ By definition, $(\phi_\infty(x))_\beta = \phi_\beta(x_{\beta'})$ As shown above, $x_{\beta'} = \pi^{\beta'}_{\beta'}\phi_\beta^{-1}(x'_\beta) = \phi_\beta^{-1}(x'_\beta)$. Therefore $\phi_\beta(x_{\beta'}) = x'_\beta$, and the proof is complete

COROLLARY 3 16 *Let M' be a cofinal subset of the directed set M. Let $\{X,\pi\}$ be an inverse system over M, $\{X',\pi'\}$ the subsystem over M' (see 2.4), and Φ the injection of $\{X,\pi\}$ into $\{X',\pi'\}$. Then, the inverse limit ϕ_∞ of Φ is a 1-1 map of X_∞ onto X'_∞*

This follows from 3 15 with $N = M'$

COROLLARY 3 17 *Let M' be cofinal in M, and let $\{X,\pi\}$ be an inverse system over M. Then two elements $x,y \in X_\infty$ coincide if and only if $x_\alpha = y_\alpha$ for each $\alpha \in M'$.*

LEMMA 3.18. *Let Φ and Ψ be maps $\{X,\pi\} \to \{X',\pi'\}$ such that, for each $\alpha \in M'$, we have $\phi(\alpha) < \psi(\alpha)$, and $\psi_\alpha = \phi_\alpha \pi^{\psi(\alpha)}_{\phi(\alpha)}$ Then $\phi_\infty = \psi_\infty$*

PROOF. By 3.11 and 3.2, we have

$$\pi'_\alpha\phi_\infty = \phi_\alpha\pi_{\phi(\alpha)} = \phi_\alpha\pi^{\psi(\alpha)}_{\phi(\alpha)}\pi_{\psi(\alpha)} = \psi_\alpha\pi_{\psi(\alpha)} = \pi'_\alpha\psi_\infty.$$

Since this holds for each $\alpha \in M'$, we have $\phi_\infty = \psi_\infty$.

4. DIRECT LIMITS

DEFINITION 4.1. Let $\{G,\pi\}$ be a direct system over the directed set M where each G^α is an R-module, and each π^β_α is homomorphic. Let $\sum G$ denote the direct sum (v,5.5) of the R-modules of $\{G,\pi\}$. If

$g \in G^\alpha$, we shall agree to identify g with its image in $\sum G$ under the map i_α of v,5 6. For each $\alpha < \beta$ in M and each $g^\alpha \in G^\alpha$ the element

$$(1) \qquad \pi_\alpha^\beta g^\alpha - g^\alpha$$

of $\sum G$ is called a *relation* Let Q be the subgroup of $\sum G$ generated by all relations The *direct limit* of $\{G,\pi\}$ is the factor group

$$G^\infty = (\sum G)/Q.$$

The natural map $\sum G \to G^\infty$ defines homomorphisms

$$\pi_\alpha. \quad G^\alpha \to G^\infty$$

called *projections*.

Since each relation (1) is mapped into zero, we have

LEMMA 4 2 *If $\alpha < \beta$, then $\pi_\beta \pi_\alpha^\beta = \pi_\alpha$.*

LEMMA 4 3 *If $u \in G^\infty$, then there exists an α and a $g^\alpha \in G^\alpha$ such that $\pi_\alpha g^\alpha = u$*

PROOF. This is the first place where the directedness of M is needed. Now u is an image of $v \in \sum G$, and $v = \sum_1^n g^{\alpha_k}$ where $g^{\alpha_k} \in G^{\alpha_k}$. Since M is directed, there exists an $\alpha > \alpha_k$ for $k = 1, \cdots, n$. Let $g_k = \pi_{\alpha_k}^\alpha g^{\alpha_k}$, and let $v' = \sum_1^n g_k$ Since $g_k - g^{\alpha_k} \in Q$, it follows that $v' - v \in Q$. Hence v' also maps on u But $v' \in G^\alpha$

LEMMA 4 4 *If $g \in G^\gamma$ and $\pi_\gamma g = 0$, then there is a $\delta > \gamma$ such that $\pi_\gamma^\delta g = 0$*

PROOF By hypothesis the element g of $\sum G$ is a linear combination of elements of the form (1) Since a multiple of (1) also has the same form, g is a sum of elements of the form (1) Choose a $\delta > \gamma$ and all α,β entering such a sum Then

$$\pi_\gamma^\delta g = (\pi_\gamma^\delta g - g) + g$$

is also such a sum and δ exceeds all α,β entering the sum Since

$$\pi_\alpha^\beta g^\alpha - g^\alpha = (\pi_\alpha^\delta g^\alpha - g^\alpha) + (\pi_\beta^\delta(-\pi_\alpha^\beta g^\alpha) - (-\pi_\alpha^\beta g^\alpha)),$$

it follows that $\pi_\gamma^\delta g$ is a sum of terms (1) where each $\beta = \delta$ All terms with a common α can be lumped into a single term. Thus

$$\pi_\gamma^\delta g = \sum_\alpha (\pi_\alpha^\delta g^\alpha - g^\alpha).$$

Since $\sum G$ is a direct sum, any relation is a consequence of relations in the individual summands It follows then that $g^\alpha = 0$ for each $\alpha \neq \delta$. Clearly $\pi_\delta^\delta g^\delta - g^\delta = 0$, so $\pi_\gamma^\delta g = 0$.

LEMMA 4 5 *$\pi_\alpha g^\alpha = \pi_\beta g^\beta$ if and only if there exists a $\gamma > \alpha,\beta$ such that $\pi_\alpha^\gamma g^\alpha = \pi_\beta^\gamma g^\beta$.*

PROOF. If such a γ exists, then $\pi_\alpha^\gamma g^\alpha - g^\alpha$ and $\pi_\beta^\gamma g^\beta - g^\beta$ are in Q. Hence their difference $g^\beta - g^\alpha$ is also in Q. But this implies $\pi_\alpha g^\alpha = \pi_\beta g^\beta$. Conversely, if $\pi_\alpha g^\alpha = \pi_\beta g^\beta$, choose a $\gamma > \alpha,\beta$ and let $g = \pi_\alpha^\gamma g^\alpha - \pi_\beta^\gamma g^\beta$. By 4 2, $\pi_\gamma g = 0$ Then by 4 4 there exists a $\delta > \gamma$ such that

$$\pi_\gamma^\delta g = \pi_\alpha^\delta g^\alpha - \pi_\beta^\delta g^\beta = 0.$$

It is to be observed that the directedness of M is not used in the definition of the direct limit As a consequence of 4.3 and 4.5, we have the following theorem, which provides an alternative definition of G^∞ using the directedness of M.

THEOREM 4 6 *Let $\{G,\pi\}$ be a direct system of R-modules We shall say that $g^\alpha \in G^\alpha$ is equivalent to $g^\beta \in G^\beta$ if there exists a $\gamma > \alpha,\beta$ such that $\pi_\alpha^\gamma g^\alpha = \pi_\beta^\gamma g^\beta$ This relation divides the elements $g^\alpha \in G^\alpha$, for all α, into disjoint equivalence classes The sum $\bar{g}_1 + \bar{g}_2 = \bar{g}$ of two equivalence classes is obtained by choosing a G^α in which $\bar{g}_1,\bar{g}_2$ have representatives g_1^α,g_2^α and defining $\bar{g}$ to be the class of $g_1^\alpha + g_2^\alpha$ The product $r\bar{g}$ is defined to be the class of rg^α where g^α is in the class $\bar{g}$ Then the equivalence classes form an R-module isomorphic to G^∞ under the correspondence which attaches to each class the image in G^∞ of its representatives.*

THEOREM 4 7 *If $\{G,\pi\}$ is a direct system over M, and, for each $\alpha < \beta$ in M, π_α^β: $G^\alpha \to G^\beta$ has kernel zero [is a homomorphism onto] then, for each α, π_α $G^\alpha \to G^\infty$ has kernel zero [is a homomorphism onto]*

PROOF The first half is an immediate consequence of 4 4. Suppose $u \in G^\infty$, and $\alpha \in M$. By 4 3, there exists a β and a g^β such that $\pi_\beta(g^\beta) = u$ Choose a $\gamma > \alpha,\beta$ Since π_α^γ is onto, there exists a g^α such that $\pi_\alpha^\gamma(g^\alpha) = \pi_\beta^\gamma(g^\beta)$ It follows that $\pi_\alpha(g^\alpha) = u$; hence π_α is onto

COROLLARY 4 8 *If each π_α^β is an isomorphism, so also is each π_α*

The results of 4 7 should be compared with the corresponding ones for inverse systems (see 3.4 and the subsequent remark; also 3 9)

EXAMPLE 4 9. Let G be a group and M the collection of all finite subsets of G. For each $\alpha \in M$, let G^α be the subgroup of G generated by the elements of α. Define $\alpha < \beta$ to mean that $G^\alpha \subset G^\beta$, and let π_α^β: $G^\alpha \to G^\beta$ be the inclusion It is easily proved that M is directed, $\{G,\pi\}$ is a direct system, and $G^\infty \approx G$ Thus, *any group is the direct limit of its subgroups on finite bases.*

EXAMPLE 4.10 Let $\{H_j\}$ be a family of groups indexed by a set J, and let H be their direct sum Let M be the collection of finite subsets of J ordered by inclusion If $\alpha \in M$, define $G^\alpha = \sum_{j\in\alpha} H_j$. If $\alpha \subset \beta$, define π_α^β: $G^\alpha \to G^\beta$ in the obvious way (the coordinates of $\pi_\alpha^\beta(g)$ are those of g for $j \in \alpha$ and are zero for $j \in \beta - \alpha$). It follows quickly that $G^\infty \approx H$ Thus, *a direct sum is a direct limit of finite direct sums*

DEFINITION 4.11. Let Φ: $\{G,\pi\} \to \{G',\pi'\}$ be a map of one direct system of R-modules into another. The homomorphisms ϕ^α: $G^\alpha \to G'^{\phi(\alpha)}$ are the components of a homomorphism $\tilde{\Phi}$ $\sum G \to \sum G'$. Because of the commutativity condition of 3 3, $\tilde{\Phi}$ maps each relation of $\sum G$ into a relation of $\sum G'$ Hence $\tilde{\Phi}$ induces a homomorphism (linear map) of the factor group

$$\phi^\infty:\ G^\infty \to G'^\infty$$

called the *direct limit* of Φ. It satisfies

$$\phi^\infty \pi_\alpha = \pi'_{\phi(\alpha)}\phi^\alpha, \qquad \alpha \in M.$$

Referring to the alternative definition of 4 6, note that $g^\alpha \sim g^\beta$ implies $\phi^\alpha g^\alpha \sim \phi^\beta g^\beta$ Hence Φ maps an equivalence class in $\{G,\pi\}$ into one in $\{G',\pi'\}$ This mapping of classes is precisely ϕ^∞.

Definition 4 11 may be applied in the special case of a map Φ $\{G,\pi\} \to G'$ of a direct system $\{G,\pi\}$ into a single R-module G'

The proof of the following theorem is straightforward and is left to the reader

THEOREM 4 12 *Let* $\mathrm{Dir}(\mathcal{G}_R)$ *be the category of direct systems and maps all of whose elements belong to* $\mathcal{G}_R$ *Then the operations of assigning a direct limit to each direct system and a direct limit to each map of one such into another is a covariant functor from* $\mathrm{Dir}(\mathcal{G}_R)$ *to* $\mathcal{G}_R$

REMARK It is to be noted that we do not attempt to assign a topology to a direct limit of topological groups This can be done in a natural way using the topology of the product space However the group Q of relations is usually not closed in $\sum G$ even in the case of compact groups As a consequence the topology induced in G^∞ is not a Hausdorff topology This means that no analog of 4 12 would hold for any reasonable category of topological groups

THEOREM 4 13 *Let* $\{G,\pi\},\{G',\pi'\}$ *be direct systems over* M,M' *respectively, and let* Φ $\{G',\pi'\} \to \{G,\pi\}$ *be a map If there exists a directed subset* N *of* M' *such that* (1°) N *is cofinal in* M', (2°) $\phi(N)$ *is cofinal in* M, *and* (3°) ϕ^β $G'^\beta \approx G^{\phi(\beta)}$ *for each* $\beta \in N$, *then* ϕ^∞ $G'^\infty \approx G^\infty$.

PROOF If $\beta \in M'$, abbreviate $\phi(\beta)$ by β' Suppose $g \in G'^\infty$ and $\phi^\infty(g) = 0$ Since N is cofinal, there exists a $\beta \in N$ and a $g^\beta \in G'^\beta$ such that $\pi'_\beta g^\beta = g$ (see 4 3) By 4 11, $\pi_{\beta'}\phi^\beta(g^\beta) = \phi^\infty(g) = 0$. By 4.4, there exists a $\delta > \beta'$ in M such that $\pi^\delta_{\beta'}\phi^\beta(g^\beta) = 0$ Since $\phi(N)$ is cofinal, there exists a $\gamma \in N$ such that $\gamma' > \delta$ Then $\pi^{\gamma'}_{\beta'}\phi^\beta(g^\beta) = 0$ By the commutativity condition of 2 3, we have $\phi^\gamma\pi'^\gamma_\beta(g^\beta) = 0$ Since ϕ^γ is isomorphic, $\pi'^\gamma_\beta(g^\beta) = 0$ Hence $g = \pi'_\beta(g^\beta) = \pi'_\gamma\pi'^\gamma_\beta(g^\beta) = 0$ This proves that ϕ^∞ has kernel zero Suppose now that $g \in G^\infty$. Since $\phi(N)$ is cofinal, there exists a $\beta \in N$ and a $g^{\beta'} \in G^{\beta'}$ such that $\pi_{\beta'}(g^{\beta'}) = g$ Since ϕ^β is onto, there exists a $g'^\beta \in G'^\beta$ such that $\phi^\beta(g'^\beta) = g^{\beta'}$. Let

$g' = \pi'_\beta(g'^\beta)$. Then $g' \in G'^\infty$, and by the commutativity condition of 4.11, $\phi^\infty(g') = g$.

COROLLARY 4.14. *Let M' be cofinal in M, $\{G,\pi\}$ a direct system over M, $\{G',\pi'\}$ the subsystem over M', and $\Phi: \{G',\pi'\} \to \{G,\pi\}$ the injection (see 2.4). Then $\phi^\infty: G'^\infty \approx G^\infty$.*

This follows from 4.13 with $N = M'$.

LEMMA 4.15. *Let Φ,Ψ be maps $\{G,\pi\} \to \{G',\pi'\}$ such that, for each $\alpha \in M$, we have $\phi(\alpha) < \psi(\alpha)$ and $\psi^\alpha = \pi'^{\psi(\alpha)}_{\phi(\alpha)}\phi^\alpha$.* Then $\phi^\infty = \psi^\infty$.

PROOF. If $g \in G^\infty$, choose an α and a $g' \in G^\alpha$ such that $\pi_\alpha g' = g$. By 4.11, we have

$$\begin{aligned}\phi^\infty g = \phi^\infty \pi_\alpha g' &= \pi'_{\phi(\alpha)}\phi^\alpha g' = \pi'_{\psi(\alpha)}\pi'^{\psi(\alpha)}_{\phi(\alpha)}\phi^\alpha g' \\ &= \pi'_{\psi(\alpha)}\psi^\alpha g' = \psi^\infty \pi_\alpha g' = \psi^\infty g.\end{aligned}$$

5. SYSTEMS OF EXACT SEQUENCES

DEFINITION 5.1. An *inverse system of lower sequences* $\{S,\pi\}$ over the directed set M is a function which attaches to each $\alpha \in M$ a lower sequence

$$S_\alpha = \{G_{\alpha,q},\phi_{\alpha,q}\}$$

(see I,2), and to each relation $\alpha < \beta$ in M, a homomorphism π^β_α: $S_\beta \to S_\alpha$ such that π^α_α = identity, and, if $\alpha < \beta < \gamma$, $\pi^\beta_\alpha\pi^\gamma_\beta = \pi^\gamma_\alpha$. Then, for any fixed q, the groups and homomorphisms $\{G_{\alpha,q},\pi^\beta_{\alpha,q}\}$ form an inverse system of groups over M. Its limit group is denoted by $G_{\infty,q}$. Again, for a fixed q, the homomorphism $\{\phi_{\alpha,q}\}$ together with the identity map of M form a map Φ_q: $\{G_{\alpha,q},\pi^\beta_{\alpha,q}\} \to \{G_{\alpha,q-1},\pi^\beta_{\alpha,q-1}\}$. The limit of Φ_q is denoted by $\phi_{\infty,q}$: $G_{\infty,q} \to G_{\infty,q-1}$. The lower sequence $S_\infty = \{G_{\infty,q},\phi_{\infty,q}\}$ so obtained is called the *inverse limit* of the system $\{S,\pi\}$. A *direct system of upper sequences* and its *limit* sequence are defined in an analogous manner.

In the above definition it is to be understood that all groups and homomorphisms in an inverse system of lower sequences belong to a fixed one of the standard categories $\mathcal{G}_R,\mathcal{G}_C$. That the limit sequence is of the same type is proved in the preceding sections. In the case of a direct system of upper sequences we shall use only the categories $\mathcal{G}_R$.

DEFINITION 5.2. A lower [upper] sequence is said to be of *order* 2 if the composition of any two successive homomorphisms of the sequence is zero: i.e. kernel $\supset$ image. Although this notion coincides with that of chain [cochain] complex (V,2), we shall prefer the "order 2" language whenever the sequences are not to be treated as chain complexes.

THEOREM 5.3. *If each sequence of an inverse [direct] system of lower [upper] sequences is of order 2, then the limit sequence is also of order 2.*

Proof The composition $\Phi_{q-1}\Phi_q$ mapping $\{G_{\alpha,q}, \pi^{\beta}_{\alpha,q}\}$ into $\{G_{\alpha,q-2}, \pi^{\beta}_{\alpha,q-2}\}$ consists of the identity map of M and the maps $\phi_{\alpha,q-1}\phi_{\alpha,q} = 0$. Hence the inverse limit of $\Phi_{q-1}\Phi_q$ is zero By 3 14, this is the composition $\phi_{\infty,q-1}\phi_{\infty,q}$. The corresponding proof for a direct system of upper sequences is left to the reader

Theorem 5 4. *If each sequence of a direct system of upper sequences is exact, then the limit sequence is also exact*

Proof One half of exactness follows from 5 3 Suppose $u \in G^{\infty,q}$ and $\phi^{\infty,q}(u) = 0$ in $G^{\infty,q+1}$. By 4 3, there exists an α and a $u^{\alpha} \in G^{\alpha,q}$ such that $\pi^{q}_{\alpha}(u^{\alpha}) = u$ By 4 9,

$$\pi^{q+1}_{\alpha}\phi^{\alpha,q}(u^{\alpha}) = \phi^{\infty,q}\pi^{q}_{\alpha}(u^{\alpha}) = 0$$

By 4 4, there exists a $\beta > \alpha$ such that

$$\pi^{\beta,q+1}_{\alpha}\phi^{\alpha,q}(u^{\alpha}) = 0.$$

Let $u^{\beta} = \pi^{\beta,q}_{\alpha}(u^{\alpha})$ It follows from the commutativity of π and ϕ that $\phi^{\beta,q}(u^{\beta}) = 0$ By exactness of S^{β}, there exists a $v^{\beta} \in G^{\beta,q-1}$ such $\phi^{\beta,q-1}(v^{\beta}) = u^{\beta}$ Let $v = \pi^{q-1}_{\beta}(v^{\beta}) \in G^{\infty,q-1}$ Then $\phi^{\infty,q-1}(v) = u$ This proves the other half of exactness

Example 5 5 The limit sequence of an inverse system of exact lower sequences need not be exact as the following example shows

Consider the diagram

$$\begin{array}{ccccc}
J & \xrightarrow{\tau} & J & \xrightarrow{\eta} & J/2J \\
\downarrow{\scriptstyle\gamma} & & \downarrow{\scriptstyle\gamma} & & \downarrow{\scriptstyle\epsilon} \\
J & \xrightarrow{\tau} & J & \xrightarrow{\eta} & J/2J
\end{array}$$

where J is the group of integers, $\tau(n) = 2n$, $\gamma(n) = 3n$, ϵ is the identity map, and η is the natural map It is easy to see that

$$S \quad 0 \to J \xrightarrow{\tau} J \xrightarrow{\eta} J/2J \to 0$$

is an exact sequence, and that the homomorphisms γ,γ,ϵ together yield a homomorphism $\phi.\ S \to S$

Let M be the set of positive integers with the usual order For each $\alpha \in M$ define $S_{\alpha} = S$ and $\pi^{\alpha}_{\alpha+1} = \phi$. Define all other π^{β}_{α} for $\beta < \alpha$ by composition $\pi^{\beta}_{\alpha} = (\phi)^{\beta-\alpha}$. Then $\{S_{\alpha},\pi^{\beta}_{\alpha}\}$ is an inverse system of exact lower sequences. Let

$$S_{\infty}:\ 0 \to G \to G \to H \to 0$$

be the limit sequence of this inverse system. Since ϵ is the identity map, it follows that $H = J/2J$. An element of G is a sequence $\{n_\alpha\}$, $\alpha \varepsilon M$ of integers such that $n_\alpha = 3n_{\alpha+1}$. It follows that n_α is divisible by arbitrarily high powers of 3; i.e. $n_\alpha = 0$ and $G = 0$. Thus S_∞ is the sequence $0 \to 0 \to 0 \to J/2J \to 0$ which is not exact.

THEOREM 5.6. *Let $\{S,\pi\}$ be an inverse system of exact lower sequences over M where all groups and homomorphisms of $\{S,\pi\}$ belong to the category $\mathcal{G}_C$ of compact groups. Then the limit sequence S_∞ of $\{S,\pi\}$ is also exact.*

PROOF. One-half of exactness follows from 5.3. Suppose $g \varepsilon G_{\infty,q}$ and $\phi_{\infty,q}(g) = 0$. Let $g_\alpha = \pi_{\alpha,q}(g)$ be the coordinate of g in $G_{\alpha,q}$. Since $\phi_{\infty,q}(g) = 0$, it follows that $\phi_{\alpha,q}(g_\alpha) = 0$ for each α. Since S_α is exact, $X_\alpha = (\phi_{\alpha,q+1})^{-1}(g_\alpha)$ is a closed nonvacuous subset of $G_{\alpha,q+1}$. From the relation $\phi_{\alpha,q+1}\pi^\beta_{\alpha,q+1} = \pi^\beta_{\alpha,q}\phi_{\beta,q+1}$, it follows that $\pi^\beta_{\alpha,q+1}$ maps X_β into X_α. Let ρ^β_α be the map so defined. Then $\{X_\alpha,\rho^\beta_\alpha\}$ is an inverse system of compact spaces. Since each space is nonvacuous, it follows from 3.6 that the limit space X_∞ is nonvacuous. It is easily seen that $X_\infty \subset G_{\infty,q+1}$ and $\phi_{\infty,q+1}$ maps X_∞ into g. This proves the other half of exactness.

THEOREM 5.7. *Let $\{S,\pi\}$ be an inverse system of exact lower sequences over M where all terms of $\{S,\pi\}$ belong to the category of finite dimensional vector spaces over a fixed field F. Then the limit sequence S_∞ of $\{S,\pi\}$ is also exact.*

PROOF. If G is a vector space over F, a *variety* V in G means a coset of some linear subspace H of G. The *dimension* of V is that of H. It is easily seen that, under a linear map, the image (inverse image) of a variety is a variety. It is also clear that the intersection of a collection of varieties, if not vacuous, is also a variety. We interrupt the proof for a basic lemma.

LEMMA 5.8. *If G is a finite dimensional vector space over F, and $\{V_j\}$ is a collection of varieties in G ($j \varepsilon J$) such that any finite number have a nonvacuous intersection, then $\bigcap_{j\varepsilon J} V_j \neq 0$.*

Since G has finite dimension, any subspace has a finite dimension, and therefore the same holds for any variety. By hypothesis, any finite subcollection of $\{V_j\}$ has an intersection of dimension $\geqq 0$. Since all such dimensions are integers, there is a smallest such integer k, and a corresponding $V = \bigcap_{i=1}^h V_{j_i}$ with $\dim V = k$. For any $j \varepsilon J$, $V \cap V_j \neq 0$ by hypothesis. Also $\dim V \cap V_j = k$ by the minimal property of k. Thus $\dim V = \dim V \cap V_j$ which implies $V = V \cap V_j$, and this in turn implies $V \subset V_j$. This proves the lemma.

We return now to the proof proper of 5.7. As before, one-half of exactness follows from 5.3. Suppose $g \varepsilon G_{\infty,q}$ and $\phi_{\infty,q}(g) = 0$. If g_α is the coordinate of g in $G_{\alpha,q}$, then $\phi_{\alpha,q}(g_\alpha) = 0$. Define $V^0_\alpha =$

$(\phi_{\alpha,q+1})^{-1}(g_\alpha)$. Since S_α is exact, V_α^0 is not vacuous As a coset of the kernel of $\phi_{\alpha,q+1}$, V_α^0 is a variety Consider now the family Ψ of all functions $v = \{V_\alpha\}$ defined for $\alpha \in M$ such that V_α is a variety in $G_{\alpha,q+1}$, $V_\alpha \subset V_\alpha^0$, and, if $\alpha < \beta$, $\pi_\alpha^\beta(V_\beta) \subset V_\alpha$. Clearly $v^0 = \{V_\alpha^0\}$ belongs to Ψ If $u,v \in \Psi$, the relation $u \subset v$ means $U_\alpha \subset V_\alpha$ for each $\alpha \in M$. This defines a partial order in Ψ By Zorn's lemma, the partially ordered set Ψ contains a maximal simply ordered subset Ψ' For each α, define V_α^1 to be the intersection of all V_α for $V \in \Psi'$. Since Ψ' is simply ordered, any finite subset has a smallest element Hence, by 5 8 and the assumption that $G_{\alpha,q+1}$ has finite dimension, V_α^1 is not vacuous It is easily proved that $\pi_\alpha^\beta(V_\beta^1) \subset V_\alpha^1$ whenever $\alpha < \beta$. Therefore $v^1 = \{V_\alpha^1\}$ lies in Ψ. Since $v^1 \subset v$ for each $v \in \Psi'$ and Ψ' is maximal, it follows that $v^1 \in \Psi'$ Thus we have found an element $v^1 \in \Psi$ such that $v \subset v^1$ in Ψ implies $v = v^1$ It remains to show that each V_α^1 is a single element of $G_{\alpha,q+1}$. Consider any fixed element $\alpha \in M$. The collection $\{\pi_\alpha^\beta(V_\beta^1)\}$ for all $\beta > \alpha$ has the property that any finite subcollection has a nonvacuous intersection (M is a directed set). Hence by 5 8 there is an element g'_α in their intersection For any $\gamma \in M$, choose a $\beta > \alpha,\gamma$ and define

$$V_\gamma = V_\gamma^1 \cap (\pi_\gamma^\beta(\pi_\alpha^\beta)^{-1}(g'_\alpha)).$$

The choice made of g'_α insures that V_γ is not vacuous Clearly V_γ is independent of the choice of β From this it follows that $V_\gamma \subset V_\gamma^1$, $V_\alpha = g'_\alpha$, and $\pi_\gamma^\delta(V_\delta) \subset V_\gamma$ if $\gamma < \delta$ Thus $v = \{V_\gamma\}$ belongs to Ψ and $v \subset v^1$ By the minimal property of v^1, we have $v = v^1$; hence $g'_\alpha = V_\alpha^1$ Thus $g = \{g'_\alpha\}$ is an element of $G_{\infty,q+1}$ and $\phi_{\infty,q+1}(g') = g$. This completes the proof of 5 7

REMARK The failure of exactness under inverse limits can be described with greater precision Let Φ. $\{G,\pi\} \to \{G',\pi'\}$ be a map of one inverse system of groups into another Then the following relations always hold

$$\text{kernel } \phi_\infty = \lim \{\text{kernel } \phi_\alpha\},$$
$$\text{image } \phi_\infty \subset \lim \{\text{image } \phi_\alpha\}.$$

In the second case equality holds if all groups are compact or finite dimensional vector spaces over a field The example 5 5 shows that equality does not hold in general In the case of direct systems equality holds in both cases.

6. COMMUTATIVITY OF THE OPERATIONS OF LIMIT AND FACTOR GROUP

DEFINITION 6 1 Let $\{G,\pi\}$ be an inverse system of groups over M. Suppose, for each $\alpha \in M$, that H_α is a subgroup of G_α, and suppose, for each $\alpha < \beta$ in M, that π_α^β maps H_β into H_α. Let ρ_α^β: $H_\beta \to H_\alpha$ be the map defined by π_α^β. Clearly $\{H,\rho\}$ is an inverse system over M. It is called a *system of subgroups* of $\{G,\pi\}$. For each $\alpha \in M$, define $K_\alpha = G_\alpha/H_\alpha$, and, for each $\alpha < \beta$ in M, define σ_α^β: $K_\beta \to K_\alpha$ to be the map induced by π_α^β. Then $\{K,\sigma\}$ is an inverse system over M called the *system of factor groups* of $\{G,\pi\}$ by $\{H,\rho\}$. The *inclusion* map Φ: $\{H,\rho\} \to \{G,\pi\}$ and the *natural* map Ψ: $\{G,\pi\} \to \{K,\sigma\}$ are defined in the obvious way. By definition, each element of the limit group H_∞ of $\{H,\rho\}$ is an element of G_∞, and ϕ_∞: $H_\infty \to G_\infty$ is the inclusion. The corresponding definitions for direct systems are left to the reader.

THEOREM 6.2. *Let $\{G,\pi\},\{H,\rho\},\{K,\sigma\}$ be an inverse system of groups, subgroups, and factor groups over M where all terms belong to the category $\mathcal{G}_C$ of compact groups or to the category of finite dimensional vector spaces over a fixed field F; then ψ_∞: $G_\infty \to K_\infty$ induces an isomorphism $G_\infty/H_\infty \approx K_\infty$.*

PROOF. For each α, adjoin an infinite set of trivial groups and maps to

$$H_\alpha \xrightarrow{\phi} G_\alpha \xrightarrow{\psi} K_\alpha$$

so as to obtain a lower sequence S_α

$$\cdots \to 0 \to H_\alpha \xrightarrow{\phi} G_\alpha \xrightarrow{\psi} K_\alpha \to 0 \to \cdots$$

It is clear that S_α is an exact sequence. For each $\alpha < \beta$ in M, adjoin to $\pi_\alpha^\beta,\rho_\alpha^\beta,\sigma_\alpha^\beta$ an infinite set of trivial maps so as to obtain a map τ_α^β: $S_\beta \to S_\alpha$. Then $\{S,\tau\}$ is an inverse system of exact sequences. It is also clear that the limit sequence S_∞ consists of

$$\cdots \to 0 \to H_\infty \xrightarrow{\phi} G_\infty \xrightarrow{\psi} K_\infty \to 0 \to \cdots$$

By 5.6 and 5.7, S_∞ is an exact sequence. It follows that ψ_∞ must be onto and its kernel is H_∞. This completes the proof.

It is to be observed that example 5.5 shows that 6.2 does not hold if it is merely required that all terms belong to the category $\mathcal{G}$.

THEOREM 6.3. *Let $\{G,\pi\},\{H,\rho\},\{K,\sigma\}$ be a direct system of groups, subgroups, and factor groups over M where all terms belong to the category $\mathcal{G}_R$; then ψ_∞: $G_\infty \to K_\infty$ induces an isomorphism $G_\infty/H_\infty \approx K_\infty$.*

The proof is analogous to that of 6.2 and uses 5.4 in place of 5.6 and 5.7.

EXERCISES

A. Inverse limits of spaces

1 Prove that every countable directed set contains a cofinal subsequence (sequence = directed set isomorphic with the set of positive integers)

2. Show that, if each $X_\alpha, \alpha \in M$, has a countable base for its open sets, and if M is countable, then $\prod_{\alpha \in M} X_\alpha$ has a countable base

3. If $\{X_\alpha, \pi\}$ is an inverse system indexed by M, each X_α has a countable base and M has a cofinal subsequence, then the limit X_α has a countable base

4 An inverse limit of compact connected spaces is connected Exhibit an inverse sequence of connected spaces with a disconnected limit.

B. Limits of groups.

1 Let G be a group and $\{G^\alpha\}$ a system of subgroups such that $\bigcup G^\alpha = G$ and for any G^α, G^β there is a G^γ with $G^\alpha \subset G^\gamma$, $G^\beta \subset G^\gamma$. Show that G is naturally isomorphic with the limit of the direct system $\{G^\alpha, \pi^\beta_\alpha\}$ where the π^β_α are inclusions

2 Let G be a compact group and $\{G_\alpha\}$ a system of (closed) subgroups such that $\bigcap G_\alpha$ is the identity and for any G_α, G_β there is G_γ with $G_\gamma \subset G_\alpha \cap G_\beta$ Let $H_\alpha = G/G_\alpha$, and for each $G_\alpha \subset G_\beta$ let π^α_β $H_\alpha \to H_\beta$ be the natural map Show that G is naturally isomorphic to the limit of the inverse system $\{H_\alpha, \pi^\alpha_\beta\}$

3 Show that the above results remain valid for nonabelian groups, provided the definitions of Chapter VIII are suitably interpreted

C Direct sums and products.

1 Let $\{G,\pi\}, \{{}^\gamma G, {}^\gamma\pi\}$, $\gamma \in \Gamma$ be direct systems of groups over the same directed set M Let ${}^\gamma\phi$ $\{{}^\gamma G, {}^\gamma\pi\} \to \{G,\pi\}$ be maps such that, for each $\alpha \in M$, the maps ${}^\gamma\phi^\alpha$ ${}^\gamma G^\alpha \to G^\alpha$ form an injective representation of G^α as a direct sum Show the same holds for the limiting maps ${}^\gamma\phi^\infty$. ${}^\gamma G^\infty \to G^\infty$.

2 In the above assume that ${}^\gamma p$ $\{G,\pi\} \to \{{}^\gamma G, {}^\gamma\pi\}$ and that the maps ${}^\gamma\phi^\alpha$: $G^\alpha \to {}^\gamma G^\alpha$ yield a projective representation of G^α as a direct product. Show that the same holds for ${}^\gamma\phi^\infty$ $G^\infty \to {}^\gamma G^\infty$.

3 Establish propositions similar to 1 and 2 for inverse systems.

D Split exact sequences.

Definition. An exact lower sequence $\{G_n, \phi_n\}$ and a sequence of homomorphisms

$$\psi_{3i+1}.\quad G_{3i+1} \to G_{3i+2}, \quad i = 0, \pm 1, \pm 2, \cdots,$$

such that $\psi_{3i+1}\phi_{3i+2}$ is the identity map of G_{3i+2} for each i, is called a *split exact lower sequence* The homomorphisms ψ_n (defined only for $n \equiv 1 \bmod 3$) are called the splitting homomorphisms of the exact sequence A map of a split exact sequence into another one is defined in the obvious fashion

1. Show that $\{\phi_n\}$ and $\{\psi_{3i+1}\}$ yield a split exact lower sequence if and only if the following conditions hold (for all i).

(a) $\phi_{3i} = 0$,

(b) $\phi_{3i+1}\phi_{3i+2} = 0$,

(c) $\psi_{3i+1}\phi_{3i+2} =$ identity,

(d) ϕ_{3i+1}: $G_{3i+1} \to G_{3i}$ and ψ_{3i+1}: $G_{3i+1} \to G_{3i+2}$ yield a projective representation of G_{3i+1} as a direct sum

2 Let $\{S,\pi\}$ be a direct [inverse] system in which each S is a split exact sequence Show that the limit is also a split exact sequence.

3. Reformulate the above results for upper sequences.

4. Let (X,A) be a pair and $r\colon X \to A$ a retraction. Show that the homology sequence of (X,A) together with the homomorphisms r_* form a split exact sequence.

E p-ADIC GROUPS AND SOLENOIDS

1 Let G_n be a cyclic group of order p^n with generator g_n Consider the inverse system

$$G_1 \overset{\phi_1}{\leftarrow} G_2 \overset{\phi_2}{\leftarrow} G_3 \leftarrow \cdot \quad \leftarrow G_n \overset{\phi_n}{\leftarrow} G_{n+1} \leftarrow \cdots$$

where $\phi_n(g_{n+1}) = g_n$ The inverse limit G_∞ is called the *p-adic group.* Show that G_∞ is compact, totally disconnected, and perfect Show that it has a countable base for its open sets, and, therefore, is homeomorphic with the Cantor set Show that for each integer n the subgroup p^nG_∞ is open and closed and yields a fundamental sequence of neighborhoods of zero

DEFINITION Let S^1 be the multiplicative group of all complex numbers z with $|z| = 1$. Consider the inverse sequence

$$S^1 \overset{\phi}{\leftarrow} S^1 \leftarrow \cdots \leftarrow S^1 \overset{\phi}{\leftarrow} S^1 \leftarrow \cdots$$

where $\phi(z) = z^p$ The inverse limit Σ_p is called the *p-adic solenoid*

2. Show that there is a continuous homomorphism $\psi\colon R \to \Sigma_p$ of the addive group R of real numbers into Σ_p such that the kernel of ψ is zero, and the image of ψ is dense in Σ_p.

3. Show that the p-adic solenoid contains a subgroup isomorphic with the p-adic group.

4. In euclidean 3-space R^3 consider the disc $D = \{(x_1 + 2)^2 +$

$x_2^2 \leq 1, x_3 = 0\}$ and the solid torus T obtained by revolving D around the x_1-axis. In D consider the disc

$$D_0 = \left\{\left(x_1 + \frac{5}{2}\right)^2 + x_2^2 \leq c^2,\ x_3 = 0\right\} \qquad \text{where} \qquad c < \frac{1}{2}\sin\frac{\pi}{p}$$

and the discs D_i obtained from D_0 by revolving D around its center by the angle $(2\pi i)/p$ ($i = 1, \ldots, p - 1$). The choice of c insures the disjointness of the discs $D_0, D_1, \ldots, D_{p-1}$. Now assume that as D revolves around the x_1-axis it also revolves around its own center in such a fashion that, as one revolution around the x_1-axis is complete, D_0 becomes D_1, D_1 becomes D_2 etc. Then the discs $D_0, D_1, \cdots, D_{p-1}$ sweep out a torus T^1 which runs p times around the inside of the torus T. To describe the situation arithmetically, represent T by means of pairs of complex numbers (z,s), $|z| \leq 1$, $|s| = 1$ (z describes a point on D and s describes the angle of revolution around the x_1-axis). Then define a mapping θ: $T \to T$ by

$$\theta(z,s) = \left(s\left(\frac{z}{c} + \frac{1}{2}\right), s^p\right)$$

where $0 < c < \frac{1}{2} \sin \pi/p$. Then $T^1 = \theta(T)$. Show that the p-adic solenoid is homeomorphic with the inverse limit of the sequence

$$T \xleftarrow{\theta} T \xleftarrow{\theta} \cdots \xleftarrow{\theta} T \leftarrow T \leftarrow \cdots$$

which in turn is homeomorphic with the intersection $\bigcap_{n=0}^{\infty} \theta^n T$ where $\theta^0 T = T$, $\theta^n T = \theta(\theta^{n-1} T)$.

F. LIMITING GROUPS FOR INFINITE COMPLEXES

1. Let K be a (possibly infinite) simplicial complex and L a subcomplex. Let $\{K_\alpha\}$ be the family of all finite subcomplexes of K ordered by inclusion. Show that the groups $H_q(K_\alpha, K_\alpha \cap L, G)$ $[H^q(K_\alpha, K_\alpha \cap L, G)]$ together with the homomorphisms of these groups induced by inclusions form a direct [inverse] system of groups. The resulting limit groups are denoted by $\overrightarrow{H}_q(K,L,G)$ $[\overleftarrow{H}^q(K,L,G)]$ and are called the *direct limit homology* [*inverse limit cohomology*] *groups of* (K,L) (cf. VI, Exer. B).

2. Show that $H_q(K,L;G)$ and $\overrightarrow{H}_q(K,L,G)$ are isomorphic. Show that $H^q(K,L;G)$ and $\overleftarrow{H}^q(K,L,G)$ are isomorphic if G is compact or a vector space over a field.

3. Let K be a locally finite simplicial complex, and L a subcomplex. A subcomplex K_α of K is called *counterfinite* if K_α contains all but a finite number of simplexes of K. Show that with a suitable ordering

of the counterfinite subcomplexes K_α, the groups $\mathcal{H}_q(K,L \cup K_\alpha;G)$ [$\mathcal{H}^q(K,L \cup K_\alpha,G)$] (see VI, Exer. B) with suitable homomorphisms induced by inclusions, form an inverse [direct] system of groups The resulting limit groups are denoted by $\overleftarrow{\mathcal{H}}_q(K,L;G)$ [$\overrightarrow{\mathcal{H}}^q(K,L,G)$] and are called the *inverse limit homology* [*direct limit cohomology*] *groups of* (K,L).

4. Show that $\mathcal{H}^q(K,L;G)$ and $\overrightarrow{\mathcal{H}}^q(K,L;G)$ are isomorphic. Show that $\mathcal{H}_q(K,L;G)$ and $\overleftarrow{\mathcal{H}}_q(K,L,G)$ are isomorphic if G is compact or a vector space over a field

CHAPTER IX

The Čech homology theory

1. INTRODUCTION

In this chapter the Čech homology and cohomology theories are defined and the axioms for such are verified The degree of generality of the results has been greatly increased by the use of modifications introduced by Dowker [Annals of Math 51 (1950), 278-292].

The Čech *cohomology* theory is defined on the category $\mathfrak{a}_1$ of arbitrary pairs (X,A) and their maps The coefficient group G is taken to be an R-module for any ring R, and the resulting $H^q(X,A)$ are in $\mathcal{G}_R$. The axioms are verified without exception Čech cohomology groups with compact coefficients are not defined

The Čech homology groups are defined under the same circumstances as the Čech cohomology groups Further, if (X,A) is a compact pair, then the homology group $H_q(X,A)$ is also defined for $G \in \mathcal{G}_C$, and is itself in $\mathcal{G}_C$ The axioms are valid without restrictions except for the Exactness axiom, which is valid only after drastic restrictions The homology sequence of any pair is defined, and it is proved that the composition of any two homomorphisms is zero To obtain the full Exactness axiom, we must restrict (X,A) to be a compact pair and G must be either compact or a vector space over a field. In case (X,A) is triangulable, exactness is proved without this restriction on G This is accomplished by showing directly that $H_q(X,A)$ is isomorphic in a natural way to the group based on the chains of a simplicial division.

The failure of the Čech homology theory to satisfy the Exactness axiom is directly traceable to the facts established in VIII,6 concerning the interchangeability of factorization and inverse limits

2. COVERINGS, NERVES, AND PROJECTIONS

DEFINITION 2 1. An *indexed family of sets in a space* X is a function α defined on a set V_α of indices such that, for each $v \in V_\alpha$, α_v (the value of α on v) is a subset of X If $X = \bigcup \alpha_v$, then α is called a *covering* of X. It is called an *open* (*closed*) covering of X if each α_v is open (closed) in X The set of all open coverings of X is denoted by $\mathrm{Cov}(X)$,

If A is a subset of X, and V_α^A is a subset of V_α such that $A \subset \bigcup \alpha_v$ for $v \in V_\alpha^A$, then we say that α is a *covering* of the pair (X,A) with (V_α,V_α^A) as indexing pair. The set of all open coverings of (X,A) is denoted by $\text{Cov}(X,A)$.

One should be careful not to confuse the sets $\text{Cov}(X)$ and $\text{Cov}(X,0)$. However one may regard $\text{Cov}(X)$ as the subset of $\text{Cov}(X,0)$ consisting of coverings α indexed by $(V_\alpha,0)$

DEFINITION 2.2. Let α be an indexed family of sets in a space X. Let s_α be the simplicial complex consisting of all simplexes whose vertices are elements of V_α (if V_α is finite, then s_α is itself a simplex). If s is a simplex of s_α, the *carrier* of s, denoted by $\text{Car}_\alpha(s)$, is the intersection of those sets α_v which correspond to vertices v of s. The *nerve* of α, denoted by X_α, is the subcomplex of s_α consisting of all simplexes with nonvacuous carriers. If α is a covering of the pair (X,A) indexed by (V_α,V_α^A), then we denote by A_α the subcomplex of X_α consisting of all simplexes s with vertices in V_α^A such that $A \cap \text{Car}_\alpha(s) \neq 0$. The pair (X_α,A_α) is then called the *nerve* of α.

Note that α_v is the carrier of the vertex v.

LEMMA 2.3. *If s' is a face of s, then* $\text{Car}_\alpha(s') \supset \text{Car}_\alpha(s)$.

PROOF. Since each vertex of s' is one of s, it follows that each term of the intersection defining $\text{Car}_\alpha(s')$ is a term of that defining $\text{Car}_\alpha(s)$.

It follows from 2.3 that each face of a simplex of the nerve X_α is also a simplex of X_α. Therefore X_α is a simplicial complex.

DEFINITION 2.4. If $f\colon (X,A) \to (Y,B)$ and β is a covering of (Y,B), then $f^{-1}\beta$ is the covering α of (X,A) with the same indexing pair, $(V_\alpha,V_\alpha^A) = (V_\beta,V_\beta^B)$, and defined by $\alpha_v = f^{-1}(\beta_v)$ for each $v \in V_\beta$. It follows from the continuity of f that, if β is an open (closed) covering, so also is $f^{-1}\beta$.

LEMMA 2.5. *If $f\colon (X,A) \to (Y,B)$ and $\alpha = f^{-1}\beta$, then the nerve X_α is a subcomplex of Y_β and A_α is a subcomplex of B_β. The inclusion map $(X_\alpha,A_\alpha) \to (Y_\beta,B_\beta)$ is denoted by f_β.*

This follows from the fact that $f(\text{Car}_\alpha(s)) \subset \text{Car}_\beta(s)$ for each simplex s of $s_\alpha = s_\beta$.

LEMMA 2.6. *If $f\colon (X,A) \to (X,A)$ is the identity, and α is a covering of (X,A), then $f^{-1}\alpha = \alpha$ and f_α is the identity map of (X_α,A_α).*

LEMMA 2.7. *If $f\colon (X,A) \to (Y,B)$, $g\colon (Y,B) \to (Z,C)$, γ is a covering of (Z,C), and $\beta = g^{-1}\gamma$, then $f^{-1}g^{-1}\gamma = (gf)^{-1}\gamma$ and $(gf)_\gamma = g_\gamma f_\beta$.*

DEFINITION 2.8. Let α and β be two coverings of (X,A). The covering β is called a *refinement* of α (notation $\alpha < \beta$) if every set of β is contained in some set of α, and every set of β indexed by an element of V_β^A is contained in some set of α indexed by V_α^A. If $\alpha < \beta$, a function $p\colon (V_\beta,V_\beta^A) \to (V_\alpha,V_\alpha^A)$ is called a *projection* if $\alpha_{pv} \supset \beta_v$ for each

$v \in V_\beta$. The vertex mapping p extends uniquely to a simplicial map $s_\beta \to s_\alpha$ which is also denoted by p

LEMMA 2 9 *The relation $<$ is a quasi-order, i e. $\alpha < \alpha$, and $\alpha < \beta < \gamma$ implies $\alpha < \gamma$*

This follows from the corresponding properties of the relation $\subset$.

Note that $\alpha < \beta$ and $\beta < \alpha$ do not imply $\alpha = \beta$ For example $\alpha < \beta$ and $\beta < \alpha$ holds if $A = V_\alpha^A = V_\beta^A = 0$ and both α and β have a set $= X$

LEMMA 2 10 *The set $\mathrm{Cov}(X,A)$ of open coverings of (X,A) is a directed set with respect to the relation $<$*

PROOF By 2 9 and VIII,2 1 it is sufficient to prove that, for any $\alpha, \beta \in \mathrm{Cov}(X,A)$, there exists a $\gamma \in \mathrm{Cov}(X,A)$ such that $\alpha < \gamma$ and $\beta < \gamma$. Define $V_\gamma = V_\alpha \times V_\beta$ and $V_\gamma^A = V_\alpha^A \times V_\beta^A$ If $v \in V_\gamma$, then $v = (v_1,v_2)$ where $v_1 \in V_\alpha$, $v_2 \in V_\beta$ Define

$$\gamma_v = \alpha_{v_1} \cap \beta_{v_2}.$$

It is clear that γ has the desired properties.

Note that the covering γ is a "smallest" common refinement of α,β. That is, if δ is any common refinement of α,β, then $\gamma < \delta$.

LEMMA 2 11. *For any α, the identity map $s_\alpha \to s_\alpha$ is a projection. If $p.\ s_\beta \to s_\alpha$, $p'.\ s_\gamma \to s_\beta$ are projections, then their composition $pp'\ s_\gamma \to s_\alpha$ is a projection*

The first statement follows from $\alpha_v \supset \alpha_v$. For the second it suffices to note that $\alpha_{pp'v} \supset \beta_{p'v} \supset \gamma_v$ implies $\alpha_{pp'v} \supset \gamma_v$.

LEMMA 2 12 *If $\alpha < \beta$ are coverings of (X,A), then a projection $p\ s_\beta \to s_\alpha$ maps (X_β, A_β) into (X_α,A_α) This simplicial map of the nerve of β into that of α is also called a projection and is denoted by the same symbol p*

PROOF Since $\alpha_{pv} \supset \beta_v$, any intersection of β_v's is contained in the intersection of the corresponding sets of α Hence, for any face s of s_β, $\mathrm{Car}_\beta(s) \subset \mathrm{Car}_\alpha(p(s))$ Consequently if the $\mathrm{Car}_\beta(s)$ is nonvacuous (meets A), so also $\mathrm{Car}_\alpha(p(s))$

THEOREM 2 13 *If $\alpha < \beta$ are two coverings of (X,A), then any two projections $p,p'\ (X_\beta,A_\beta) \to (X_\alpha,A_\alpha)$ are contiguous simplicial maps (see VI,3 1).*

PROOF Let s be a simplex of X_β, and let $x \in \mathrm{Car}_\beta(s)$ Then, for any vertex v of s, we have $x \in \beta_v$. Since $\beta_v \subset \alpha_{pv}$ and $\beta_v \subset \alpha_{p'v}$, x lies in both α_{pv} and $\alpha_{p'v}$ Hence the simplex s' of s_α spanned by the two images of the vertices of s has a nonvacuous carrier. Thus, $p(s)$ and $p'(s)$ are faces of s' in X_α If s is in A_β, choose $x \in A \cap \mathrm{Car}_\beta(s')$. The same argument shows that s' is in A_α

COROLLARY 2 14. *Using homology and cohomology groups of com-*

plexes in the sense of VI,3.9, *we have, for any coefficient group G, that the homomorphisms*

$$H_q(X_\beta,A_\beta;G) \to H_q(X_\alpha,A_\alpha;G), \qquad H^q(X_\alpha,A_\alpha;G) \to H^q(X_\beta,A_\beta;G)$$

induced by a projection $(X_\beta,A_\beta) \to (X_\alpha,A_\alpha)$ *are independent of the choice of the projection and are therefore uniquely associated with the relation* $\alpha < \beta$.

LEMMA 2.15 *If* $f\colon (X,A) \to (Y,B)$, *and* $\alpha < \beta$ *are coverings of* (Y,B), *then, if* $\alpha' = f^{-1}\alpha$, $\beta' = f^{-1}\beta$, *we have* $\alpha' < \beta'$. *If* $p\colon (Y_\beta,B_\beta) \to (Y_\alpha,B_\alpha)$ *is a projection, then p maps* $(X_{\beta'},A_{\beta'})$ *into* $(X_{\alpha'},A_{\alpha'})$. *If* p' *is the map so defined by p, then* p' *is a projection and commutativity holds in the diagram*

$$\begin{array}{ccc} (X_{\alpha'},A_{\alpha'}) & \xleftarrow{\;p'\;} & (X_{\beta'},A_{\beta'}) \\ \downarrow{\scriptstyle f_\alpha} & & \downarrow{\scriptstyle f_\beta} \\ (Y_\alpha,B_\alpha) & \xleftarrow{\;p\;} & (Y_\beta,B_\beta) \end{array}$$

PROOF. By definition, $V_{\beta'} = V_\beta$ and, for each $v \in V_\beta$, $p(v) = p'(v)$. Therefore $\alpha_{pv} \supset \beta_v$ implies $f^{-1}\alpha_{pv} \supset f^{-1}\beta_v$, i e $\alpha'_{pv} \supset \beta'_v$ Thus, p' is a projection. It follows from now 2 14, that p' maps $(X_{\beta'},A_{\beta'})$ into $(X_{\alpha'},A_{\alpha'})$ Commutativity in the diagram follows from the fact that f_α,f_β are inclusion maps (2.5), and p defines p'. This completes the proof.

Beginning with 2 4 the discussion of this chapter was limited to coverings of pairs. However the definitions and results may be duplicated for coverings of spaces (without a distinguished subset) and to indexed families of sets which are not necessarily coverings.

3. THE ČECH GROUPS

We are now prepared to define the Čech homology and cohomology groups of an arbitrary pair (X,A) over any coefficient group G in $\mathcal{G}_R$. The homology groups of complexes, used in the definition, are those defined in VI,3 9

DEFINITION 3.1 Let Cov(X,A) be the directed set of all open coverings of (X,A) (see 2.10). For each $\alpha \in$ Cov(X,A), let (X_α,A_α) be its nerve (2.2), and let

$$H_{q,\alpha} = H_q(X_\alpha,A_\alpha;G), \qquad H^q_\alpha = H^q(X_\alpha,A_\alpha,G).$$

For each relation $\alpha < \beta$ in $\operatorname{Cov}(X,A)$, let

$$\pi_\alpha^\beta\colon\ H_{q,\beta} \to H_{q,\alpha}, \qquad \pi_\beta^\alpha\colon\ H_\alpha^q \to H_\beta^q$$

be the homomorphisms induced by any projection $(X_\beta,A_\beta) \to (X_\alpha,A_\alpha)$ (see 2 14). The collections $\{H_{q,\alpha},\pi_\alpha^\beta\},\{H_\alpha^q,\pi_\beta^\alpha\}$ are called *the qth Čech homology and cohomology systems of (X,A) over G*

THEOREM 3 2. *The qth Čech homology [cohomology] system of (X,A) over G is an inverse [direct] system of groups defined on the directed set* $\operatorname{Cov}(X,A)$.

PROOF. By 2.11, the identity map of (X_α,A_α) is a projection; hence π_α^α is the identity map of $H_{q,\alpha}$. Again, by 2 11, if $\alpha < \beta < \gamma$, the projection $(X_\gamma,A_\gamma) \to (X_\alpha,A_\alpha)$ can be chosen as the composition of prescribed projections $(X_\gamma,A_\gamma) \to (X_\beta,A_\beta) \to (X_\alpha,A_\alpha)$ From this it follows that $\pi_\alpha^\beta\pi_\beta^\gamma = \pi_\alpha^\gamma$ for homology, and $\pi_\gamma^\beta\pi_\beta^\alpha = \pi_\gamma^\alpha$ for cohomology.

DEFINITION 3.3. The inverse [direct] limit of the qth Čech homology [cohomology] system of (X,A) over G is denoted by $H_q(X,A;G)$ $[H^q(X,A,G)]$ and is called *the qth Čech homology [cohomology] group of (X,A) over G.* The group G may belong to any one of the categories $\mathcal{G}_R$, and by VIII,3 14 and VIII,4 12, the groups $H_q(X,A;G)$ and $H^q(X,A,G)$ are in the same category as G

For $G \in \mathcal{G}_C$ the situation is as follows Each of the cohomology groups $H^q(X_\alpha,A_\alpha,G)$ is in $\mathcal{G}_C$ and the groups form a direct system. Since the limit of a direct system of compact groups is not defined (see Remark VIII,4), the cohomology groups for $G \in \mathcal{G}_C$ are not defined. For homology, the Čech system is an inverse system and the passage to the limit is permissible. However the groups $H_q(X_\alpha,A_\alpha,G)$ themselves are not defined for $G \in \mathcal{G}_C$ since in general the complex X_α is infinite One could try to avoid this difficulty by replacing the directed set $\operatorname{Cov}(X,A)$ by its subset $\operatorname{Cov}^f(X,A)$ consisting only of finite coverings (i e. coverings α with V_α finite) The definitions and results of the preceding section remain valid for the directed set $\operatorname{Cov}^f(X,A)$ so that 3.1-3.3 could be repeated with $\operatorname{Cov}(X,A)$ replaced by $\operatorname{Cov}^f(X,A)$ Of course, the resulting limiting group $H_q^f(X,A;G)$ may not be isomorphic with $H_q(X,A,G)$ (see X,9) However, if the pair (X,A) is compact, then (see below) $\operatorname{Cov}^f(X,A)$ is a cofinal subset of $\operatorname{Cov}(X,A)$ and therefore by VIII,3 16 the limits H_q^f and H_q are isomorphic. Thus, for compact pairs, we may limit our attention to finite coverings, and thereby define the homology groups $H_q(X,A;G)$ with $G \in \mathcal{G}_C$ Then the group $H_q(X,A,G)$ is also in $\mathcal{G}_C$. The situation resembles the one of Chapter VI where the group $H_q(K,L;G)$ for $G \in \mathcal{G}_C$ was defined only when K is finite.

To justify the above discussion we need

LEMMA 3.4. *If the pair (X,A) is compact, then the set* $\mathrm{Cov}'(X,A)$ *consisting of finite coverings is a cofinal subset of* $\mathrm{Cov}(X,A)$.

PROOF. Let $\alpha \in \mathrm{Cov}(X,A)$ be a covering indexed by the pair (V_α,V_α^A). Since both X and A are compact, there exist finite subsets $U \subset V_\alpha$, $W \subset V_\alpha^A$ such that

$$\bigcup_{v \in U} \alpha_v = X, \qquad \bigcup_{v \in W} \alpha_v \supset A.$$

Then the covering α defines a covering β with indexing set $(U \cup W, W)$. Since β is finite and $\alpha < \beta$, the proposition follows.

In the definition of $\mathrm{Cov}(X,A)$, the indexing pair (V_α,V_α^A) for a covering α is any pair of sets Since the set of "all" sets involves one of the usual logical difficulties, the same is true of $\mathrm{Cov}(X,A)$. We shall now show how this difficulty can be avoided.

Let M be any infinite set. Consider the subset $\mathrm{Cov}_M(X,A)$ of $\mathrm{Cov}(X,A)$ consisting of all coverings α with $V_\alpha \subset M$ We now observe that the set $M \times M$ has the same cardinal as M, and therefore, as in 2.10, we can prove that $\mathrm{Cov}_M(X,A)$ is a directed set.

LEMMA 3.5. *Let $\omega(X)$ be the least cardinal such that the space X has a base for open sets of cardinal power $\omega(X)$ If $M \subset N$ and M has power at least $\omega(X)$, then the set* $\mathrm{Cov}_M(X,A)$ *is cofinal in* $\mathrm{Cov}_N(X,A)$

PROOF. Let $\alpha \in \mathrm{Cov}_N(X,A)$, and let β be a family of sets in X, indexed by the set M, and such that β_m, $m \in M$, runs through a base for the open sets of X. Let V_γ (V_γ^A) be the subset of M consisting of all $m \in M$ such that $\beta_m \subset \alpha_v$ for some $v \in V_\alpha$ $(v \in V_\alpha^A)$ Then the indexed family β defines a covering γ with (V_γ,V_γ^A) as indexing pair and $\alpha < \gamma$. Since $\gamma \in \mathrm{Cov}_M(X,A)$, it follows that the latter set is cofinal in $\mathrm{Cov}_N(X,A)$.

The homology and cohomology groups of (X,A) may now be defined as limits of appropriate directed systems defined on $\mathrm{Cov}_M(X,A)$ where M is a set of cardinal power at least $\omega(X)$ This definition seems to depend on the choice of M. However, if N is another such set, then by 3 5 both sets Cov_M and Cov_N are cofinal in the set $\mathrm{Cov}_{M \cup N}$. This yields an isomorphism of the limit groups defined using M and those defined using the set N. It is easy to see that these isomorphisms define a transitive system of groups as defined in I,6. Consequently the limit groups may be regarded as independent of the choice of the set M.

We have carried out the discussion for a single pair (X,A); however the same applies to any admissible category $\mathfrak{a}$ such that the cardinals $\omega(X)$, $X \in \mathfrak{a}$ have an upper bound.

THEOREM 3.4. (*Dimension axiom*). *If P is a single point, then $H_q(P;G) = 0 = H^q(P,G)$ for $q \neq 0$ and $H_0(P,G) \approx G \approx H^0(P;G)$.*

PROOF. Let α be the covering of P consisting of the single set P itself. Then α is a refinement of any covering of P. This means that the set $C' \subset \mathrm{Cov}(P)$ consisting of the single element α is cofinal in $\mathrm{Cov}(P)$. The inverse system $H_{q\alpha}$, $\alpha \in C'$, has but a single term. By the definition of inverse limit (VIII,3.1), the inverse limit of the single term $H_{q\alpha}$ is itself $H_{q\alpha}$. By VIII,3.16, it follows that $H_{q\alpha}$ is isomorphic to $H_q(P;G)$. The nerve P_α consists of a single vertex. The desired result follows now from VI,3.8 (see end of VI,4).

4. INDUCED HOMOMORPHISMS

THEOREM 4.1. *Let* $f\colon (X,A) \to (Y,B)$, *let* $f^{-1}\colon \mathrm{Cov}(Y,B) \to \mathrm{Cov}(X,A)$ *be the associated map of the coverings* (2.1, 2.4), *and, for each* $\alpha \in \mathrm{Cov}(Y,B)$ *let* $f_\alpha\colon (X_{\alpha'},A_{\alpha'}) \to (Y_\alpha,B_\alpha)$ *be the inclusion map of the nerve of* $\alpha' = f^{-1}\alpha$ *into that of* α (see 2.5). *Then the induced homomorphisms*

$$f_{\alpha *}\colon\ H_q(X_{\alpha'},A_{\alpha'};G) \to H_q(Y_\alpha,B_\alpha;G),$$
$$[f^*_\alpha\colon\ H^q(Y_\alpha,B_\alpha;G) \to H^q(X_{\alpha'},A_{\alpha'};G)]$$

for all $\alpha \in \mathrm{Cov}(Y,B)$ *together with* f^{-1} *form a map* $\Phi(f)$ *of the* q^{th} *Čech homology [cohomology] system of* (X,A) $[(Y,B)]$ *over* G *into that of* (Y,B) $[(X,A)]$.

PROOF. By 2.15, f^{-1} is order preserving. Therefore, according to VIII,2.3, $\Phi(f)$ is a map if commutativity holds in the diagram

$$\begin{array}{ccc}
H_q(X_{\alpha'},A_{\alpha'}) & \xleftarrow{\ \pi^{\beta'}_{\alpha'}\ } & H_q(X_{\beta'},A_{\beta'}) \\
\downarrow{\scriptstyle f_{\alpha *}} & & \downarrow{\scriptstyle f_{\beta *}} \\
H_q(Y_\alpha,B_\alpha) & \xleftarrow{\ \pi^{\beta}_{\alpha}\ } & H_q(Y_\beta,B_\beta)
\end{array}$$

where $\alpha < \beta$ in $\mathrm{Cov}(Y,B)$. By 2.15, a projection $p\colon (Y_\beta,B_\beta) \to (Y_\alpha,B_\alpha)$ defines a projection $p'\colon (X_{\beta'},B_{\beta'}) \to (X_{\alpha'},A_{\alpha'})$. Hence $f_\alpha p' = pf_\beta$. Therefore, by VI,3.8, $f_{\alpha *}p'_* = p_* f_{\beta *}$. By 3.1, $\pi^\beta_\alpha = p_*$ and $\pi^{\beta'}_{\alpha'} = p'_*$.

DEFINITION 4.2. The limit of the map $\Phi(f)$ of the qth Čech homology [cohomology] system of (X,A) $[(Y,B)]$ into that (Y,B) $[(X,A)]$ is denoted by

$$f_*\colon\ H_q(X,A;G) \to H_q(Y,B;G), \qquad [f^*\colon\ H^q(Y,B;G) \to H^q(X,A;G)]$$

and is called *the homomorphism induced by* f. The coefficient group G belongs to one of the categories $\mathcal{G}_R$; then f_* and f^* belong to the same category (see VIII,3.14,4.12). If the pair (X,A) is compact, then we replace Cov by Cov′ throughout, and as above derive the definition of f_* (homology only) for G in the category $\mathcal{G}_c$

THEOREM 4.3 (*Axiom* 1). *If* f: $(X,A) \to (X,A)$ *is the identity, then* f_* *and* f^* *are identities*

It is only necessary to observe that $\Phi(f)$ is the identity.

THEOREM 4.4 (*Axiom* 2). *If* f. $(X,A) \to (Y,B)$, *and* g. $(Y,B) \to (Z,C)$, *then* $(gf)_* = g_*f_*$ *and* $(gf)^* = f^*g^*$

PROOF. It is clear that $(gf)^{-1}$. Cov $(Z,C) \to$ Cov(X,A) is the composition of g^{-1} and f^{-1} If $\alpha \in$ Cov(Z,C), $\beta = g^{-1}\alpha$, and $\gamma = f^{-1}\beta = (gf)^{-1}\alpha$, then $g_\alpha f_\beta = (gf)_\alpha$ because all three maps g_α: $(Y_\beta,B_\beta) \to (Z_\alpha,C_\alpha)$, f_β. $(X_\gamma,A_\gamma) \to (Y_\beta,B_\beta)$ and $(gf)_\alpha$ $(X_\gamma,A_\gamma) \to (Z_\alpha,C_\alpha)$ are inclusion maps (2.5) Therefore $\Phi(gf)$ is the composition of $\Phi(g)$ and $\Phi(f)$ (see VIII,2.3). The desired result follows now from the functional properties of the limit homomorphism (VIII,3.14,4.12).

5. THE HOMOTOPY AXIOM

It will be convenient to establish the Homotopy axiom in the alternative form (Axiom 5′) given in I,3.

THEOREM 5.1 (*Homotopy axiom*) *Let* g_0,g_1: $(X,A) \to (X,A) \times I$ *be defined by* $g_0(x) = (x,0)$, $g_1(x) = (x,1)$; *then* $g_{0*} = g_{1*}$ *and* $g_0^* = g_1^*$ *for any coefficient group G for which the appropriate Čech groups are defined*

The proof will be preceded by several definitions and lemmas.

LEMMA 5.2 *If* α *is a finite open covering of I by connected sets, then its nerve* I_α *is acyclic* (*see* VI,5.6)

PROOF. We shall first reduce the general case to the case when no inclusion holds between the sets of the covering. Indeed suppose $\alpha_{v_1} \subset \alpha_{v_2}$ for some two indices $v_1,v_2 \in V_\alpha$ Let β be the covering obtained by removing the element v_1 from V_α. Then $\alpha < \beta$ and $\beta < \alpha$. Let π_1. $I_\beta \to I_\alpha$ and π_2: $I_\alpha \to I_\beta$ be projections. Then $\pi_2\pi_1$: $I_\beta \to I_\beta$ and $\pi_2\pi_1$ $I_\alpha \to I_\alpha$ are projections and therefore induce identity maps on the homology groups It follows that I_α and I_β have isomorphic homology groups. Since the covering β has fewer sets than α, it follows by induction that we can limit our attention to coverings for which no inclusion holds If α is such a covering, an order $v_0 < v_1 < \cdots v_n$ of the elements of V_α can be chosen so that the left end points l_i and the right end points r_i of α_{v_i} satisfy

$$0 = l_0 \leqq l_1 < l_2 < \cdots < l_n, \qquad r_0 < r_1 < \cdots < r_{n-1} \leqq r_n = 1.$$

Consider the simplicial maps f_i: $I_\alpha \to I_\alpha$ defined for $i = 0, \cdots, n$ by

$$f_i(v_j) = v_j \qquad \text{for} \quad j \leqq i,$$
$$f_i(v_j) = v_i \qquad \text{for} \quad j \geqq i.$$

We assert that f_{i+1} and f_i are contiguous. Indeed let s be any simplex of I_α. If all the vertices of s are less than v_{i+1}, then $f_{i+1}(s) = f_i(s)$. If some of the vertices of s are larger than v_i, then $f_i(s)$ has the form $s'v_i$, and $f_{i+1}(s)$ is either of the form $s'v_iv_{i+1}$ or $s'v_{i+1}$ where s' is a simplex with vertices $<v_i$. In the first case $f_i(s)$ is a face of $f_{i+1}(s)$. Thus we may assume that $f_i(s) = s'v_i, f_{i+1}(s) = s'v_{i+1}$. Since all the vertices of s' are $<v_i$, it follows that $\mathrm{Car}_\alpha(s')$ is a connected set with end points l,r satisfying $l \leqq l_i$, $r \leqq r_i$. Since $\mathrm{Car}_\alpha(s') \cap \alpha_{i,i+1} \neq 0$, it follows that we have

$$l \leqq l_i \leqq l_{i+1} < r.$$

Consequently $\mathrm{Car}_\alpha(s') \cap \alpha_i \cap \alpha_{i+1} \neq 0$ so that $s'v_iv_{i+1}$ is a simplex containing both $f_i(s)$ and $f_{i+1}(s)$. Thus f_{i+1} and f_i are contiguous. It follows from VI,3.2 that $f_{i+1*} = f_{i*}$. However f_n is the identity map of I_α while f_0 maps I_α into the single vertex v_0. Since $f_{n*} = f_{0*}$, it follows that I_α is acyclic.

DEFINITION 5.3. A covering α of I indexed by the set $N = (0,1, \cdots, n)$ $(n \geqq 0)$ is called *regular* if the sets α_i are open and connected, and if

$$0 \in \alpha_0, \qquad 0 \text{ non } \in \alpha_1, \qquad 1 \in \alpha_n, \qquad 1 \text{ non } \in \alpha_{n-1},$$
$$\alpha_i \cap \alpha_{i+1} \neq 0 \quad \text{for} \quad i = 0, \cdots, n-1,$$
$$\alpha_i \cap \alpha_j = 0 \quad \text{for} \quad i < j - 1.$$

LEMMA 5.4. *The regular coverings form a cofinal subset of* $\mathrm{Cov}(I)$.

PROOF. Let α be any open covering of I. Consider the family Φ of components of the sets of α. Then Φ is a family of connected open sets covering I, and by compactness Φ contains a finite subfamily Φ_0 covering I. In Φ_0 select a minimal subset Φ_1 such that no proper subset of Φ_1 covers I. Then Φ_1 after suitable indexing constitutes a regular covering which is a refinement of α.

DEFINITION 5.5. Let $\alpha \in \mathrm{Cov}(X,A)$ be a covering indexed by the pair (V_α, V_α^A). Suppose to each $v \in V_\alpha$ there corresponds a regular covering β^v of I indexed by $N^v = (0,1, \cdots, n^v)$. Consider the set W of all pairs (v,i), $v \in V_\alpha$, $i \in N^v$ and let W' be the subset consisting of pairs (v,i) with $v \in V_\alpha^A$. The covering $\gamma \in \mathrm{Cov}(X \times I, A \times I)$ indexed by (W,W') and defined by

$$\gamma_{v,i} = \alpha_v \times \beta_i^v$$

is called a *stacked covering over* α. The coverings β^v are called the *stacks* of γ.

LEMMA 5.6. *Stacked coverings form a cofinal subset of* $\mathrm{Cov}(X \times I, A \times I)$.

PROOF. Let $\delta \in \text{Cov}(X \times I, A \times I)$ be indexed by (V_δ, V_δ^A). For each (x,t) select open sets $U(x,t) \subset X$, $V(x,t) \subset I$ such that $U(x,t) \times V(x,t)$ contains (x,t) and is contained in one of the sets δ_v, and if $x \in A$, then $U(x,t) \times V(x,t)$ is contained in one of the sets δ_v with $v \in V_\delta^A$. For each fixed $x \in X$, the sets $V(x,t)$ constitute a covering of I, which therefore by 5.3 has a regular refinement β^x. For each set β_i^x there is an open set $U(x,i)$ in X such that $U(x,i) \times \beta_i^x$ is in one of the sets of δ_v, and if $x \in A$, then $U(x,i) \times \beta_i^x$ is in one of the sets δ_v with $v \in V_\delta^A$. Define $\alpha_x = \bigcap_i U(x,i)$. Then α is a covering of (X,A) indexed by (X,A), and the covering γ stacked over α with β^x as stacks is a refinement of δ.

LEMMA 5.7. *Let γ be a stacked covering over $\alpha \in \text{Cov}(X)$. If the nerve X_α is a (finite) simplex, then the nerve $(X \times I)_\gamma$ is acyclic.*

PROOF. Without loss of generality we may assume that none of the sets of the covering of α is empty. Let V and W be the indexing sets of α and γ, and let β^v be the stacks. Define a covering δ of I indexed by W as follows:

$$\delta_{v,i} = \beta_i^v.$$

Let s be any simplex with vertices $(v_0,i_0), \cdots, (v_n,i_n)$ in W. Then

$$\begin{aligned}\text{Car}_\gamma(s) &= \bigcap_j \gamma_{(v_j,i_j)} = \bigcap_j \alpha_{v_j} \times \bigcap_j \beta_{i_j}^{v_j} \\ &= \bigcap_j \alpha_{v_j} \times \bigcap_j \delta_{v_j,i_j} = \bigcap_j \alpha_{v_j} \times \text{Car}_\delta(s).\end{aligned}$$

Since X_α is a simplex, the set $\bigcap_j \alpha_{v_j}$ is not empty. It follows that $\text{Car}_\gamma(s)$ is empty if and only if $\text{Car}_\delta(s)$ is empty. Thus $(X \times I)_\gamma = I_\delta$. Since I_δ is acyclic by 5.2, it follows that $(X \times I)_\gamma$ is acyclic.

We shall abbreviate $(X \times I)_\gamma$ by $X \times I_\gamma$.

LEMMA 5.8. *Let γ be a covering stacked over the covering α of (X,A) indexed by (V_α, V_α^A). Consider the simplicial maps*

$$l,u.\quad (X_\alpha, A_\alpha) \to (X \times I_\gamma, A \times I_\gamma)$$

defined for $v \in V_\alpha$ by

$$l(v) = (v,0), \qquad u(v) = (v,n^v).$$

Then $l_* = u_*$ and $l^* = u^*$.

PROOF. For each simplex s of X_α consider the subcomplex $C(s)$ of $X \times I_\gamma$ consisting of all simplexes whose vertices have the form (v,i) where v is a vertex of s. The simplex s is the nerve of a covering α' of a subset X' of X and $C(s)$ is the nerve of a covering γ' stacked over α'. Thus by 5.7, $C(s)$ is acyclic. If s' is a face of s, then $C(s') \subset C(s)$, and if s is in A_α, then $C(s)$ is in $A \times I_\gamma$. If c is a chain in the simplex s, then $l(c)$ and $u(c)$ are chains in $C(s)$. Thus $C(s)$ is a common acyclic carrier for the maps l and u and by VI,5.8 we have $l_* = u_*$ and $l^* = u^*$.

PROOF OF 5.1. Let D be the subset of $\mathrm{Cov}(X \times I, A \times I)$ consisting of stacked coverings Since D is cofinal, we can use the set D to define the Čech groups of $(X \times I, A \times I)$. Let $\gamma \varepsilon D$ be stacked over α. Consider the coverings $\gamma_0 = g_0^{-1}\gamma$, $\gamma_1 = g_1^{-1}\gamma$ of (X,A) and the inclusion maps

$$g_{0\gamma}\colon (X_{\gamma_0},A_{\gamma_0}) \to (X \times I_\gamma, A \times I_\gamma),$$
$$g_{1\gamma}\colon (X_{\gamma_1},A_{\gamma_1}) \to (X \times I_\gamma, A \times I_\gamma).$$

The map u defined in 5 8 may be factored into $u = g_{1\gamma}u'$ where u': $(X_\alpha,A_\alpha) \to (X_{\gamma_1},A_{\gamma_1})$ is an isomorphism defined by $u'(v) = (v,n^v)$.

Observe that $\gamma_0 < \gamma_1$ and that the map

$$\pi\colon (X_{\gamma_1},A_{\gamma_1}) \to (X_{\gamma_0},A_{\gamma_0})$$

defined by $\pi(v,i) = (v,0)$ is a projection. Further observe that $l = g_{0\gamma}\pi u'$ It follows that

$$l_* = g_{0\gamma*}\pi_* u'_*, \qquad u_* = g_{1\gamma*}u'_*.$$

Since by 5 8, $l_* = u_*$, and since u'_* is an isomorphism, it follows that

$$g_{1\gamma*} = g_{0\gamma*}\pi_*.$$

Thus by VIII,3 18 we have $g_{0*} = g_{1*}$. For cohomology we prove similarly $g_{1\gamma}^* = \pi^* g_{0\gamma}^*$ and apply VIII,4 15.

6. INVARIANCE UNDER EXCISION

THEOREM 6 1 (*Excision axiom*) *If U is open in X and its closure $\bar{U}$ is contained in the interior of $A \subset X$, then the inclusion map $f\colon (X - U, A - U) \to (X,A)$ induces isomorphisms*

$$f_*\colon H_q(X - U, A - U) \approx H_q(X,A),$$
$$f^*\colon H^q(X,A) \approx H^q(X - U, A - U),$$

for any coefficient group G for which the respective Čech groups are defined.

PROOF. For notational brevity let $X' = X - U$, $A' = A - U$. Let D be the subset of $\mathrm{Cov}(X,A)$ consisting of all coverings α indexed by (V_α,V_α^A) such that

(1) if $\alpha_v \cap U \neq 0$, then $v \varepsilon V_\alpha^A$ and $\alpha_v \subset A$.

By VIII,3.15, the conclusion of the theorem is a consequence of the following three propositions.

(2) D is cofinal in $\mathrm{Cov}(X,A)$.

(3) $f^{-1}(D)$ is cofinal in $\mathrm{Cov}(X',A')$.

(4) If $\alpha \in D$ and $\beta = f^{-1}\alpha$, then

$$f_{\alpha *}: \; H_q(X'_\beta, A'_\beta) \approx H_q(X_\alpha, A_\alpha), \qquad f^*_\alpha: \; H^q(X_\alpha, A_\alpha) \approx H^q(X'_\beta, A'_\beta).$$

To prove (2) consider any covering α of (X,A) with indexing pair (V_α, V^A_α). Let V' be a set disjoint with V_α and in a 1-1 correspondence with V^A_α. For each $v \in V^A_\alpha$, the corresponding element of V' will be denoted by v' Consider the covering γ of (X,A) indexed by $(V_\alpha \cup V', V^A_\alpha \cup V')$ and defined as follows

$$\begin{aligned} \gamma_v &= \alpha_v - \overline{U} && \text{for} \quad v \in V_\alpha, \\ \gamma_{v'} &= \alpha_v \cap \operatorname{Int} A && \text{for} \quad v' \in V'. \end{aligned}$$

Since $\overline{U} \subset \operatorname{Int} A$, it follows that γ is a covering of (X,A) Clearly $\alpha < \gamma$ and $\gamma \in D$

To prove (3) consider any covering β of (X',A') indexed by $(V_\beta, V^{A'}_\beta)$ Define $\alpha \in \operatorname{Cov}(X,A)$ indexed by the same pair $(V_\beta, V^{A'}_\beta)$ as follows:

$$\alpha_v = \beta_v \cup U$$

Then $\beta = f^{-1}\alpha$. Choose $\gamma \in D$ so that $\alpha < \gamma$. Then $\beta = f^{-1}\alpha < f^{-1}\gamma$, so that $f^{-1}D$ is cofinal in $\operatorname{Cov}(X',A')$.

To prove (4), it suffices, in view of VI,3 5, 3.6, to prove

$$X_\alpha = X'_\beta \cup A_\alpha, \qquad A'_\beta = X'_\beta \cap A_\alpha. \tag{5}$$

Since (X'_β, A'_β) is a subcomplex of (X_α, A_α) (see 2 5), we have the inclusions

$$X_\alpha \supset X'_\beta \cup A_\alpha, \qquad A'_\beta \subset X'_\beta \cap A_\alpha.$$

It thus remains to prove

$$X_\alpha \subset X'_\beta \cup A_\alpha, \qquad A'_\beta \supset X'_\beta \cap A_\alpha. \tag{6}$$

Let s be a simplex of X_α which is not in X'_β Then $\operatorname{Car}_\alpha(s) \neq 0$ and $\operatorname{Car}_\alpha(s) \cap X' = \operatorname{Car}_\beta(s) = 0$. Consequently $0 \neq \operatorname{Car}(s) \subset U$. This implies that, for every vertex v of s, we have $\alpha_v \cap U \neq 0$, and therefore, since $\alpha \in D$, that $v \in V^A_\alpha$ Since $U \subset A$, it follows that $\operatorname{Car}_\alpha(s) \cap A \neq 0$ so that s is a simplex of A_α. This proves the first part of (6).

Let s be a simplex of $X'_\beta \cap A_\alpha$ It follows that the vertices of s are in V^A_α and that

$$\operatorname{Car}_\alpha(s) \cap X' = \operatorname{Car}_\beta(s) \neq 0, \qquad \operatorname{Car}_\alpha(s) \cap A \neq 0.$$

If $\operatorname{Car}_\alpha(s) \subset X'$, then

$$\operatorname{Car}_\beta(s) \cap A' = \operatorname{Car}_\alpha(s) \cap X' \cap A = \operatorname{Car}_\alpha(s) \cap A \neq 0$$

and s is in A'_β. If $\mathrm{Car}_\alpha(s) \cap U \neq 0$, then, since α is in D, it follows that $\alpha_v \subset A$ for every vertex v of s. Thus $\mathrm{Car}_\alpha(s) \subset A$ and

$$\mathrm{Car}_\beta(s) \cap A' = \mathrm{Car}_\alpha(s) \cap A \cap X' = \mathrm{Car}_\alpha(s) \cap X' \neq 0,$$

so that s is in A'_β. This concludes the proof of 6.1.

7. THE BOUNDARY OPERATOR AND EXACTNESS

The Čech homology and cohomology groups of (X,A), A, and X are defined as limits of suitable systems of groups defined over the directed sets $\mathrm{Cov}(X,A)$, $\mathrm{Cov}(A,0)$, and $\mathrm{Cov}(X,0)$ respectively. In order to define the boundary operator and discuss exactness, it will be convenient to have equivalent definitions in which all these systems are defined over the same directed set. It appears that the directed set $\mathrm{Cov}(X,A)$ is most suitable for this purpose.

DEFINITION 7.1. If $\alpha \in \mathrm{Cov}(X,A)$, let S_α $[S^\alpha]$ be the homology [cohomology] sequence of (X_α,A_α) over G. If $\alpha < \beta$ in $\mathrm{Cov}(X,A)$, let $\pi^\beta_\alpha\colon S_\beta \to S_\alpha$ $[\pi^\beta_\alpha\colon S^\alpha \to S^\beta]$ be a map induced by a projection $(X_\beta,A_\beta) \to (X_\alpha,A_\alpha)$. The resulting limit sequence is called the *adjusted homology* [*cohomology*] *sequence of* (X,A). The groups and homomorphisms of the adjusted sequences are written

$$\cdots \leftarrow H_q(X,A) \xleftarrow{j'_*} H_q(X)_{(X,A)} \xleftarrow{i'_*} H_q(A)_{(X,A)} \xleftarrow{\partial'} H_{q+1}(X,A) \leftarrow \cdots$$

$$\cdots \to H^q(X,A) \xrightarrow{j'^*} H^q(X)_{(X,A)} \xrightarrow{i'^*} H^q(A)_{(X,A)} \xrightarrow{\delta'} H^{q+1}(X,A) \to \cdots$$

To compare the groups with the subscript (X,A) with the groups without this subscript, we introduce two maps,

$$\phi\colon \mathrm{Cov}(X,A) \to \mathrm{Cov}(A,0),$$
$$\psi\colon \mathrm{Cov}(X,A) \to \mathrm{Cov}(X,0).$$

Let $\alpha \in \mathrm{Cov}(X,A)$ be indexed by the pair (V_α,V^A_α). Then $\phi\alpha$ is indexed by $(V^A_\alpha,0)$ and $(\phi\alpha)_v = A \cap \alpha_v$ for $v \in V^A_\alpha$. The covering $\psi\alpha$ is indexed by $(V_\alpha,0)$ and satisfies $(\psi\alpha)_v = \alpha_v$. Observe that

$$A_\alpha = A_{\phi\alpha}, \qquad X_\alpha = X_{\psi\alpha}.$$

The maps ϕ and ψ and the appropriate identity maps of the homology groups yield maps of inverse systems

$$\Phi\colon \{H_q(A_\alpha;G),\pi^\alpha_\beta\}_{(A,0)} \to \{H_q(A_\alpha;G),\pi^\alpha_\beta\}_{(X,A)}$$
$$\Psi\colon \{H_q(X_\alpha;G),\pi^\alpha_\beta\}_{(X,0)} \to \{H_q(X_\alpha;G),\pi^\alpha_\beta\}_{(X,A)}$$

where the subscript indicates the directed set on which these inverse systems are defined The inverse limits of the maps Φ and Ψ are homomorphisms

$$\phi_\infty\colon\ H_q(A;G) \to H_q(A;G)_{(X,A)}, \qquad \psi_\infty\colon\ H_q(X;G) \to H_q(X;G)_{(X,A)}.$$

For cohomology, Φ and Ψ are maps of direct systems and their limits are

$$\phi^\infty\colon\ H^q(A;G)_{(X,A)} \to H^q(A,G), \qquad \psi^\infty\colon\ H^q(X;G)_{(X,A)} \to H^q(X;G).$$

LEMMA 7 2. *The homomorphisms ϕ_∞, ψ_∞, ϕ^∞, and ψ^∞ are isomorphisms.*

PROOF. In view of VIII,3.15 and VIII,4.13, it suffices to show that the image of ϕ is a cofinal subset of Cov(A,0) and that the image of ψ is a cofinal subset of Cov(X,0). Let $\alpha \in$ Cov(A,0) be indexed by the pair (V,W) Let V^+ be a set consisting of V and a single element v_0 not in V. For each $v \in V$ select an open set β_v in X so that $\alpha_v = A \cap \beta_v$ Define $\beta_{v_0} = X$. Then $\beta \in$ Cov(X,A) and has (V^+,V) as indexing set. The covering $\phi\beta$ agrees with α but has $(V,0)$ as indexing set, thus $\alpha < \phi\beta$ and the image of ϕ is cofinal

Let $\alpha \in$ Cov(X,0) be indexed by (V,W) Then α defines a covering $\beta \in$ Cov(X,A) indexed by (V,V) The covering $\psi\beta$ then agrees with α but is indexed by $(V,0)$. Thus $\alpha < \psi\beta$ and the image of ψ is cofinal

DEFINITION 7.3 The homomorphisms

$$\partial\colon\ H_q(X,A;G) \to H_{q-1}(A;G), \qquad \delta\ \ H^q(A,G) \to H^{q+1}(X,A,G)$$

are defined from the diagrams

$$H_q(X,A;G) \xrightarrow{\partial'} H_{q-1}(A;G)_{(X,A)} \xleftarrow{\phi_\infty} H_{q-1}(A,G)$$

$$H^q(A;G) \xleftarrow{\phi_\infty} H^q(A;G)_{(X,A)} \xrightarrow{\delta'} H^{q+1}(X,A,G)$$

as $\partial = \phi_\infty^{-1}\partial'$ and $\delta = \delta(\phi^\infty)^{-1}$.

THEOREM 7 4. (*Axiom* 3) *Let* $f.\ (X,A) \to (Y,B)$. *Then* $(f|A)_*\partial = \partial f_*$ *and* $\delta(f|A)^* = f^*\delta$

PROOF. The formula for homology is an immediate consequence of the commutativity relations in the diagram

$$\begin{array}{ccccc} H_q(X,A;G) & \xrightarrow{\partial'} & H_{q-1}(A,G)_{(X,A)} & \xleftarrow{\phi_\infty} & H_{q-1}(A;G) \\ \downarrow f_* & & \downarrow (f|A)'_* & & \downarrow (f|A)_* \\ H_q(Y,B;G) & \xrightarrow{\partial'} & H_{q-1}(B,G)_{(Y,B)} & \xleftarrow{\phi_\infty} & H_{q-1}(B;G) \end{array}$$

where $(f|A)'_*$ is defined as the appropriate limit map. The commutativity relations in each square are immediate consequences of the definitions. The proof for cohomology is similar.

Theorem 7.5. *The homology [cohomology] sequence of a pair (X,A) over any coefficient group G is isomorphic with the adjusted homology [cohomology] sequence. The isomorphism is given by the maps ϕ_∞,ψ_∞ [ϕ^∞,ψ^∞] and the identity map of $H_q(X,A,G)$ [$H^q(X,A,G)$] onto itself.*

Proof. We shall only carry out the proof for homology, the proof for cohomology is quite similar. In view of 7.2 we need only verify commutativity relations in the diagram

$$\begin{array}{ccccccc} & & H_q(A) & \xrightarrow{i_*} & H_q(X) & & \\ & \overset{\partial}{\nearrow} & \downarrow \phi_\infty & & \downarrow \psi_\infty & \overset{j_*}{\searrow} & \\ H_{q+1}(X,A) & & & & & & H_q(X,A) \\ & \underset{\partial'}{\searrow} & & & & \underset{j'_*}{\nearrow} & \\ & & H_q(A)_{(X,A)} & \xrightarrow{i'_*} & H_q(X)_{(X,A)} & & \end{array}$$

Commutativity in the left-hand triangle is a direct consequence of the definition of ∂. To prove commutativity in the middle square, we discuss the maps $\phi_\infty i'_*$ and $i_*\psi_\infty$ in greater detail. These maps are limits of maps

$$\tau,\eta\colon\ \{H_q(A_\alpha),\pi^\alpha_\beta\}_{(A,0)} \to \{H_q(X_\alpha),\pi^\alpha_\beta\}_{(X,A)}$$

where the subscript is used to indicate the directed set over which the inverse systems are defined. The map τ carries each covering $\alpha \in \text{Cov}(X,A)$ indexed by (V_α,V^A_α) into the covering $\tau\alpha = \phi\alpha \in \text{Cov}(A,0)$ indexed by the set $(V^A_\alpha,0)$ and satisfying $(\tau\alpha)_v = A \cap \alpha_v$. Further τ_α is the map $H_q(A_{\tau\alpha},G) \to H_q(X_\alpha,G)$ induced by the inclusion $A_{\tau\alpha} \subset X_\alpha$. The map η assigns to each covering $\alpha \in \text{Cov}(X,A)$ indexed by (V_α,V^A_α) the covering $\eta\alpha = i^{-1}\psi\alpha \in \text{Cov}(A,0)$ indexed by $(V,0)$ and satisfying $(\eta\alpha)_v = A \cap \alpha_v$. Again η_α is the map $H_q(A_{\eta\alpha};G) \to H_q(X_\alpha;G)$ induced by inclusion $A_{\eta\alpha} \subset X_\alpha$. Clearly $\eta\alpha < \tau\alpha$; furthermore the inclusion map $A_{\tau\alpha} \subset A_{\eta\alpha}$ is a projection. It follows that $\pi^{\tau\alpha}_{\eta\alpha}$ is the map $H_q(A_{\tau\alpha},G) \to H_q(A_{\eta\alpha};G)$ induced by inclusion. Thus we have

$$\eta\alpha < \tau\alpha, \qquad \tau_\alpha = \eta_\alpha\pi^{\tau\alpha}_{\eta\alpha}.$$

Thus by VIII,3.18 we have $\tau_\infty = \eta_\infty$, i.e. $\phi_\infty i'_* = i^*\psi_\infty$.

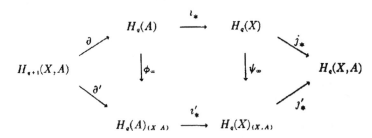

The situation in the right-hand triangle is similar. The maps $j'_*\psi_\infty$ and j_* are limits of maps

$$\tau,\eta: \quad \{H_q(X_\alpha;G),\pi^\alpha_\beta\}_{(X,0)} \rightarrow \{H_q(X_\alpha,A_\alpha,G),\pi^\alpha_\beta\}_{(X,A)}.$$

For each $\alpha \in \mathrm{Cov}(X,A)$ indexed by (V_α,V^A_α) the coverings $\tau\alpha$, $\eta\alpha \in \mathrm{Cov}(X,0)$ coincide with α but are indexed by $(V_\alpha,0)$ and (V_α,V^A_α) respectively. We have $X_{\tau\alpha} = X_{\eta\alpha} = X_\alpha$ and both τ_α and η_α are the map $H_q(X_\alpha,G) \rightarrow H_q(X_\alpha,A_\alpha;G)$ induced by inclusion Again we have $\eta\alpha < \tau\alpha$ and the identity map $X_\alpha \rightarrow X_\alpha$ is a projection of $\tau\alpha$ into $\eta\alpha$. Thus again

$$\eta\alpha < \tau\alpha, \qquad \tau_\alpha = \eta_\alpha\pi^{\tau\alpha}_{\eta\alpha}$$

so that $\tau_\infty = \eta_\infty$.

With the proof of 7.5 concluded, the question of the exactness of the homology and cohomology sequence is replaced by the question of the exactness of the adjusted sequences The adjusted sequences are however limits of systems of exact sequences defined over the directed set $\mathrm{Cov}(X,A)$. Thus the results of VIII,5 may be applied

THEOREM 7 6 (*Exactness axiom*) *For any pair* (X,A) *and any* G *in a category* $\mathcal{G}_R$, *the cohomology sequence is exact while the homology sequence is a sequence of order* 2 (*see* VIII,5 2). *If* (X,A) *is compact and* G *is either in* $\mathcal{G}_C$ *or in* $\mathcal{G}_F$ (*the category of vector spaces over a field* F), *then the homology sequence is also exact*

PROOF. The first part of the theorem is a consequence of VIII,5 3 and VIII,5.4. If (X,A) is compact, then, in defining the groups occurring in the homology sequence, we may limit our attention to finite coverings. If G is compact, then, for each finite covering α, the homology sequence of (X_α,A_α) over G is composed of compact groups and therefore by VIII,5 6 the limit sequence is exact. If G is in $\mathcal{G}_F$ and is finite dimensional (over F), then the groups of the homology sequence of (X_α,A_α) over F are all finite dimensional and the exactness of the limit follows from VIII,5 7 If $G \in \mathcal{G}_F$ is not finite dimensional, then G may be represented as a direct sum $\sum G_\beta$ where each G_β is finite dimensional. This decomposition yields a decomposition of the homology sequence of (X,A) over G into a direct sum of the homology sequences over G_β Since each of these is exact, it follows that the homology sequence over G is exact

It will be shown later (9.4) that the homology sequence of a triangulable pair is exact without restriction on the coefficient group In X,4 we shall construct a compact pair for which the homology sequence with integer coefficients is not exact.

8. CLOSED SUBSETS

It will be shown here that the reason for using coverings $\alpha \in \mathrm{Cov}(X,A)$ indexed by pairs (V_α,V_α^A) in the definition of $H_e(X,A)$ is due to the fact that A was not assumed to be closed. If A is closed, we can use the set $\mathrm{Cov}(X)$ consisting of coverings α indexed by a single set V_α.

DEFINITION 8.1. A covering $\alpha \in \mathrm{Cov}(X,A)$ indexed by (V_α,V_α^A) is called *proper* if V_α^A is the set of all $v \in V$ with $\alpha_v \cap A \neq 0$.

LEMMA 8.2. *For each covering $\alpha \in \mathrm{Cov}(X)$ indexed by a set V_α, consider the covering $\rho\alpha \in \mathrm{Cov}(X,A)$ defined by α and indexed by (V_α,V') where V' is the set of all $v \in V_\alpha$ with $\alpha_v \cap A \neq 0$. The map ρ: $\mathrm{Cov}(X) \to \mathrm{Cov}(X,A)$ is then a 1-1 order preserving correspondence between $\mathrm{Cov}(X)$ and the set of all proper coverings in $\mathrm{Cov}(X,A)$.*

The proof is obvious.

LEMMA 8.3. *If A is a closed subset of X, then the proper coverings form a cofinal subset of $\mathrm{Cov}(X,A)$. If X is a T_1-space, then the converse is also true.*

PROOF. Let $\alpha \in \mathrm{Cov}(X,A)$ be a covering indexed by (V_α,V_α^A). Let V' be the set of all $v \in V_\alpha^A$ with $\alpha_v \cap A \neq 0$ and consider the covering $\beta \in \mathrm{Cov}(X,A)$ indexed by (V_α,V') and defined by

$$\beta_v = \alpha_v - A \qquad \text{for} \qquad v \in V_\alpha - V',$$
$$\beta_v = \alpha_v \qquad \text{for} \qquad v \in V'.$$

Then β is proper and $\alpha < \beta$.

Suppose now that X is a T_1-space and A is not closed. Let $x \in \bar{A} - A$. Consider the covering α consisting of the sets $\alpha_1 = X$, $\alpha_2 = X - (x)$ and indexed by $V_\alpha = (1,2)$, $V_\alpha^A = (2)$. Let $\beta \in \mathrm{Cov}(X,A)$ be any refinement of α. Then for $v \in V_\beta^A$ we must have $\beta_v \subset X - (x)$. Thus $x \in \beta_v$ for some $v \in V_\beta - V_\beta^A$. Since β_v is open, it follows that $\beta_v \cap A \neq 0$ so that β is not a proper covering. Thus the proper coverings are not cofinal.

These lemmas show that, if A is closed in X, then, in defining the adjusted homology sequence of (X,A), we may replace the set $\mathrm{Cov}(X,A)$ on which the direct and inverse systems are defined by the subset consisting only of proper coverings. By 8.2, this set may in turn be replace by the set $\mathrm{Cov}(X)$ of coverings of X. For each covering $\alpha \in \mathrm{Cov}(X)$ we then consider the homology [cohomology] sequence of the pair (X_α,A_α) where A_α consists of those simplexes of X_α whose carriers meet A. This is precisely the nerve of the covering $\rho\alpha \in \mathrm{Cov}(X,A)$. The groups of the thus modified adjusted homology sequence will be indicated by the subscript X.

If the pair (X,A) is compact, we can, in addition to the above changes, restrict ourselves to finite coverings Thus the homology [cohomology] sequence can be defined as a limit taken over the directed set $\mathrm{Cov}'(X)$ of finite coverings of X.

9. ČECH GROUPS OF TRIANGULABLE SPACES

DEFINITION 9 1 Let (X,A) be a pair with a triangulation $T = \{t,(K,L)\}$ For every vertex A of K consider the open star $\mathrm{st}(A)$ of A in K as defined in II,3 6 and define

$$\tau_A = t(\mathrm{st}(A)).$$

There results a covering τ of (X,A) with $V_\tau = \{A\} = |K^0|$, $V_\tau^A = |L^0|$. We shall say that τ is the covering associated with the triangulation T.

LEMMA 9 2. *The nerve of (X,A) in the covering τ is (K,L), i e.* $X_\tau = K$ and $A_\tau = L$

PROOF The vertices of K and of X_τ are the same; thus K and X_τ are subcomplexes of the same simplex Let $A^0, \cdots, A^n$ be distinct vertices of K Then, since t is a homeomorphism,

$$\bigcap_{i=0}^{n} \tau_{A^i} = \bigcap_{i=0}^{n} t(\mathrm{st}(A^i)) = t\Big(\bigcap_{i=0}^{n} \mathrm{st}(A^i)\Big). \tag{1}$$

It follows that $\bigcap \tau_{A^i} \neq 0$ if and only if $\bigcap \mathrm{st}(A^i) \neq 0$, and, by II,3.7, this holds if and only if $A^0, \cdots, A^n$ are vertices of a simplex s of K. Thus $X_\tau = K$ From (1) we deduce that

$$\Big(\bigcap_{i=0}^{n} \tau_{A^i}\Big) \cap A = t\Big(\bigcap_{i=0}^{n} |L| \cap \mathrm{st}(A^i)\Big)$$

so that the simplex s with vertices $A^0, \cdots, A^n$ is in A_τ if and only if $A^0, \cdots, A^n$ are vertices of L, and

$$\bigcap_{i=0}^{n} |L| \cap \mathrm{st}(A^i) \neq 0. \tag{2}$$

Now it follows directly from the definition of $\mathrm{st}(A^i)$

$$|L| \cap \mathrm{st}(A^i) = \mathrm{st}_L(A^i)$$

where $\mathrm{st}_L(A^i)$ is the open star of the vertex A^i constructed relative to the complex L Thus, by II,3.7, condition (2) is equivalent to s being a simplex of L Thus $A_\tau = L$

THEOREM 9 3 *Let $T = \{t,(K,L)\}$ be a triangulation of the pair (X,A). The homology [cohomology] sequence $S_{(K,L)}$ of (K,L) (in the simplicial theory) and the adjusted homology [cohomology] sequence S of (X,A) (in the Čech theory) are isomorphic The isomorphism is obtained* *by taking the covering τ associated with T and regarding $S_{(K,L)}$ as the homology [cohomology] sequence of the nerve of τ. This yields maps*

$$\pi_\tau: S \to S_\tau, \qquad [\pi^\tau: S^\tau \to S]$$

which are isomorphisms.

This result is valid for all coefficient groups G for which the appropriate Čech groups are defined

COROLLARY 9.4. *The Čech homology sequence of a triangulable pair (X,A) is exact for any coefficient group G.*

The proof of 9.3 will be preceded by a definition and two lemmas.

DEFINITION 9.5. The *mesh* of a covering α of a metric space X is the maximum of the diameters of the sets α.

LEMMA 9.6 *A subset C of the directed set $\mathrm{Cov}(X)$ of all open coverings of a compact metric space X is cofinal if and only if for every $\epsilon > 0$ the set C contains a covering of mesh $< \epsilon$*

PROOF The condition is clearly necessary. To prove that it also is sufficient, consider a covering α of X. Let ϵ be the Lebesgue number of this covering (see II,8.5) and let $\beta \in C$ be a covering with mesh $\beta < \epsilon$. Then $\alpha < \beta$ and C is cofinal

LEMMA 9.7 *Let $T' = \{t',(K',L')\}$ be the barycentric subdivision of the triangulation $T = \{t,(K,L)\}$ of (X,A) Then the covering τ' associated with T' is a refinement of the covering τ associated with T, and the projections*

$$\pi_\tau^{\tau'}: S_{\tau'} \approx S_\tau, \qquad [\pi_\tau^{\tau'}: S^\tau \approx S^{\tau'}]$$

are isomorphisms

PROOF We recall that $(K',L') = (\mathrm{Sd}\, K, \mathrm{Sd}\, L)$ Let $l: (K',L') \to (K,L)$ be the linear map connected with this subdivision. Then $t' = tl$. Any vertex B of K' is the barycenter of some simplex s of K Let $A^0, \ldots, A^n$ be the vertices of s Then, from the definition of l and the open stars, it follows that

$$l(\mathrm{st}(B)) = \bigcap_{i=0}^{n} \mathrm{st}(A^i).$$

Consequently

$$(1) \qquad \tau'_B = t'\mathrm{st}(B) = tl\mathrm{st}(B) = \bigcap_{i=0}^{n} t\mathrm{st}(A^i) = \bigcap_{i=0}^{n} \tau_{A^i}.$$

This implies that τ' is a refinement of τ

Let $\pi: (K',L') \to (K,L)$ be a projection as defined in VI,7,6, i.e. π is a simplicial map which to every vertex B of K' assigns a vertex of the simplex of K whose barycenter is B. Hence (1) implies that π

also is a projection of the covering τ' into τ. Thus the last part of the lemma is a consequence of VI,7.2.

PROOF OF 9.3. Let ${}^nT = \{{}^nt,({}^nK,{}^nL)\}$ be the nth barycentric subdivision of the triangulation T, and let $\{{}^n\tau\}$ be the corresponding sequence of coverings of X. It follows from II,7.5 that

$$\lim_{n\to\infty} \text{mesh } {}^n\tau = 0.$$

Thus, if we regard $\{{}^n\tau\}$ as coverings of X, they form a cofinal subset of Cov(X). Consequently, if we regard $C = \{{}^n\tau\}$ as coverings of (X,A), they will form a cofinal subset of the set R of proper coverings of (X,A). By 8 3, the set R is cofinal in Cov(X,A) Thus C is cofinal in Cov(X,A) Let S' be the inverse [direct] limit of the system $\{S_\tau,\pi_\alpha^\beta\}$ $[\{S^\tau,\pi_\alpha^\beta\}]$ defined over the directed set C. The homomorphism π_τ $[\pi^\tau]$ can then be represented as the composition

$$S \xrightarrow{\psi_\infty} S' \xrightarrow{\pi'_\tau} S_\tau \qquad [S \xleftarrow{\psi_\infty} S' \xleftarrow{\pi'^\tau} S^\tau]$$

where ψ_∞ $[\psi^\infty]$ is the limit map of the injection of the system defined over Cov(X,A) into the subsystem defined over C. Since C is cofinal in Cov(X,A), it follows that ψ_∞ $[\psi^\infty]$ is an isomorphism (VIII,3 16, 4 14) In the system over C, all the maps are isomorphisms by 9.8 Thus VIII,3 4 implies that π'_τ. $S' \to S_\tau$ $[\pi'^\tau$. $S^\tau \to S']$ is an isomorphism This shows that the map in 9 3 is an isomorphism

10. PARTIALLY EXACT HOMOLOGY AND COHOMOLOGY THEORIES

As we have seen, the Čech homology groups satisfy all the axioms except for the Exactness axiom, which holds only in very special circumstances. Nevertheless the Čech groups are very useful (particularly for compact pairs) because of various features that will be discussed in the following chapters. In later chapters we shall encounter systems other than the Čech homology theory which fail to satisfy the Exactness axiom in its present form. It is therefore useful to generalize the concept of a homology theory so as to include the Čech theory as well as the other theories that we shall encounter

We begin by modifying the axioms for an admissible category $\mathfrak{A}$ (see I,1) by replacing condition (5) by the following

(5′) If (X,A) is triangulable, then $(X,A) \in \mathfrak{A}$. If f: $(X,A) \to (Y,B)$ and (X,A), (Y,B) are triangulable, then $f \in \mathfrak{A}$

This axiom asserts that the category $\mathfrak{T}$ of triangulable pairs is a subcategory of $\mathfrak{A}$: this is satisfied by all the admissible categories that we have encountered.

Now we replace the Exactness axiom by a weaker one:

AXIOM 4′ (*Partial Exactness axiom*). *If (X,A) is admissible, the homology [cohomology] sequence of (X,A) is a sequence of order* 2 (*see* VIII,5 2) *If (X,A) is triangulable, then the sequence is exact*

A system $H = \{H_q(X,A),f_*,\partial\}$ $[H = \{H^q(X,A),f^*,\delta\}]$ satisfying axioms 1-3, 5-7, and 4′ is called a *partially exact* homology [cohomology] theory. The Čech homology theory (on the category $\mathfrak{A}_1$) is a typical example of a partially exact homology theory Others will appear later.

In applying the results of Chapter I to a partially exact homology theory, we must check the extent to which the exactness axiom is used in the proof. For instance in proving that $H_q(X,X) = 0$ (I,8 1) the first proof makes full use of exactness: however, the alternative proof uses only the order 2 property of the homology sequence Thus this proposition remains valid for partially exact theories An inspection of the proof of I,10 2 shows that the homology and cohomology sequences of a triple are of order 2 even for a partially exact theory Similarly, inspection of the proof of I,8.6 shows that the reduced homology and cohomology sequences of a pair are of order 2.

NOTE

The development of the Čech theory The first definition of homology groups of the Čech type was made by Vietoris [Math Ann. 97 (1927), 454-472] He restricted himself to compact metric spaces and used a specific metric to define his cycles About the same time Alexandroff [Annals of Math 30 (1928), 101-187] introduced the concept of approximating a compact metric space by an inverse sequence of complexes (called a projection spectrum), and successfully defined Betti numbers Pontrjagin [Math Ann 105 (1931), 165-205] added to this the notion of an inverse sequence of groups, and obtained homology groups It was Čech [Fund Math 19 (1932), 149-183] who first defined the nerve of a finite covering by open sets, and used such complexes as *approximations* to a space By using inverse systems instead of sequences, he defined homology groups of arbitrary spaces. When it became clear that the Čech definitions for noncompact spaces did not give a fully satisfactory theory, Dowker [Annals of Math. 51 (1950), 278-292] found the satisfactory modification based on pairs of infinite coverings.

A parallel development has taken place based on an idea of Alexander [Proc Nat Acad Sci 21 (1935), 509-512] that a q-cochain may be defined as a function from ordered sets of $q + 1$ points in the space to the coefficient group He subsequently modified this approach by introducing the notion of a *grating* [Bull Amer Math. Soc 53 (1947), 201-233]. Using another modification suggested by Wallace, Spanier [Annals of Math. 49 (1948), 407-427] showed that the cohomology theory obtained from such cochains satisfies our axioms on compact spaces and is isomorphic there to the Čech theory The advantage of this approach lies in the simplicity of the definition of cochain A disadvantage is that there appears to be no equally simple and dual construction of chains and their homology groups.

EXERCISES

A. Connectedness and quasi-components.

Definition. Let X be a topological space and let $x_0 \in X$ The *component* of x_0 in X is the union of all connected subsets of X containing x_0 The *quasi-component* of x_0 is the intersection of all the simultaneously open and closed subsets of X containing x_0.

1. Show that. (1) components are closed, connected, and disjoint, (2) quasi-components are closed and disjoint, (3) eash quasi-component is a union of components; (4) a quasi-component is a component if and only if it is connected; (5) if X is compact or if the number of quasi-components is finite, then each quasi-component is a component. Show by an example that a quasi-component need not be connected.

2. Let $x_0,x_1 \in X$, $g \in G$, and $g \neq 0$ Show that the elements $(gx_0)_x$ and $(gx_1)_x$ of the Čech group $H_0(X,G)$, as defined in I,7.1, are equal if and only if x_0 and x_1 belong to the same quasi-component.

3. For each element h of the Čech group $H^0(X,G)$ where G is in the category $\mathcal{G}_R$, consider the function $X \to G$ as defined in I,7.1c. Show that this establishes an isomorphism between $H^0(X;G)$ and the R-module of all *continuous* functions $X \to G$, where G is taken with the discrete topology. Describe the group $\tilde{H}^0(X;G)$ in a similar fashion

4. Show that $H_0(X;G)$ is isomorphic with $\text{Hom}(H^0(X;J);G)$ where J is the group of integers.

B. 0-dimensional sets.

Definition. A space X is called *0-dimensional* if every open covering of X has a refinement consisting of disjoint sets

1 Show that a compact space is 0-dimensional if and only if it is totally disconnected (i e each component reduces to a single point).

2 Show that every compact 0-dimensional space is the limit space of an inverse system of finite sets.

3 Show that, if X is 0-dimensional, then $H_q(X;G) = 0$, $H^q(X;G) = 0$ for $q > 0$ (Čech groups).

4. Let (X,X_1,X_2) be a proper triad (with respect to the Čech groups) where $X = X_1 \cup X_2$, and let $A = X_1 \cap X_2$ be 0-dimensional. Show that $H_q(X;G) \approx H_q(X_1;G) + H_q(X_2;G)$ for $q > 1$. If X_1 and X_2 are connected and $H_1(X_1,G) = 0 = H_1(X_2,G)$, then $H_1(X,G) \approx \tilde{H}_0(A,G)$.

C. LIMITING GROUPS.

We shall consider here the category $\mathfrak{a}_H$ of pairs (X,A), where X is a Hausdorff space and A is a subspace of X, and of all maps of such pairs.

1. Given a (partially exact) homology [cohomology] theory H on the category $\mathfrak{a}_C$ of compact pairs, define an extension $\vec{H}$ [$\overleftarrow{H}$] of H to the category $\mathfrak{a}_H$ using direct limits of the groups $H_q(X_\alpha,A_\alpha)$ [$H^q(X_\alpha,A_\alpha)$] where (X_α,A_α) is any compact pair contained in (X,A). Verify the axioms for the limiting theory $\vec{H}$ [$\overleftarrow{H}$].

2. Assume in 1 that H is given on the category $\mathfrak{a}_H$. Define a natural map $\vec{H} \to H$ [$H \to \overleftarrow{H}$].

3. Show that, if H is the singular homology theory on the category $\mathfrak{a}_H$, then $\vec{H} \to H$ is an isomorphism. Show that, if the coefficient group is compact or a vector space over a field, then, for singular cohomology, $H \to \overleftarrow{H}$ is an isomorphism.

4. Show that, if (X,A) is in $\mathfrak{a}_H$ and A is closed, the groups $\vec{H}_q(X,A)$ [$\overleftarrow{H}^q(X,A)$] may be defined equivalently as limits of the groups $H_q(X_\alpha,A \cap X_\alpha)$ [$H^q(X_\alpha,A \cap X_\alpha)$] where X_α is any compact subset of X.

5. Let K be a (possibly infinite) simplicial complex and L a subcomplex of K. Show that the inverse limiting cohomology group $\overleftarrow{H}^q(K,L,G)$ (see VIII, Exer. F) is naturally isomorphic with the limiting group $\overleftarrow{H}^q(|K|,|L|,G)$ using singular cohomology and using the weak topology in $|K|$.

D. HOMOLOGY WITH COMPACT CARRIERS.

DEFINITION. A partially exact homology theory H defined on the category $\mathfrak{a}_H$ is said to have *compact carriers* if, for each $u \in H_q(X,A)$, there is a compact pair $(X',A') \subset (X,A)$ such that u is in the image of the homomorphism $H_q(X',A') \to H_q(X,A)$ induced by inclusion.

1. Assume that H is exact and has compact carriers. Let (X,A) be a pair in $\mathfrak{a}_H$ and (X',A') a compact pair contained in (X,A). If $u \in H_q(X',A')$ is in the kernel of the homomorphism $H_q(X',A') \to H_q(X,A)$ induced by inclusion, show that there exists a compact pair (X'',A'') such that $(X',A') \subset (X'',A'') \subset (X,A)$, and that u is the kernel of $H_q(X',A') \to H_q(X'',A'')$. (Hint: consider the triple (X,A,A') to reduce the general case to the case $A = A'$, then consider the triple (X,X',A').)

2 Show that for any exact homology theory H on $\mathfrak{A}_H$ the following conditions are equivalent.

(a) H has compact carriers,

(b) the map $\vec{H} \to H$ of C,2 is an isomorphism,

(c) the map $\vec{H} \to H$ of C,2 is onto

3. Show that singular homology has compact carriers

E. ADMISSIBLE AND UNIQUENESS CATEGORIES.

1. Let H be a partially exact homology [cohomology] theory on an admissible category $\mathfrak{A}$. Let $\mathfrak{A}'$ be the (full) subcategory of $\mathfrak{A}$ determined by those pairs (X,A) for which the homology [cohomology] sequence is exact Show that $\mathfrak{A}'$ is an admissible category and that H is an exact homology [cohomology] theory on $\mathfrak{A}'$.

2. Let $\mathcal{J}_H$ be the category consisting of all compact pairs (X,A) which have the homotopy type of a triangulable pair and of all maps of such pairs Show that $\mathcal{J}_H$ is an admissible category.

3 Show that any partially exact homology [cohomology] theory on the category $\mathfrak{A}_C$ is an exact theory on the subcategory $\mathcal{J}_H$.

4. Show that $\mathcal{J}_H$ is a uniqueness category for homology and cohomology. (This result will be generalized in Chapter XII, of the second volume.)

CHAPTER X

Special features of the Čech theory

1. INTRODUCTION

The discussion in this chapter is centered around the continuity property of the Čech groups Roughly speaking this property may be stated as follows. If the compact pair (X,A) is the inverse limit of the compact pairs (X_α,A_α), then the Čech groups of (X,A) are limits (inverse for homology, and direct for cohomology) of the groups of (X_α,A_α) At the end of the chapter it is shown that among all the partially exact theories (IX,10) on the category $\mathfrak{a}_C$, the Čech theory is essentially the only one satisfying this "continuity axiom." Thus a complete axiomatic description of the Čech groups is obtained (for compact pairs) despite the lack of exactness

As an application of the continuity property, we show that the Čech groups satisfy a much stronger form of the Excision axiom. In particular, in the Čech theory, every compact triad is a proper triad (see I,14 1) This fact is very useful in applications.

Sections 6-9 deal with homology theories for noncompact spaces obtained by compactifying the spaces (in some standard fashion), and then taking the Čech groups of the compactified space Two methods of compactification are discussed The first is the compactification of a locally compact space by adding a point at infinity; the other one is the Tychonoff compactification of a normal space The first yields a theory on the category $\mathfrak{a}_{LC}$ of locally compact pairs This theory is especially interesting for two reasons (1) In the case of a locally finite infinite simplicial complex, the groups are isomorphic with those based on finite cochains and infinite chains (2) The theory does not admit a reduced theory because in the category $\mathfrak{a}_{LC}$ only compact spaces are collapsible (in the sense of I,7 3) The theory for normal spaces obtained by using the Tychonoff compactification turns out to be isomorphic with the Čech theory based on *finite* open coverings.

Throughout the chapter, the Čech groups will only be applied to compact pairs (X,A) and therefore (see IX,8) we may assume that the set $\mathrm{Cov}'(X)$ of finite open coverings of X is the directed set used in the definition of the Čech groups of (X,A).

2. FORMULATION OF CONTINUITY

Let $\mathfrak{A}$ denote a category of pairs (X,A) and maps, and let Inv$\mathfrak{A}$ denote the category of inverse systems having values in $\mathfrak{A}$. Elements of Inv$\mathfrak{A}$ will be denoted by boldface symbols. Thus, $(\mathbf{X},\mathbf{A}) \in$ Inv$\mathfrak{A}$ means

$$(\mathbf{X},\mathbf{A}) = \{(X_m,A_m),\pi_{m_2}^{m_1}\}$$

is an inverse system over some directed set M with values $(X_m,A_m),\pi_{m_2}^{m_1}$ in $\mathfrak{A}$. Similarly, if $(\mathbf{Y},\mathbf{B}) = \{(Y_n,B_n),\rho_{n_2}^{n_1}\}$ is an inverse system over N and belongs to Inv$\mathfrak{A}$, a map

$$\mathbf{f}: (\mathbf{X},\mathbf{A}) \to (\mathbf{Y},\mathbf{B})$$

of Inv$\mathfrak{A}$ is composed of a map $f: N \to M$ and maps $f_n: (X_{fn},A_{fn}) \to (Y_n,B_n)$ as in the definition VIII,2.3.

The operation of taking the inverse (or direct) limit will be denoted by lim. Thus lim $(\mathbf{X},\mathbf{A})$ is a pair (X,A) where X is the limit of the inverse system $\{X_m,\pi_{m_2}^{m_1}\}$, and A is the limit of the system of subspaces $\{A_m,\pi_{m_2}^{m_1}|A_{m_1}\}$. Likewise, if $\mathbf{f} \in$ Inv$\mathfrak{A}$, then lim $\mathbf{f}$: lim $(\mathbf{X},\mathbf{A}) \to$ lim $(\mathbf{Y},\mathbf{B})$ is defined as VIII,3.10.

We shall assume that the category $\mathfrak{A}$ is such that the operation of inverse limit maps Inv$\mathfrak{A}$ into $\mathfrak{A}$:

$$\lim: \text{Inv}\mathfrak{A} \to \mathfrak{A}.$$

For example, this is true if $\mathfrak{A}$ is the category of compact pairs. As observed in VIII,3.14, lim is a covariant functor.

Suppose now that a partially exact homology [cohomology] theory H is defined on $\mathfrak{A}$ with values in $\mathcal{G}_R$ or $\mathcal{G}_C$ [$\mathcal{G}_R$]. Then H can be applied to $(\mathbf{X},\mathbf{A}) \in$ Inv$\mathfrak{A}$ to yield inverse [direct] systems

$$H_q(\mathbf{X},\mathbf{A}) = \{H_q(X_m,A_m),\pi_{m_2*}^{m_1}\}, \qquad [H^q(\mathbf{X},\mathbf{A}) = \{H^q(X_m,A_m),\pi_{m_2}^{m_1*}\}],$$

and maps

$$\partial: H_q(\mathbf{X},\mathbf{A}) \to H_{q-1}(\mathbf{A}), \qquad [\delta: H^{q-1}(\mathbf{A}) \to H^q(\mathbf{X},\mathbf{A})].$$

defined in the obvious way. Similarly, if $\mathbf{f}$: $(\mathbf{X},\mathbf{A}) \to (\mathbf{Y},\mathbf{B})$, then the maps f_{m*} [f_m^*] form a map

$$\mathbf{f}_*: H_q(\mathbf{X},\mathbf{A}) \to H_q(\mathbf{Y},\mathbf{B}), \qquad [\mathbf{f}^*: H^q(\mathbf{Y},\mathbf{B}) \to H^q(\mathbf{X},\mathbf{A})].$$

We have thus a kind of "homology theory" on the category Inv$\mathfrak{A}$. Its values are, corresponding to those of H, in Inv$\mathcal{G}_R$ or Inv$\mathcal{G}_C$ [Dir$\mathcal{G}_R$].

It follows that the two composite functors $H_q \lim$ and $\lim H_q$ are both defined on Inv$\mathfrak{a}$ and both have values in $\mathcal{G}_R$ or $\mathcal{G}_C$. They are comparable as follows

THEOREM 2.1. *Let* $(X,A) = \lim (\mathbf{X},\mathbf{A})$ *and let* $\pi_m\colon (X,A) \to (X_m,A_m)$ *be the projections. Then* $\pi_m = \pi_m^{m'}\pi_{m'}$ *for* $m < m'$ *and* $\pi_{m*} = \pi^{m'}_{m*}\pi_{m'*}$. *Thus the homomorphisms* $\{\pi_{m*}\}$ *constitute a homomorphism*

$$H_q(\lim (\mathbf{X},\mathbf{A})) \to H_q(\mathbf{X},\mathbf{A}).$$

This defines, by VIII,3.10,3.13, *a limit homomorphism*

$$l(q,\mathbf{X},\mathbf{A})\colon\ H_q(\lim (\mathbf{X},\mathbf{A})) \to \lim H_q(\mathbf{X},\mathbf{A}).$$

This is a natural transformation of $H \lim$ *into* $\lim H$ *in the following sense. If* $(\mathbf{X},\mathbf{A}) \in$ Inv$\mathfrak{a}$ *then commutativity holds in the diagram*

$$\begin{array}{ccc}
H_q(\lim (\mathbf{X},\mathbf{A})) & \xrightarrow{\ l\ } & \lim H_q(\mathbf{X},\mathbf{A}) \\
\downarrow{\scriptstyle\partial} & & \downarrow{\scriptstyle\lim \partial} \\
H_{q-1}(\lim (\mathbf{A})) & \xrightarrow{\ l\ } & \lim H_{q-1}(\mathbf{A})
\end{array}$$

Likewise, if $\mathbf{f}\colon\ (\mathbf{X},\mathbf{A}) \to (\mathbf{Y},\mathbf{B})$ *is in* Inv$\mathfrak{a}$, *then commutativity holds in the diagram*

$$\begin{array}{ccc}
H_q(\lim (\mathbf{X},\mathbf{A})) & \xrightarrow{\ l\ } & \lim H_q(\mathbf{X},\mathbf{A}) \\
\downarrow{\scriptstyle(\lim \mathbf{f})_*} & & \downarrow{\scriptstyle\lim (\mathbf{f}_*)} \\
H_q(\lim (\mathbf{Y},\mathbf{B})) & \xrightarrow{\ l\ } & \lim H_q(\mathbf{Y},\mathbf{B})
\end{array}$$

The proofs follow from VIII,3.11. A similar result holds for cohomology.

THEOREM 2.2. *Let* $(X,A) = \lim (\mathbf{X},\mathbf{A})$ *and* $\pi_m\colon\ (X,A) \to (X_m,A_m)$ *be the projections. Then the maps* $\{\pi_m^*\}$ *constitute a homomorphism*

$$H^q(\mathbf{X},\mathbf{A}) \to H^q(\lim (\mathbf{X},\mathbf{A}))$$

thereby defining a limit homomorphism

$$l(q,\mathbf{X},\mathbf{A})\colon\ \lim H^q(\mathbf{X},\mathbf{A}) \to H^q(\lim (\mathbf{X},\mathbf{A})).$$

The homomorphism l so defined is a natural transformation of lim H *into* H lim *in the sense that commutativity holds in the two diagrams*

$$\begin{array}{ccc}
\lim H^q(\mathbf{X},\mathbf{A}) & \xrightarrow{l} & H^q(\lim (\mathbf{X},\mathbf{A})) \\
\uparrow \lim \delta & & \uparrow \delta \\
\lim H^{q-1}(\mathbf{A}) & \xrightarrow{l} & H^{q-1}(\lim (\mathbf{A}))
\end{array}$$

$$\begin{array}{ccc}
\lim H^q(\mathbf{X},\mathbf{A}) & \xrightarrow{l} & H^q(\lim (\mathbf{X},\mathbf{A})) \\
\uparrow \lim (\mathbf{f}^*) & & \uparrow (\lim \mathbf{f})^* \\
\lim H^q(\mathbf{Y},\mathbf{B}) & \xrightarrow{l} & H^q(\lim (\mathbf{Y},\mathbf{B}))
\end{array}$$

DEFINITION 2.3. A partially exact homology [cohomology] theory H with values in $\mathcal{G}_R$ or $\mathcal{G}_C$ [$\mathcal{G}_R$] is said to be *continuous* on the category $\mathcal{a}$ if the transformation l of 2.1 [2.2] is a natural equivalence, i e for each $(\mathbf{X},\mathbf{A}) \varepsilon \operatorname{Inv}\mathcal{a}$

$$l(q,\mathbf{X},\mathbf{A}): \quad H_q(\lim (\mathbf{X},\mathbf{A})) \approx \lim H_q(\mathbf{X},\mathbf{A})$$

$$[l(q,\mathbf{X},\mathbf{A}): \quad \lim H^q(\mathbf{X},\mathbf{A}) \approx H^q(\lim (\mathbf{X},\mathbf{A}))].$$

We emphasize that the concept of continuity is not defined for *cohomology* theories with values in the category $\mathcal{G}_C$

DEFINITION 2.4. An inverse system $\{(X_\alpha,A_\alpha),\pi_\alpha^\beta\}$ is called a *nested* system if each X_α is a subspace of some fixed space Z and each π_β^α is an inclusion map $(X_\beta,A_\beta) \subset (X_\alpha,A_\alpha)$ The intersection $(X,A) = (\bigcap X_\alpha,\bigcap A_\alpha)$ is called the *intersection* of the nested system

THEOREM 2 5 *If (X,A) is the intersection of the nested system $\{(X_\alpha,A_\alpha),\pi_\alpha^\beta\}$, then each point $x \varepsilon X$ regarded as an element of each X_α yields an element $\phi(x)$ of the inverse limit (X_∞,A_∞). The mapping ϕ established a homeomorphism ϕ. $(X,A) \approx (X_\infty,A_\infty)$.*

Obviously ϕ is a 1-1 correspondence The verification that ϕ and ϕ^{-1} are continuous is left to the reader.

THEOREM 2 6 *Let H be a continuous and partially exact homology [cohomology] theory with values in $\mathcal{G}_R$ or $\mathcal{G}_C$ [$\mathcal{G}_R$] Let (X,A) be the intersection of the nested system $\{(X_\alpha,A_\alpha),\pi_\alpha^\beta\}$ of compact pairs in a compact space Z Let i_α. $(X,A) \subset (X_\alpha,A_\alpha)$. Then each system $\{u_\alpha\}$ where $u_\alpha \varepsilon H_q(X_\alpha,A_\alpha)$ and $\pi_{\alpha *}^\beta u_\beta = u_\alpha$ determines a unique element $u \varepsilon H_q(X,A)$ such that $u_\alpha = i_{\alpha *}u$, and vice versa. [Each $u \varepsilon H^q(X,A)$ is of the form*

$i_\alpha^* u_\alpha$ *for some* α *and some* $u_\alpha \in H_q(X_\alpha,A_\alpha)$. *If* $i_\alpha^* u_\alpha = 0$, *then* $\pi_\alpha^{\beta *} u_\alpha = 0$ *for some* $\beta > \alpha$]

PROOF If we identify (X,A) with the inverse limit of $\{(X_\alpha,A_\alpha),\pi_\alpha^\beta\}$ using the mapping ϕ of 2.5, the conclusion of 2 6 becomes a direct consequence of the continuity of the homology [cohomology] theory H

3. CONTINUITY OF THE ČECH THEORY

THEOREM 3.1. *The Čech homology theory based on a coefficient group which is in* $\mathcal{G}_R$ *or* $\mathcal{G}_C$ *is continuous on the category of compact pairs. The same conclusion holds for the Čech cohomology theory based on a coefficient group in* $\mathcal{G}_R$

A number of lemmas, useful later, will precede the proof.

Since we deal with compact pairs, we can limit our attention to finite coverings only Since the subset A of a compact pair (X,A) is always closed, we may assume (see IX,8) that the Čech groups of (X,A) are defined as limits of inverse or direct systems of groups defined over the directed set $\mathrm{Cov}^f(X)$ of finite open coverings of X rather then on on the set $\mathrm{Cov}^f(X,A)$

DEFINITION 3 2 If α,β are families of sets in X defined on the same indexing set $V_\alpha = V_\beta$, and, for each $v \in V_\alpha$, $\alpha_v \supset \beta_v$, then α is called an *enlargement* of β, and β is called a *reduction* of α If α is an indexed family of sets in X, $\bar{\alpha}$ denotes the family of closures of sets of α.

LEMMA 3 3 *If* X *is a normal space and* α *is a finite open covering of* X, *then there exists a closed covering* β *of* X *which is a reduction of* α

PROOF If V_α has but one element v, then $\alpha_v = X$. Let $\beta_v = X$. Then β is the desired covering Proceeding by induction, assume the lemma holds in any normal space if V_α has k elements Let α be a covering of X such that V_α has $k + 1$ elements Select a fixed $v_0 \in V_\alpha$ and let $A = X - \bigcup \alpha_v$ for $v \neq v_0$ Since X is normal, there is an open set U such that $A \subset U$ and $\bar{U} \subset \alpha_{v_0}$ Let $X' = X - U$, and let $\alpha'_v = \alpha_v \cap X'$ for $v \neq v_0$ Then α' is an open covering of the normal space X' defined on a set of k elements Let β' be a closed reduction of α'. Define $\beta_{v_0} = \bar{U}$, and $\beta_v = \beta'_v$ for $v \neq v_0$. Then β is clearly a closed reduction of α.

LEMMA 3.4 *Let* X *be a normal space,* $\alpha \in \mathrm{Cov}^f(X)$, *and let* β *be a finite indexed family of closed sets in* X *such that* $\alpha < \beta$ *Then there exists an open enlargement* γ *of* β *such that* $\alpha < \bar{\gamma} < \gamma < \beta$ *and the nerves of* $\bar{\gamma},\gamma$ *and* β *coincide* $X_{\bar{\gamma}} = X_\gamma = X_\beta$.

PROOF. For convenience, we can suppose that V_β consists of the integers $1,2, \cdots, k$. Choose a $v \in V_\alpha$ such that $\beta_1 \subset \alpha_v$ Let B be the union of those intersections of sets of $\beta_2, \cdots, \beta_k$ which do not meet β_1. Since X is normal, there is an open set γ_1 such that $\beta_1 \subset \gamma_1$, $\bar{\gamma}_1 \subset \alpha_{v_1}$, and $\bar{\gamma}_1 \cap B = 0$. (If $\beta_1 = 0$, we must choose $\gamma_1 = 0$.) It follows that the indexed family $\bar{\gamma}_1, \beta_2, \cdots, \beta_k$ is $> \alpha$, and its nerve coincides with X_β. Repeat the foregoing process on the element β_2 of the family $\bar{\gamma}_1, \beta_2, \cdots, \beta_k$ obtaining γ_2 and a new indexed family $\bar{\gamma}_1, \bar{\gamma}_2, \beta_3, \cdots, \beta_k$. An obvious induction completes the proof.

DEFINITION 3.5. If α is an indexed family of sets in X and $A \subset X$, then α is said to be *regular relative to* A if the following two conditions are satisfied: (1) if $\alpha_v \cap A = 0$, $v \in V_\alpha$, then $\bar{\alpha}_v \cap A = 0$, and (2) if $\alpha_{v_i} \cap A \neq 0$ for $i = 1, \cdots, k$, and $D = \alpha_{v_1} \cap \cdots \cap \alpha_{v_k} \neq 0$, then $D \cap A \neq 0$.

The two conditions can be restated in terms of nerves as follows: (1°) if v is a vertex of $A_{\bar{\alpha}}$, then v is a vertex of A_α, and (2°) if a simplex s of X_α has all its vertices in A_α then it lies in A_α, or, using the terminology of II,9.1, A_α is a full subcomplex of X_α.

LEMMA 3.6. *If X is a normal space, and A',X' are closed sets in X with $A' \subset X'$, then the finite open coverings of X which are regular relative to both A' and X' form a cofinal family in* $\mathrm{Cov}^f(X)$.

PROOF. Let $\alpha \in \mathrm{Cov}^f(X)$. By 3.3, there is a closed covering β of A' (consisting of closed sets of A') such that $\alpha < \beta$. By 3.4, there is an enlargement γ of β consisting of open sets of X' such that $\alpha < \bar{\gamma}$ and $X_{\bar{\gamma}} = X_\gamma = X_\beta$. Let U be the union of the sets of γ. Then $B = X' - U$ is a closed set. By 3.3, there is a covering δ of B by closed sets of B such that $\alpha < \delta$. Form the composite covering $\epsilon = \{\bar{\gamma},\delta\}$ of X' by closed sets of X'. Clearly $\alpha < \epsilon$. By 3.4, there is an enlargement η of ϵ consisting of open sets of X such that $\alpha < \eta$ and $X_\eta = X_\epsilon$. We require in addition that $\bar{\eta}_v \cap A' = 0$ for each η_v which is an enlargement of a set of δ. If this is not the case, choose an open set $W \supset B$ such that $\overline{W} \cap A' = 0$ and replace η_v by $\eta_v \cap W$. The properties $\alpha < \eta$ and $X_\eta = X_\epsilon$ still hold. In addition, condition (1) holds for regularity relative to A' of η.

Suppose s is a simplex of X_η and its vertices lie in A'_η. Then, for each vertex v of s, $\eta_v \cap A' \neq 0$. Thus η_v is an enlargement of a set of $\bar{\gamma}$. Since $X_\eta = X_\epsilon$, it follows that s belongs to $X_{\bar{\gamma}}$. Since $X_\gamma = X_\beta$, s belongs to X_β. Therefore $\mathrm{Car}_\beta(s) \neq 0$. But $\mathrm{Car}_\beta(s) \subset A' \cap \mathrm{Car}_\eta(s)$. It follows that $\mathrm{Car}_\eta(s)$ meets A'. Therefore s lies in A'_η.

Suppose now that s is a simplex of X_η and its vertices are in X'_η. Since $X_\eta = X_\epsilon$, s belongs to X_ϵ. Therefore $\mathrm{Car}_\epsilon(s) \neq 0$. But $\mathrm{Car}_\epsilon(s) \subset X' \cap \mathrm{Car}_\eta(s)$. It follows that $\mathrm{Car}_\eta(s)$ meets X'. Therefore s lies in X'_η.

Thus η is regular relative to A' and satisfies condition (2) for regularity relative to X'. Let U' be the union of the open sets of η. Let W' be an open set such that $W' \supset X - U'$ and $\overline{W}' \cap X' = 0$. Let

α' be the covering α cut down to W'. If we adjoin α' to η, we obtain a composite covering $>\alpha$ which is regular relative to A' and X'.

LEMMA 3.7. *Let $\{X_m,\pi^{m_1}_{m_2}\}$ be an inverse system of compact spaces, and let $X = \lim \{X_m,\pi^{m_1}_{m_2}\}$. Then the elements of $\mathrm{Cov}^f(X)$ of the form $\pi_m^{-1}(\beta)$, where $m \in M$, $\beta \in \mathrm{Cov}^f(X_m)$, form a cofinal family.*

PROOF. Suppose $\alpha \in \mathrm{Cov}^f(X)$. Since sets of the form $\pi_m^{-1}(U)$, where U is open in X_m, form a base for the open sets of X (VIII,3.12), it follows from the compactness of X (VIII,3.6) that there is a covering α' consisting of sets $\pi_{m_i}^{-1}(U_i)$ $(i = 1, \cdots, k)$ such that $\alpha < \alpha'$. Since M is directed, there is an $m > m_i$ for $i = 1, \cdots, k$. Define a covering β of X_m by $\beta_i = (\pi^m_{m_i})^{-1}(U_i)$ for $i = 1, \cdots, k$, and $\beta_{k+1} = X_m - \pi_m(X)$. Then $\pi_m^{-1}(\beta_i) = \pi_{m_i}^{-1}(U_i)$ $(i = 1, \cdots, k)$ and $\pi_m^{-1}(\beta_{k+1}) = 0$. It follows that $\pi_m^{-1}(\beta) > \alpha$.

LEMMA 3.8. *Let $\{(X_m,A_m),\pi^{m_1}_{m_2}\}$ be an inverse system of compact pairs, and let $(X,A) = \lim \{(X_m,A_m),\pi^{m_1}_{m_2}\}$. Then elements of $\mathrm{Cov}^f(X)$ of the form $\alpha = \pi_m^{-1}(\beta)$ $(m \in M,\ \beta \in \mathrm{Cov}^f(X_m))$ and such that $(X_\alpha,A_\alpha) = (X_{m\beta},A_{m\beta})$ form a cofinal family.*

PROOF. Suppose $\gamma \in \mathrm{Cov}^f(X)$. By 3.7, there is an m_0 and a $\delta \in \mathrm{Cov}(X_{m_0})$ such that $\pi_{m_0}^{-1}(\delta) > \gamma$. Let $X' = \pi_{m_0}(X)$ and $A' = \pi_{m_0}(A)$. By 3.6, there is an $\epsilon \in \mathrm{Cov}(X_{m_0})$ which is regular relative to both X' and A' and $\delta < \epsilon$. Let U (W) be the union of the sets of ϵ which do not meet X' (A'), and let U',W' be their complements in X_{m_0}. Then U',W' are open, $X' \subset U'$, $A' \subset W'$, and, if a set of ϵ meets U' (W'), it also meets X' (A'). By two applications of VIII,3.7, there is an $m > m_0$ such that $\pi^m_{m_0}(X_m) \subset U'$ and $\pi^m_{m_0}(A_m) \subset W'$. Define $\beta = (\pi^m_{m_0})^{-1}(\epsilon)$. Then

$$\alpha = \pi_m^{-1}(\beta) = \pi_{m_0}^{-1}(\epsilon) > \pi_{m_0}^{-1}(\delta) > \gamma.$$

Since $\alpha = \pi_m^{-1}(\beta)$, we have, by IX,2.5, that $(X_\alpha,A_\alpha) \subset (X_{m\beta},A_{m\beta})$. Suppose s is a simplex of $X_{m\beta}$. Then $\mathrm{Car}_\beta(s) \neq 0$. Since $\mathrm{Car}_\epsilon(s) \supset \pi^m_{m_0}(\mathrm{Car}_\beta(s)) \subset U'$, we have that $\mathrm{Car}_\epsilon(s)$ meets U'. Hence for each vertex v of s, ϵ_v meets X'. Since ϵ is regular relative to X', $\mathrm{Car}_\epsilon(s)$ meets X'. But $\mathrm{Car}_\alpha(s) = \pi_{m_0}^{-1}(\mathrm{Car}_\epsilon(s))$. It follows that $\mathrm{Car}_\alpha(s) \neq 0$. Therefore s belongs to X_α. This proves that $X_\alpha = X_{m\beta}$. The same argument with X, X_m, X_{m_0}, U' replaced by A, A_m, A_{m_0}, W' shows that $A_\alpha = A_{m\beta}$. This proves the lemma.

PROOF OF 3.1. The proof will be given only for the case of homology; the case of cohomology is left to the reader.

Suppose then that $(\mathbf{X},\mathbf{A}) = \{(X_m,A_m),\pi^{m_1}_{m_2}\}$ is an inverse system of compact pairs. Let $(X,A) = \lim (\mathbf{X},\mathbf{A})$. We must prove two propositions concerning the homomorphism $l(q,\mathbf{X},\mathbf{A})$ of 2.1. First, its kernel is zero, and, second, it is onto.

Suppose $u \in H_q(X,A)$ lies in the kernel of l Since $\pi_{m*}(u)$ is the coordinate of $l(u)$ in $H_q(X_m,A_m)$, it follows that $\pi_{m*}(u) = 0$ for each $m \in M$ If $\beta \in \mathrm{Cov}(X_m)$, then the coordinate of $\pi_{m*}(u)$ in $H_q(X_{m\beta},A_{m\beta})$ is zero If $\alpha = \pi_m^{-1}(\beta)$ and $(X_\alpha,A_\alpha) = (X_{m\beta},A_{m\beta})$, it follows from the definition of π_{m*} (IX,4.2) that the coordinate of u in $H_q(X_\alpha,A_\alpha)$ is zero. By 3.8, such α's form a cofinal family in $\mathrm{Cov}'(X)$. Therefore, $u = 0$

Suppose now that $v \in \lim H_q(\mathbf{X},\mathbf{A})$ Let v_m be the coordinate of v in $H_q(X_m,A_m)$ For any $\beta \in \mathrm{Cov}'(X_m)$, let $v_{m\beta}$ be the coordinate of v_m in $H_q(X_{m\beta},A_{m\beta})$. We must find a $u \in H_q(X,A)$ such that $l(u) = v$ It suffices to construct the coordinates u_α of u for α in some cofinal family D in $\mathrm{Cov}'(X)$ Let D be the cofinal family described in 3 8 For each such α choose an $m \in M$ and a $\beta \in \mathrm{Cov}'(X_m)$ such that $\alpha = \pi_m^{-1}(\beta)$ and $(X_\alpha,A_\alpha) = (X_{m\beta},A_{m\beta})$ Define $u_\alpha = v_{m\beta}$

Suppose $\alpha_1 < \alpha_2$ in D, and m_1,β_1 and m_2,β_2 are the choices for α_1 and α_2. Choose $m_3 > m_i$ and define $\beta'_i = (\pi^{m_3}_{m_i})^{-1}(\beta_i)$ $(i = 1,2)$. By 3 6, there exists an $\epsilon \in \mathrm{Cov}'(X_{m_3})$ which is a refinement of β'_1 and β'_2 and is regular relative to both $\pi_{m_3}(X)$ and $\pi_{m_3}(A)$ It follows, as in the proof of 3.8, that there is an $m_4 > m_3$ such that, if $\beta = (\pi^{m_4}_{m_3})^{-1}(\epsilon)$, and $\alpha = \pi^{-1}_{m_4}(\beta)$, then $(X_\alpha,A_\alpha) = (X_{m_4\beta},A_{m_4\beta})$ Let $\gamma_i = (\pi^{m_4}_{m_i})^{-1}(\beta_i)$ for $i = 1,2$. Then $(X_{\alpha_i},A_{\alpha_i}) = (X_{m_4\gamma_i},A_{m_4\gamma_i}) = (X_{m_i\beta_i},A_{m_i\beta_i})$ for $i = 1,2$, because each is contained in the following and the first equals the third. Therefore $u_{\alpha_i} = v_{m_4\gamma_i}$ Since $\epsilon > \beta'_1,\beta'_2$, we have $\beta > \gamma_1,\gamma_2$ and $\alpha > \alpha_1,\alpha_2$ Let $u'_\alpha = v_{m_4\beta}$ Since commutativity holds in the diagram

$$\begin{array}{ccc} H_q(X_\alpha,A_\alpha) & = & H_q(X_{m_4\beta},A_{m_4\beta}) \\ \downarrow & & \downarrow \\ H_q(X_{\alpha_i},A_{\alpha_i}) & = & H_q(X_{m_4\gamma_i},A_{m_4\gamma_i}) \end{array}$$

it follows that u'_α projects into u_{α_i} $(i = 1,2)$ Therefore u_{α_2} projects into u_{α_1} This proves that the elements u_α for $\alpha \in D$ are the coordinates of an element $u \in H_q(X,A)$. It shows also that u_α is independent of the choice of m and β

It remains to show that $l(u) = v$, or that $\pi_{m*}(u) = v_m$ for each $m \in M$. This follows if $(\pi_{m*}(u))_\epsilon = v_{m\epsilon}$ for ϵ in a cofinal family in $\mathrm{Cov}(X_m)$. Let ϵ be regular relative to $\pi_m(X)$ and $\pi_m(A)$. By 3 6, such coverings form a cofinal family. As in the proof of 3.8, there is an $m_1 > m$ such that, if $\beta = (\pi^{m_1}_m)^{-1}(\epsilon)$ and $\alpha = \pi^{-1}_{m_1}(\beta)$, then $(X_\alpha,A_\alpha) = (X_{m_1\beta},A_{m_1\beta})$. Because u_α is independent of the choice of the m,β used in its definition, it follows that $u_\alpha = v_{m\beta}$ Therefore u_α and $v_{m_1\beta}$ have the same image in $H_q(X_{m\epsilon},A_{m\epsilon})$ under the inclusion map $(X_{m_1\beta},A_{m_1\beta}) \to (X_{m\epsilon},A_{m\epsilon})$. These images are $(\pi_{m*}(u))_\epsilon$ and $v_{m\epsilon}$ respectively. This completes the proof of 3.1.

4. CONTINUITY VERSUS EXACTNESS

It will be shown that the Čech "homology theory" with integer coefficients is not exact on all compact pairs. The proof is based on the continuity of the Čech theory. This and Theorem 12.2 below imply that the (essentially unique) homology theory with integer coefficients on triangulable pairs does not admit an extension to compact pairs which is simultaneously continuous and exact.

EXAMPLE 4.1. Let (E,S) be the 2-cell and its boundary defined in the complex plane by the conditions $|z| \leq 1$ and $|z| = 1$ respectively. Consider the maps

$$\phi\colon\ S \to S, \qquad f\colon\ E \to E$$

defined by

$$\phi(z) = -z, \qquad f(z) = z^3.$$

If for each $z \in S$ we identity z and $\phi(z)$, the pair (E,S) becomes a pair (P,C) where P is the projective plane and C is a projective line in P. Since $f\phi(z) = (-z)^3 = -z^3 = \phi f(z)$, it follows that the map f induces a map $\bar{f}\colon\ (P,C) \to (P,C)$ In XIV, Example 9 7, the reduced homology sequence with integer coefficients of (P,C) will be calculated, and also the endomorphism of this homology sequence induced by $\bar{f}$. It will be shown that the only nontrivial groups are $H_1(P)$, $H_1(C)$, and $H_2(P,C)$ and that the diagram

$$\begin{array}{ccccc} H_2(P,C) & \xrightarrow{\partial} & H_1(C) & \xrightarrow{i_*} & H_1(P) \\ \downarrow{\bar{f}_*} & & \downarrow{\bar{f}_*} & & \downarrow{\bar{f}_*} \\ H_2(P,C) & \xrightarrow{\partial} & H_1(C) & \xrightarrow{i_*} & H_1(P) \end{array}$$

is isomorphic with the diagram of VIII,5 5.

Let M be the set of positive integers with the usual order For each $\alpha \in M$ define $(P_\alpha,C_\alpha) = (P,C)$ and $\pi^\alpha_{\alpha+1} = \bar{f}$ Define all other π^β_α for $\beta < \alpha$ by transitivity $\pi^\beta_\alpha = \pi^\beta_{\alpha-1}\pi^{\alpha-1}_\alpha$. Then $\{(P_\alpha,C_\alpha),\pi^\beta_\alpha\}$ is an inverse system of spaces indexed by M. Let (P_∞,C_∞) be the limit pair By VIII,3 6, (P_∞,C_∞) is a compact pair. If we employ integer coefficients and the Čech homology theory (or any other continuous homology theory), then the (reduced) homology sequence of (P_∞,C_∞) is isomorphic with the inverse limit of the homology sequences of (P,C) under the maps $(\pi_\alpha^\beta)_{**}$. This limit sequence is then precisely the sequence obtained as the limit in VIII,5 5 Thus in the reduced homology sequence of (P_∞,C_∞) the group $H_1(P_\infty)$ is cyclic of order 2 while all the other groups are trivial. Thus the reduced homology sequence of (P_∞,C_∞) is not exact. The nonreduced homology sequence differs only in the positions $H_0(P_\infty)$ and $H_0(C_\infty)$ (which are cyclic infinite) and also is not exact

The pair (P_∞,C_∞) may be described alternatively as follows Consider the pair $(\tilde{E},\tilde{S})$ where $\tilde{S}$ is the 3-adic solenoid and $\tilde{E}$ is the join of $\tilde{S}$ with a point Using the antipodism of S^1, define an antipodism on $\tilde{S}$ by a passage to the limit Identifying antipodic pairs of points on $\tilde{S}$, one obtains a pair homeomorphic with (P_∞,C_∞).

5. RELATIVE HOMEOMORPHISMS AND EXCISIONS

DEFINITION 5 1 A map $f.$ $(X,A) \to (Y,B)$ is called a *relative homeomorphism* if f maps $X - A$ homeomorphically onto $Y - B$

LEMMA 5 2 *If f $(X,A) \to (Y,B)$ is a map of compact pairs and f maps $X - A$ in a 1-1 fashion onto $Y - B$, then f is a relative homeomorphism.*

PROOF We need only verify the continuity of the map g $Y - B \to X - A$ defined by $g(y) = f^{-1}(y)$ Let U be any open subset of $X - A$ Then

$$g^{-1}(U) = f(U) = Y - B - f(X - U).$$

Since $X - U$ is compact, $f(X - U)$ is closed, and therefore $g^{-1}(U)$ is open. Thus g is continuous

DEFINITION 5 3. A homology [cohomology] theory H (exact or inexact) is said to be *invariant under relative homeomorphisms* if for every admissible relative homeomorphism f $(X,A) \to (Y,B)$ the homomorphisms f_* [f^*] are isomorphisms in all dimensions

THEOREM 5 4 *The Čech homology and cohomology theories on the category $\mathfrak{a}_C$ of compact pairs are invariant under relative homeomorphisms*

PROOF. Let f: $(X,A) \to (Y,B)$ be a relative homeomorphism of compact pairs, and let $\{B_\alpha\}$ be the collection of all closed sets of Y such that $B \subset \operatorname{Int} B_\alpha$ If B_α and B_β are in the collection, then $B_\alpha \cap B_\beta$ is also in the collection Therefore $\{(Y,B_\alpha),\rho_\alpha^\beta\}$ where ρ_α^β $(Y,B_\beta) \subset (Y,B_\alpha)$ form a nested system with (Y,B) as intersection Let $A_\alpha = f^{-1}(B_\alpha)$ Then $\{(X,A_\alpha),\tau_\alpha^\beta\}$, where τ_α^β: $(X,A_\beta) \subset (X,A_\alpha)$, is also a nested system with (X,A) as intersection. Let f_α: $(X,A_\alpha) \to (Y,B_\alpha)$ be the map defined by f. We show first that $f_{\alpha*}$ is an isomorphism.

Let V be an open set of Y such that $B \subset V \subset \overline{V} \subset \operatorname{Int} B_\alpha$. Let $U = f^{-1}(V)$ Then $A \subset U \subset \overline{U} \subset \operatorname{Int} A_\alpha$ Consider the diagram

$$\begin{array}{ccc} H_q(X - U, A_\alpha - U) & \xrightarrow{i_*} & H_q(X, A_\alpha) \\ \downarrow{\scriptstyle g_*} & & \downarrow{\scriptstyle f_{\alpha *}} \\ H_q(Y - V, B_\alpha - V) & \xrightarrow{i'_*} & H_q(Y, B_\alpha) \end{array}$$

where i and i' are excision maps, and g is defined by f. Since f is a relative homomorphism and $A \subset U$, $B \subset V$, it follows that g is a homeomorphism Thus g_* is an isomorphism The maps i_* and i'_* are isomorphisms by the excision axiom Thus commutativity in the diagram implies that $f_{\alpha *}$ is an isomorphism

The commutativity relations in the diagram

$$\begin{array}{ccccc} H_q(X,A) & \longrightarrow & H_q(X,A_\beta) & \longrightarrow & H_q(X,A_\alpha) \\ \downarrow{\scriptstyle f_*} & & \downarrow{\scriptstyle f_{\beta *}} & & \downarrow{\scriptstyle f_{\alpha *}} \\ H_q(Y,B) & \longrightarrow & H_q(Y,B_\beta) & \longrightarrow & H_q(Y,B_\alpha) \end{array}$$

where the horizontal maps are induced by inclusions, together with 2 6 imply that f_* is an isomorphism

An analogous proof applies to cohomology

Remark The above proof did not make any appeal to the specific definition of the Čech groups and remains valid for any partially exact homology or cohomology theory satisfying the continuity axiom This gain in generality is however to a large extent illusory, since it will be shown in 12.2 that any such theory is isomorphic with the Čech theory.

Theorem 5 4 is of course a generalization of the excision axiom for compact pairs This theorem becomes the excision axiom itself if we redefine an excision to be a relative homeomorphism

In general in the categories $\mathfrak{A}_1$ or $\mathfrak{A}_C$ or $\mathfrak{A}_{LC}$ we may consider the following types of excisions

(E) Inclusion maps $f.$ $(X - U, A - U) \subset (X,A)$ where U is open in X and $\overline{U} \subset \operatorname{Int} A$.

(E_1) Inclusion maps f $(X - U, A - U) \subset (X,A)$ where U is open in X and $U \subset A$

(E_2) Inclusion maps $f.$ $(X - U, A - U) \subset (X,A)$ where $\overline{U} \subset \operatorname{Int} A$

(E_3) Relative homeomorphisms

Type (E) consists of the excisions introduced in Chapter I. Type

(E_2) in the categories $\mathcal{a}_C$ and $\mathcal{a}_{LC}$ coincides with type (E) since the map f is not admissible unless U is open. Excisions of type (E_2) in the category $\mathcal{a}_1$ are applicable in the singular homology and cohomology theory (see VII,9.1). Theorem 5.4 asserts that excisions of type (E_3) are applicable in the Čech theories on the category $\mathcal{a}_C$.

The following example shows that excisions of type (E_1) are not applicable in the singular homology theory (even in the case of compact metric spaces)

EXAMPLE 5.5. Let R be the region in the (x,y)-plane defined by

$$\sin\frac{1}{x} < y < 2 \qquad \text{for} \qquad 0 < |x| < \frac{2}{3\pi},$$
$$1 < y < 2 \qquad \text{for} \qquad x = 0.$$

Let X be the rectangle given by

$$|x| \leqq \frac{2}{3\pi}, \qquad -1 \leqq y \leqq 2,$$

and let C be the boundary of X Clearly $R \subset X$. Set $A = X - R$ It is easy to see that C is a strong deformation retract of A This implies $H_q(C) \approx H_q(A)$ (induced by the inclusion map), and therefore by I,10 5 $H_q(X,C) \approx H_q(X,A)$. Since (X,C) is a 2-cell, it follows that $H_2(X,A) \approx G$.

Let $B = \bar{R} \cap A$ be the boundary of R We shall show that, in the singular theory, we have $H_2(\bar{R},B) = 0$; since $(\bar{R},B) = (X - U, A - U)$ where $U = A - B = \text{Int } A$, it will follow that excisions of type (E_1) are not applicable in the singular theory

Since $\bar{R}$ is contractible over itself to a point, we have $H_1(\bar{R}) = H_2(\bar{R}) = 0$ Therefore, by exactness, ∂ $H_2(\bar{R},B) \approx H_1(B)$. Thus it remains to show that $H_1(B) = 0$ (in the singular theory.)

Let D be the part of B on the y-axis with $-1 \leqq y \leqq 1$, and let $D' = B - D$ By a simple argument involving local connectedness, any singular simplex in B lies wholly in D or wholly in D'. Thus the total singular complex of B divides into two disjoint singular complexes—that of D and that of D'. Hence $H_1(B)$ decomposes into the direct sum $H_1(D) + H_1(D')$ Since D is a closed line segment, it is contractible to a point So $H_1(D) = 0$. Similarly D' is homeomorphic to an open line segment; so it too is contractible to a point, $H_1(D') = 0$. Thus $H_1(B) = 0$ as required.

The above analysis of the singular complex of B implies also that $H_0(B) \approx G + G$, $\tilde{H}_0(B) \approx G$. Thus we have

$$\tilde{H}_0(B) \approx G, \qquad H_1(B) = 0$$

in the singular theory. In order to compute the groups of B in the Čech theory we assume that G is a field, thus insuring exactness. We have $H_1(B) \approx H_2(\bar{R},B)$ by exactness and $H_2(\bar{R},B) \approx H_2(X,A)$ by 5 4. Thus $H_1(B) \approx G$. Since B is connected, the nerve B_α for any finite open covering α of B also will be connected Thus $\tilde{H}_0(B_\alpha) = 0$ and $\tilde{H}_0(B) = 0$ This yields

$$\tilde{H}_0(B) \approx 0, \qquad H_1(B) \approx G.$$

This exhibits the difference between the Čech and singular theories. Incidentally it shows that the singular theory is not continuous on the category $\mathfrak{a}_C$. In the above discussion we assumed that G was a field in order to compute the Čech group $H_1(B)$ using exactness. A direct argument using continuity could be used to establish $H_1(B) \approx G$ for any coefficient group

6. HOMOLOGY THEORIES FOR LOCALLY COMPACT SPACES

We recall here that in I,2, the category $\mathfrak{a}_{LC}$ was defined to consist of pairs (X,A) such that X is a locally compact (Hausdorff) space, and A is closed in X, and to consist of maps f such that the inverse image of any compact set under f is compact.

DEFINITION 6.1. A subset A of a locally compact space (abbreviated LC-space) X is called *bounded* if $\bar{A}$ is compact A subset U of X is called *countercompact* if $X - U$ is compact.

DEFINITION 6.2. Let X be an LC-space and let ω be a point not in the space X We define a topological space $X + \omega$ to consist of the set $X \cup \omega$ and the open sets (i) the open sets of X, and (ii) the union of ω with the countercompact sets of X It is easy to see that $X + \omega$ is a compact (Hausdorff) space.

LEMMA 6.3 *Suppose (X,A), Y, and f: $X - A \to Y$ are in $\mathfrak{a}_{LC}$, and ω is not in Y Set $g(x) = f(x)$ for $x \in X - A$ and $g(x) = \omega$ for $x \in A$. Then g: $X \to Y + \omega$ is a continuous extension of f. If f is a homeomorphism, then g is a relative homeomorphism. If f is a homeomorphism, X is compact, and A is a single point, then g is a homeomorphism.*

PROOF To prove that g is continuous observe that $g^{-1}(U) = f^{-1}(U)$ for $U \subset Y$, and if $U = W \cup \omega$, then

$$g^{-1}(U) = f^{-1}(W) \cup A = X - f^{-1}(Y - W).$$

Thus, if W is countercompact, $f^{-1}(Y - W)$ is compact and $g^{-1}(U)$ is open.

The second part of the lemma is obvious To prove the last part observe that, if f is a homeomorphism and A is a single point, then g is $1 - 1$; thus, if X is compact, g is a homeomorphism.

It will be convenient to pick a fixed point ω and assume that ω is not contained in any space of the category $\mathcal{A}_{LC}$. This actually amounts to replacing $\mathcal{A}_{LC}$ by a subcategory. With this assumption it is possible to "compactify" all the spaces of $\mathcal{A}_{LC}$ by adjoining ω. We shall abbreviate $X + \omega$ by $\dot{X}$. If (X,A) is a pair in $\mathcal{A}_{LC}$, then $(\dot{X},\dot{A})$ is a pair in $\mathcal{A}_C$. If $f\colon (X,A) \to (Y,B)$ is in $\mathcal{A}_{LC}$, then by 6.3 the map $\dot{f}\colon (\dot{X},\dot{A}) \to (\dot{Y},\dot{B})$, defined by $\dot{f}(x) = f(x)$ for $x \in X$ and $\dot{f}(\omega) = \omega$, is in $\mathcal{A}_C$. It is clear that $\cdot$ is a covariant functor on $\mathcal{A}_{LC}$ with values in $\mathcal{A}_C$. We shall refer to this functor as "the single point compactification." Note that $0 = \omega$, so that $\dot{A}$ is never vacuous in the symbol $(\dot{X},\dot{A})$.

To show that $\cdot$ is an h-functor (IV,9), we convert $\mathcal{A}_C$ and $\mathcal{A}_{LC}$ into h-categories $\mathcal{A}_C'$ and $\mathcal{A}_{LC}'$ in the following way. *Couples* (i, j) are inclusions

$$(A,B) \xrightarrow{i} (X,B) \xrightarrow{j} (X,A)$$

where (X,A,B) is a triple, i.e. where X is in $\mathcal{A}_C$ (or $\mathcal{A}_{LC}$), $B \subset A \subset X$, and B and A are closed in X. *Excisions* are defined to be the relative homeomorphisms. Two maps $f_0,f_1\colon (X,A) \to (Y,B)$ are *homotopic* if there is a map $F\colon (X \times I, A \times I) \to (Y,B)$ in $\mathcal{A}_C$ (or $\mathcal{A}_{LC}$) such that $F(x,0) = f_0(x)$, $F(x,1) = f_1(x)$. *Points* are defined to be pairs (X,A) where $X - A$ is a single point.

THEOREM 6.4. *Let H be a homology [cohomology] theory on $\mathcal{A}_C$ or $\mathcal{A}_{LC}$ which is invariant with respect to relative homeomorphisms. If we regard the boundary [coboundary] operator of triples as the basic boundary [coboundary] operator, then H becomes a homology [cohomology] theory on $\mathcal{A}_C'$ or $\mathcal{A}_{LC}'$ respectively. The same holds for partially exact homology and cohomology theories.*

PROOF. The validity of Axioms 1, 2, 3 is clear. The Exactness axiom is the statement of the exactness of the homology [cohomology] sequence of a triple. If the given theory is partially exact, i.e. the homology [cohomology] sequence of a pair is a sequence of order 2, then the same holds for the sequence of a triple (see IX,10). The Homotopy axiom is satisfied since the homotopies in $\mathcal{A}_C$ (or $\mathcal{A}_{LC}$) coincide with those of $\mathcal{A}_C'$ (or $\mathcal{A}_{LC}'$). The Excision axiom holds since, by assumption, the given theory is invariant under relative homeomorphisms. Finally, if the pair (X,A) is a "point", then $X - A$ is a single point, and the inclusion map $i\colon X - A \subset (X,A)$ is a relative homeomorphism. Thus i_* $[i^*]$ are isomorphisms, which implies the Dimension axiom.

THEOREM 6.5. *The single point compactification is a covariant h-functor $\cdot\colon \mathcal{A}_{LC}' \to \mathcal{A}_C'$.*

PROOF. It is clear that · carries couples into couples and thus is a c-functor. Let f_0, f_1: $(X,A) \to (Y,B)$ be in $\mathcal{A}'_{LC}$ and let F: $(X \times I, A \times I) \to (Y,B)$ be a map in $\mathcal{A}'_{LC}$ such that $F(x,0) = f_0(x)$, $F(x,1) = f_1(x)$. Extend F to a map $\dot{F}$: $(\dot{X} \times I, \dot{A} \times I) \to (\dot{Y},\dot{B})$ be setting $\dot{F}(\omega,t) = \omega$. Then $\dot{F}$ is continuous by 6.3, and $\dot{F}(x,0) = \dot{f}_0(x)$, $\dot{F}(x,1) = \dot{f}_1(x)$. Thus $\dot{f}_0$ and $\dot{f}_1$ are homotopic. If f: $(X,A) \to (Y,B)$ is a relative homomorphism, then, since $X - A = \dot{X} - \dot{A}$ and $Y - B = \dot{Y} - \dot{B}$, it follows that $\dot{f}$ is also a relative homeomorphism. If (X,A) is a "point" in $\mathcal{A}'_{LC}$, then $\dot{X} - \dot{A} = X - A$ is a single point and $(\dot{X},\dot{A})$ is a "point" in $\mathcal{A}'_C$.

DEFINITION 6.6. Let H be a partially exact homology [cohomology] theory on the category $\mathcal{A}_C$ invariant with respect to relative homeomorphisms. Regard H as a theory on $\mathcal{A}'_C$ and denote by $\dot{H}$ the composition of H with the h-functor ·: $\mathcal{A}'_{LC} \to \mathcal{A}'_C$. There results a partially exact homology [cohomology] theory on $\mathcal{A}'_{LC}$ or, what amounts to the same, a theory on $\mathcal{A}_{LC}$ invariant with respect to relative homeomorphisms. The theory $\dot{H}$ will be called the *LC-theory associated with H.* If H is exact, so also is $\dot{H}$.

Note that, if (X,A) in $\mathcal{A}_{LC}$ is a compact pair, then the inclusion map i: $(X,A) \subset (\dot{X},\dot{A})$ is admissible and is a relative homeomorphism. Thus we have

$$\dot{H}_q(X,A) = H_q(\dot{X},\dot{A}) \overset{i_*}{\approx} H_q(X,A).$$

Thus $\dot{H} \approx H$ on $\mathcal{A}_C$, and $\dot{H}$ may be regarded as an "extension" of H.

From now on we shall limit ourselves to the case when H is the Čech theory on $\mathcal{A}_C$. The objective will be to give a direct description of the associated LC-theory $\dot{H}$ in terms of coverings.

DEFINITION 6.7. Let (X,A) be a pair in $\mathcal{A}_{LC}$ and $\alpha \in \mathrm{Cov}^f(X)$ a finite open covering of X. Let Ω_α be the subcomplex of the nerve X_α consisting of simplexes s whose carrier $\mathrm{Car}_\alpha(s)$ is not bounded (i.e. the closure of $\mathrm{Car}_\alpha(s)$ is not compact). We shall consider the groups of $(X_\alpha, A_\alpha \cup \Omega_\alpha)$. If β is a refinement of α and π: $X_\beta \to X_\alpha$ is a projection, then π maps $A_\beta \cup \Omega_\beta$ into $A_\alpha \cup \Omega_\alpha$.

DEFINITION 6.8. Let $\mathrm{Cov}_1(X)$ denote the subset of $\mathrm{Cov}^f(X)$ consisting of those coverings α such that each set α_v is either bounded or countercompact. Clearly $\mathrm{Cov}_1(X)$ is a directed subset of $\mathrm{Cov}^f(X)$. The inverse [direct] limit over the directed set $\mathrm{Cov}_1(X)$ of the groups $H_q(X_\alpha, A_\alpha \cup \Omega_\alpha)$ $[H^q(X_\alpha, A_\alpha \cup \Omega_\alpha)]$ with the homomorphisms induced by projections is denoted by $H^{\blacktriangle}_q(X,A)$ $[H^q_{\blacktriangle}(X,A)]$. The provision concerning the categories in which the coefficient group may lie are the same as for the Čech groups of compact pairs (see IX,3.3 and the subsequent remarks).

THEOREM 6.9. *For every locally compact pair (X,A) we have*

$$H_q(\dot{X},\dot{A}) \approx H_q^{\blacktriangle}(X,A), \qquad H^q(\dot{X},\dot{A}) \approx H_{\blacktriangle}^q(X,A),$$

where the groups of $(\dot{X},\dot{A})$ are the Čech groups

PROOF. In $\mathrm{Cov}^f(\dot{X})$ we consider the subsets $\mathrm{Cov}^r(X)$ of coverings α which are regular at ω, i.e. coverings for which either $\omega \in \alpha_v$ or ω non $\in \bar{\alpha}_v$ for each v (see 3.5) By 3.6, $\mathrm{Cov}^r(X)$ is a cofinal subset of $\mathrm{Cov}^f(\dot{X})$, and therefore we may assume that the groups of $(\dot{X},\dot{A})$ are defined as limits over $\mathrm{Cov}^r(X)$. Next we define an order preserving map ϕ. $\mathrm{Cov}^r(\dot{X}) \to \mathrm{Cov}_1(X)$ as follows:

$$(\phi\alpha)_v = \alpha_v \cap X = \alpha_v - \omega.$$

For the nerves we then have the inclusion maps

$$\phi_\alpha\colon\ (X_{\phi\alpha},A_{\phi\alpha} \cup \Omega_{\phi\alpha}) \subset (\dot{X}_\alpha,\dot{A}_\alpha).$$

The system $\Phi = \{\phi,\phi_{\alpha *}\}$ constitutes a map of the inverse system of groups $H_q(\dot{X}_\beta,A_\beta \cup \Omega_\beta)$ indexed by $\mathrm{Cov}_1(X)$ into the system of groups $H_q(\dot{X}_\alpha,\dot{X}_\alpha)$ indexed by $\mathrm{Cov}^r(\dot{X})$. The limit map

$$\phi_\infty\colon\ H_q^{\blacktriangle}(X,A) \to H_q(\dot{X},\dot{A})$$

is thus defined Similarly for cohomology $\{\phi,\phi_\alpha^*\}$ is a map of direct systems, yielding a limit map

$$\phi^\infty\colon\ H^q(\dot{X},\dot{A}) \to H_{\blacktriangle}^q(X,A)$$

We assert that ϕ_∞ and ϕ^∞ are isomorphisms The proof breaks up into two cases according as the pair (X,A) is compact or not.

If the pair (X,A) is compact, then $\mathrm{Cov}^f(\dot{X}) = \mathrm{Cov}^r(\dot{X})$. Further the inclusion map $i\colon\ (X,A) \subset (\dot{X},A)$ is a relative homeomorphism and therefore i_* and i^* are isomorphisms Comparing the respective definitions shows that $i_* = \phi_\infty$ and $i^* = \phi^\infty$, which concludes the proof in this case

If (X,A) is not a compact pair, then it is easy to see that ϕ maps $\mathrm{Cov}^r(X)$ onto $\mathrm{Cov}_1(X)$ and the inclusion maps ϕ_α become identities. Thus $\phi_{\alpha *}$ and ϕ_α^* are isomorphisms, and it follows from VIII,3.15 and VIII,4.13 that ϕ_∞ and ϕ^∞ also are isomorphisms

REMARK 1. The discussion of the $^\blacktriangle$-groups was incomplete since we did not define induced homomorphisms and boundary [coboundary]; the missing definitions are quite analogous to those made for the Čech groups. After these definitions are made it is trivial to check that the isomorphisms ϕ_∞ obtained above commute with f_* and ∂, thus yielding an isomorphism between homology theories.

REMARK 2. Let X be locally compact but not compact. Let

$\alpha \in \mathrm{Cov}_1(X)$. Since every finite number of countercompact sets has an unbounded intersection, it follows that Ω_α is a simplex and consequently is homologically trivial It follows then from the exactness of the reduced homology sequence that $H_q(X_\alpha,\Omega_\alpha) \approx \tilde{H}_q(X_\alpha)$. Thus in defining the (absolute) groups $H_q^\blacktriangle(X)$ for a locally compact, non-compact, space we can use the groups $\tilde{H}_q(X_\alpha)$ rather then $H_q(X_\alpha,\Omega_\alpha)$. Note that, if X is compact, then $\Omega_\alpha = 0$ and $H_q(X_\alpha,\Omega_\alpha) = H_q(X_\alpha)$. Thus in this case $H_q^\blacktriangle(X_\alpha)$ uses the groups $H_q(X_\alpha)$ instead of $\tilde{H}_q(X_\alpha)$. The difference between these two cases is of course limited to the dimension zero A similar remark applies to cohomology.

Remark 3 It is easily shown by examples that the functor $\cdot$ does not preserve excisions of type (E). Thus the singular theory is not suitable for composition with this functor

7. LC-THEORIES IN TERMS OF A SINGLE SPACE

Let H be a homology theory on $\mathfrak{A}_C$ which is invariant with respect to relative homeomorphisms Given a compact pair (X,A), the identity map α of $X - A$ has an extension

$$\alpha^\cdot \quad (X,A) \to ((X - A)^\cdot,\omega)$$

which is a relative homeomorphism. Consequently

$$H_q(X,A) \overset{\alpha_*}{\approx} H_q((X - A)^\cdot,\omega) = H_q(X - A).$$

This suggests that one can describe the original homology theory using nonrelative homology groups alone, but defined over a suitable LC category. This leads to a new type of axiomatic system which we shall proceed to describe

Let $\mathfrak{B}_{LC}$ be the subcategory of $\mathfrak{A}_{LC}$ of all LC-spaces X and all maps f: $X \to Y$ such that for every compact subset C of Y the set $f^{-1}(C)$ is compact.

By a "single space" homology theory on $\mathfrak{B}_{LC}$ we shall mean a system $H = \{H_q(X),f_*,\tau,\partial\}$ where the primitive terms are as follows:

$H_q(X)$ is an abelian group defined for each $X \in \mathfrak{B}_{LC}$ and each integer q.

If f: $X \to Y$ is in $\mathfrak{B}_{LC}$, then f_* $H_q(X) \to H_q(Y)$ is a homomorphism

If U is an open subset of the LC-space X, then τ $H_q(X) \to H_q(U)$ is a homomorphism. We shall sometimes use the notation $\tau_{(X,U)}$ to indicate the spaces X and U

If U is an open subset of the LC-space X, then ∂: $H_q(U) \to H_{q-1}(X - U)$ is a homomorphism.

The system $H = \{H_q(X), f_*, \tau, \partial\}$ is subjected to the following axioms:

AXIOM 1. *If f: $X \to X$ is the identity, then f_* is the identity.*

AXIOM 1'. *$\tau_{(X,X)}$ is the identity.*

AXIOM 2. *If f: $X \to Y$, g: $Y \to Z$, then $(gf)_* = g_* f_*$.*

AXIOM 2'. *If $V \subset U \subset X$ and V and U are open in X, then $\tau_{(U,V)}\tau_{(X,U)} = \tau_{(X,V)}$*

AXIOM 3. *If $V \subset U \subset X$ and V and U are open in X, then commutativity holds in the diagram*

$$\begin{array}{ccc} H_q(U) & \xrightarrow{\tau} & H_q(V) \\ \downarrow \partial & & \downarrow \partial \\ H_{q-1}(X - U) & \xrightarrow{h_*} & H_{q-1}(X - V) \end{array}$$

where h: $X - U \subset X - V$.

AXIOM 3'. *If f: $X \to Y$, and U and V are open subsets of X and Y, respectively, such that $f(U) \subset V$ and $f(X - U) \subset (Y - V)$, then commutativity holds in the diagram*

$$\begin{array}{ccc} H_q(X) & \xrightarrow{f_*} & H_q(Y) \\ \downarrow \tau & & \downarrow \tau \\ H_q(U) & \xrightarrow{f_{1*}} & H_q(V) \end{array}$$

where f_1: $U \to V$ is defined by f.

AXIOM 3''. *Under the same conditions as in Axiom 3', commutativity holds in the diagram*

$$\begin{array}{ccc} H_q(U) & \xrightarrow{f_{1*}} & H_q(V) \\ \downarrow \partial & & \downarrow \partial \\ H_{q-1}(X - U) & \xrightarrow{f_{2*}} & H_{q-1}(Y - V) \end{array}$$

where f_2: $X - U \to Y - V$ is defined by f.

AXIOM 4 (*Exactness axiom*). *If U is an open subset of X and j: $X - U \subset X$, then the homology sequence*

$$\cdots \leftarrow H_{q-1}(X - U) \xleftarrow{\partial} H_q(U) \xleftarrow{\tau} H_q(X) \xleftarrow{j_*} H_q(X - U) \leftarrow \cdot$$

is exact (Note: for partially exact homology theories, we assume only that the homology sequence is of order 2, and is exact whenever $(X, X - U)$ is triangulable)

AXIOM 5 (*Homotopy axiom*). *If g_0, g_1: $X \to X \times I$ are defined by $g_0(x) = (x,0)$, $g_1(x) = (x,1)$, then $g_{0*} = g_{1*}$.*

AXIOM 7 (*Dimension axiom.*) *If P is a point, then $H_q(P) = 0$ for $q \neq 0$.*

Note the total absence of any analog of the excision axiom.

The main results about the connection between ordinary and "single space" homology theories is incorporated in the following four theorems:

THEOREM 7.1. *Let H be a homology theory on the category $\mathfrak{A}_C$ invariant with respect to relative homeomorphisms. For each $X \in \mathfrak{B}_{LC}$ define $H_q^\circ(X) = H_q(X,\omega)$ If f: $X \to Y$ is in $\mathfrak{B}_{LC}$ and $\dot{f}$. $(X,\omega) \to (Y,\omega)$ is an extension of f, then define f_*° $H_q^\circ(X) \to H_q^\circ(Y)$ by setting $f_*^\circ = \dot{f}_*$. Let U be an open subset of the LC-space X, and let f. $(\dot{X},\omega) \to (\dot{U},\omega)$ be defined by $f(x) = x$ for $x \in U$, $f(x) = \omega$ for $x \in X - U$ Then define τ° $H_q^\circ(X) \to H_q^\circ(U)$ by $\tau^\circ = f_*$. Finally to define ∂°. $H_q(U) \to H_{q-1}(X - U)$ consider the map g $(\dot{X},(X - U)^\cdot) \to (U,\omega)$ defined by f Then g is a relative homeomorphism and g_* is an isomorphism Define $\partial^\circ = \partial g_*^{-1}$ where ∂: $H_q(X,(X - U)^\cdot) \to H_{q-1}((X - U)^\cdot,\omega)$ is the boundary operator in the triple $(\dot{X},(X - U)^\cdot,\omega)$ Then the system $H^\circ = \{H_q^\circ(X), f_*^\circ, \tau^\circ, \partial^\circ\}$ is a "single space" homology theory defined over the category $\mathfrak{B}_{LC}$*

THEOREM 7.2 *Let H be a "single space" homology theory defined over the category $\mathfrak{B}_{LC}$ For each $(X\ A)$ in $\mathfrak{A}_C$ define $H_q^+(X,A) = H_q(X - A)$. Let f: $(X,A) \to (Y,B)$ (in $\mathfrak{A}_C$), let $U = f^{-1}(Y - B)$, and let f_1 $U \to Y - B$ be defined by f Then f_1 is a map in $\mathfrak{B}_{LC}$ and we define f_*^+. $H_q^+(X,A) \to H_q^+(Y,B)$ to be the composition*

$$H_q(X - A) \xrightarrow{\tau} H_q(U) \xrightarrow{f_{1*}} H_q(Y - B).$$

If (X,A) is an $\mathfrak{A}_C$, then we define ∂^+: $H_q^+(X,A) \to H_{q-1}^+(A)$ to be the homomorphism ∂. $H_q(X - A) \to H_{q-1}(A)$. Then the system $H^+ = \{H_q^+(X,A), f_*^+, \partial^+\}$ is a homology theory on $\mathfrak{A}_C$ invariant with respect to relative homeomorphisms

THEOREM 7.3. *Let H be a homology theory on $\mathfrak{A}_C$ invariant with respect to relative homeomorphisms. Then $H^{\circ +}$ is again such a theory, and*

$$H_q^{\circ +}(X,A) = H_q^{\circ}(X - A) = H_q((X - A)^{\cdot},\omega).$$

Let ϕ: $(X,A) \to ((X - A)^{\cdot},\omega)$ be defined by $\phi(x) = x$ for $x \in X - A$ and $\phi(x) = \omega$ for $x \in A$. Then ϕ is a relative homeomorphism and ϕ_*: $H_q(X,A) = H_q^{\circ +}(X,A)$ yields an isomorphism $H = H^{\circ +}$ of the homology theories.

THEOREM 7.4. *Let H be a "single space" homology theory on $\mathfrak{B}_{LC}$. Then $H^{+\circ}$ coincides with H.*

The proof of the above four theorems is straightforward but lengthy in view of the large number of propositions that have to be checked. We therefore leave these proofs as an exercise for the reader. The results remain valid for partially exact homology theories, and with suitable reformulations also for cohomology. If H is the Čech theory on $\mathfrak{A}_C$ then the "single space" theory H° on $\mathfrak{B}_{LC}$ may be given a direct description using finite open coverings.

8. TYCHONOFF IMBEDDING AND COMPACTIFICATION

DEFINITION 8.1. Given any set A consider the space I^A of all functions α: $A \to I$, where $I = [0,1]$ is the closed unit interval, with the product topology defined as in V,5.2. The space I^A is called *the cube indexed by the set A*. By V,5.4, I^A is compact. If B is a subset of A, then we identify the cube I^B with a subset of I^A by extending each function α: $B \to I$ to a function α': $A \to I$ such that $\alpha'(x) = 0$ for $x \in A - B$. With this convention we call I^B a *subcube* of I^A. The mapping p_B: $I^A \to I^B$ which to each α: $A \to I$ assigns $\alpha | B$: $B \to I$ is called the *projection* of I^A onto I^B. If the set A is finite, then I^A is called a *finite* cube.

DEFINITION 8.2. A T_1-space X is called *completely regular* if, for every closed set $A \subset X$ and every point x_0 not in A, there is a continuous function f: $X \to I$ such that $f(x_0) = 0$, $f(x) = 1$ for $x \in A$.

It follows from Urysohn's lemma that normal spaces are completely regular.

LEMMA 8.3. *Every subspace of a completely regular space is completely regular.*

PROOF. Let $Y \subset X$, where X is completely regular. Let A be a closed subset of Y and $y_0 \in Y - A$. Then the closure $\bar{A}$ of A in X is a closed subset of X and $y_0 \in X - \bar{A}$. Thus there is an f: $X \to I$ such that $f(y_0) = 0$ and $f(\bar{A}) = 1$. The function $g = f | Y$: $Y \to I$ satisfies $g(y_0) = 0$ and $g(A) = 1$, thus showing that Y is completely regular.

DEFINITION 8.4 Given a topological space X consider the set ξ of all maps f. $X \to I$. Define the *Tychonoff map* T. $X \to I^\xi$ by setting

$$T(x)(f) = f(x), \qquad \text{for } x \in X, f \in \xi.$$

The continuity of T follows from v,5 3.

LEMMA 8 5 *If X is completely regular and $A \subset X$ is closed, then there is a subcube Q of I^ξ such that $T(A) = Q \cap T(X)$.*

PROOF Let α be the subset of ξ consisting of all maps f. $X \to I$ which are not identically zero on A Let $x \in X$, the condition $T(x) \in I^\alpha$ is equivalent to $f(x) = 0$ for all f $X \to I$ such that $f(A) = 0$ By complete regularity this is equivalent to $x \in A$. Thus $T(A) = I^\alpha \cap T(X)$

THEOREM 8 6 *For every topological space X the following three conditions are equivalent.*

(i) *X is completely regular,*

(ii) *the Tychonoff map T $X \to I^\xi$ defines a homeomorphism of X with $T(X)$,*

(iii) *X is homeomorphic with a subset of a compact space*

PROOF. Suppose X is completely regular If x_0, x_1 are two distinct points of X, then there is an f $X \to I$ with $f(x_0) = 0$, $f(x_1) = 1$ It follows that $T(x_0)(f) \neq T(x_1)(f)$ so that $T(x_0) \neq T(x_1)$ Thus T is 1-1 If A is any closed subset of X, it follows from 8 5 that $T(A)$ is a closed subset of $T(X)$. Thus the inverse of T is continuous and T is a homeomorphism Thus (i) $\to$ (ii) Since I^ξ is compact, it follows that (ii) $\to$ (iii) Finally, since a compact space is completely regular, it follows from 8 3 that (iii) $\to$ (i)

LEMMA 8 7 *If X is any space, then any map f. $T(X) \to I$ can be extended to a map f' $I^\xi \to I$*

PROOF. Clearly $g = fT$: $X \to I$ is an element of ξ. For each $y \in I^\xi$ define $f'(y) = y(g)$. Then f' $I^\xi \to I$ If $y \in T(X)$, then $y = T(x)$ for some $x \in X$, and we have

$$f'(y) = y(g) = T(x)(g) = g(x) = fT(x) = f(y).$$

Hence f' is an extension of f

DEFINITION 8 8. Let $X^\sim$ be a compact space and $X \subset X^\sim$ a subspace of $X^\sim$. We shall say that $X^\sim$ is a *Tychonoff compactification of X* if X is dense in $X^\sim$ (i e. $\overline{X} = X^\sim$), and if every map f: $X \to I$ admits an extension $f^\sim$: $X^\sim \to I$.

LEMMA 8.9. *Let $X^\sim$ and $Y^\sim$ be Tychonoff compactifications of X and Y respectively. Then every map f: $X \to Y$ admits a unique extension $f^\sim$: $X^\sim \to Y^\sim$.*

PROOF. In view of 8.6, we may assume without loss of generality that $Y^{\sim}$ is a subset of some cube I^A. Then the map f is described by means of coordinate maps f_a: $X \to I$ for $a \in A$. Each f_a admits an extension $f_a^{\sim}$: $X^{\sim} \to I$, and together they yield an extension $f^{\sim}$: $X^{\sim} \to I^A$ of f. Since $\overline{X} = X^{\sim}$, it follows that $f^{\sim}(X^{\sim}) = \overline{f^{\sim}(X)} = \overline{f(X)} \subset \overline{Y} = Y^{\sim}$. Thus $f^{\sim}$: $X^{\sim} \to Y^{\sim}$. The uniqueness of $f^{\sim}$ also follows from $\overline{X} = X^{\sim}$.

THEOREM 8.10. *Every completely regular space X admits a Tychonoff compactification $X^{\sim}$. The compactification is unique in the following sense: If $X^{\sim}$ and X^{+} are two Tychonoff compactifications of X, then there is a unique homeomorphism h: $X^{\sim} \approx X^{+}$ which is the identity on X.*

PROOF. Let T: $X \to I^{\xi}$ be the Tychonoff map of X. It follows from 8.7 that $\overline{T(X)}$ is a Tychonoff compactification of $T(X)$. Since X is completely regular, $T(X)$ and X are homeomorphic, so that X also has a Tychonoff compactification.

Let $X^{\sim}$ and X^{+} be two Tychonoff compactifications of X. By 8.9, the identity map i: $X \to X$ admits extensions h: $X^{\sim} \to X^{+}$, g: $X^{+} \to X^{\sim}$. Then gh: $X^{\sim} \to X^{\sim}$ and hg: $X^{+} \to X^{+}$ are both the identity on X. Thus gh = identity and hg = identity. Hence $g = h^{-1}$ and h is a homeomorphism. This completes the proof.

In view of 8.10, we may regard the Tychonoff compactification $X^{\sim}$ of a completely regular space X as a well defined space. Then it follows from 8.9 that the functions $X^{\sim}, f^{\sim}$ form a covariant functor $\sim$ on the category of completely regular spaces and continuous maps with values in the category of compact spaces and continuous maps.

LEMMA 8.11. *If X is a normal space and $A \subset X$ is closed, then the closure $\overline{A}$ of A in $X^{\sim}$ is the Tychonoff compactification $A^{\sim}$ of A.*

PROOF. Let f: $A \to I$. By Tietze's extension theorem, there is an extension g: $X \to I$ of f. The map g in turn admits an extension $g^{\sim}$: $X^{\sim} \to I$. This extends f to $\overline{A}$ so that $\overline{A} = A^{\sim}$, and the proof is complete.

If X is normal and $A \subset X$, then we shall use the notation $\overline{A}$ for the closure of A in X and the notation $\overline{A}^{\sim}$ for the closure of A in $X^{\sim}$, which, in view of 8.11, coincides with the Tychonoff compactification $(\overline{A})^{\sim}$. Clearly

$$\overline{A} = X \cap \overline{A}^{\sim}.$$

If A is a closed subset of the normal space X, then the pair (X,A) will be called a *normal* pair. The category of normal pairs and their continuous maps is denoted by $\mathfrak{A}_N$. If (X,A) is a normal pair, then by 8.11, $(X^{\sim},A^{\sim})$ is a pair. If f: $(X,A) \to (Y,B)$ is a map of normal pairs, then also $f^{\sim}$: $(X^{\sim},A^{\sim}) \to (Y^{\sim},B^{\sim})$. Thus $\sim$ may be regarded as a covariant functor $\sim$: $\mathfrak{A}_N \to \mathfrak{A}_C$.

9. HOMOLOGY THEORIES FOR NORMAL SPACES

Our next objective is to convert the category $\mathfrak{A}_N$ of normal pairs into an h-category in such a manner that the Tychonoff compactification becomes an h-functor (see IV,9). Several preliminary results are needed.

LEMMA 9 1. *If A and B are closed subsets of the normal space X, then $A^\sim \cap B^\sim = (A \cap B)^\sim$ and $A^\sim \cup B^\sim = (A \cup B)^\sim$.*

PROOF. Since $A \cap B \subset A^\sim \cap B^\sim$, it follows that $(A \cap B)^\sim \subset A^\sim \cap B^\sim$ Suppose x_0 is not in $(A \cap B)^\sim$ Let V be a closed neighborhood of x_0 in $X^\sim$ such that $V \cap (A \cap B)^\sim = 0$ Then $V \cap A$ and B are disjoint closed sets in X Let f. $X \to I$ be a Urysohn function such that $f(V \cap A) = 0$ and $f(B) = 1$ Then $f^\sim\cdot\ X^\sim \to I$ satisfies $f^\sim((V \cap A)^\sim) = 0$ and $f^\sim(B^\sim) = 1$ If $x_0 \in A^\sim$, then $x_0 \in (V \cap A)^\sim$ and $f^\sim(x_0) = 0$, thus x_0 is not in $B^\sim$, and consequently x_0 is not in $A^\sim \cap B^\sim$. This proves $A^\sim \cap B^\sim \subset (A \cap B)^\sim$. The second part of the lemma is a general property of the closure operation

THEOREM 9 2 *If the inclusion map i: $(X_1,A_1) \subset (X,A)$ of normal pairs is an excision (type (E) of §5), then the same is true of $i^\sim$*

PROOF By definition, X_1,A_1 are closed in X, and

$$A_1 = X_1 \cap A, \qquad X = \text{Int } X_1 \cup \text{Int } A$$

The last condition is equivalent to

$$\overline{X - X_1} \cap \overline{X - A} = 0$$

We will show that the same relations hold for the compactified sets. By 9 1, we have

$$A_1^\sim = X_1^\sim \cap A^\sim, \qquad \overline{X - X_1^\sim} \cap \overline{X - A^\sim} = 0.$$

Since A is closed in X, we have $X \cap A^\sim = A$, and this implies $X - A \subset X^\sim - A^\sim$ Since X is dense in $X^\sim$, we have that $X - A$ is dense in $X^\sim - A^\sim$ Therefore

$$\overline{X - A^\sim} = \overline{X^\sim - A^\sim}$$

Since the same holds with A replaced by X_1, we have

$$\overline{X^\sim - X_1^\sim} \cap \overline{X^\sim - A^\sim} = 0.$$

Thus $i^\sim$ is an excision.

LEMMA 9 3 *If X is normal and i: $X \subset X^\sim$, then the map $i^{\sim 1}$: $\text{Cov}^f(X^\sim) \to \text{Cov}^f(X)$ is onto*

PROOF. Let α be a finite open covering of X indexed by the set V. For each $v \in V$ define $\beta_v = X^\sim - (X - \alpha_v)^\sim$. Then, by 9 1,

$$\bigcup_v \beta_v = X^\sim - \bigcap_v (X - \alpha_v)^\sim = X^\sim - (\bigcap_v (X - \alpha_v))^\sim$$
$$= X^\sim - (X - \bigcup \alpha_v)^\sim = X^\sim.$$

Therefore β is an open covering of $X^\sim$ indexed by V Since $X \cap \beta_v = X - X \cap (X - \alpha_v)^\sim = X - (X - \alpha_v) = \alpha_v$, it follows that $i^{-1}\beta = \alpha$

DEFINITION 9.4. If $\alpha \in \mathrm{Cov}(X)$ and $\beta \in \mathrm{Cov}(Y)$ are coverings indexed by the set V and W respectively, then we define the covering $\alpha \times \beta \in \mathrm{Cov}(X \times Y)$ indexed by the set $V \times W$ as follows.

$$(\alpha \times \beta)_{(v, w)} = \alpha_v \times \beta_w.$$

LEMMA 9 5. *If X and Y are compact, then coverings of the form $\alpha \times \beta$, $\alpha \in \mathrm{Cov}'X$, $\beta \in \mathrm{Cov}'Y$ form a cofinal subset of* $\mathrm{Cov}(X \times Y)$.

PROOF. Let $\gamma \in \mathrm{Cov}(X \times Y)$ Since $X \times Y$ is compact, and sets of the form $U \times W$, $U \subset X$, $W \subset Y$, U and W open, form a base for the open sets in $X \times Y$, there is a finite covering δ of $X \times Y$ such that $\gamma < \delta$ and each set δ_v is of the form $\delta_v = U_v \times W_v$. For each $x \in X$ and $y \in Y$ define

$$U(x) = \bigcap U_v \qquad \text{where} \qquad x \in U_v,$$
$$W(y) = \bigcap W_v \qquad \text{where} \qquad y \in W_v.$$

The distinct sets among the $U(x)$ and $W(y)$ form finite coverings α and β of X and Y respectively. Each set $U(x) \times W(y)$ is contained in some set $U_v \times W_v = \delta_v$ so that $\delta < \alpha \times \beta$. Thus $\gamma < \alpha \times \beta$

LEMMA 9.6. *Let i. $A \subset X$ be an inclusion map where A is a dense subset of the space X Let f. $A \to Y$ where Y is compact The map f has an extension F. $X \to Y$ if and only if, for every covering $\beta \in \mathrm{Cov}'(Y)$, there is a covering $\alpha \in \mathrm{Cov}'(X)$ such that $f^{-1}\beta < i^{-1}\alpha$.*

PROOF. Assume that the extension F: $X \to Y$ of f exists. Given any $\gamma \in \mathrm{Cov}'(Y)$, define $\alpha = F^{-1}(\gamma)$ Then, since $f = Fi$, we have $f^{-1}\gamma = i^{-1}F^{-1}\gamma = i^{-1}\alpha$. Thus the condition is necessary.

To prove the sufficiency of the condition, consider for each $x \in X$ the family $N(x)$ of all neighborhoods of x For each $V \in N(x)$ define $S_V = \overline{f(V \cap A)}$. If $V_1, \cdots, V_n \in N(x)$, then $V = V_1 \cap \cdots \cap V_n \in N(x)$, and $S_V \subset S_{V_1} \cap \cdots \cap S_{V_n}$. Since $S_V \neq 0$, it follows that $S_{V_1} \cap \cdots \cap S_{V_n} \neq 0$. Thus any finite number of subsets of $\{S_V\}$ has a nonvacuous intersection Since the sets S_V are closed in Y, and Y is compact, it follows that the intersection $F(x) = \cap S_V$ for $V \in N(x)$ is nonvacuous. We shall prove that $F(x)$ consists of a single point. Indeed let $y_1, y_2 \in F(x)$. Select a covering $\beta \in \mathrm{Cov}'(Y)$ consisting of two sets β_1,β_2 such that $y_1 \in \beta_1 - \bar{\beta}_2$, $y_2 \in \beta_2 - \bar{\beta}_1$. Let $\alpha \in \text{Cov}^f(X)$ be such that $f^{-1}\beta < \imath^{-1}\alpha$ Let $x \in \alpha_v$; then $\alpha_v = V$ is an element of the family $N(x)$, and

$$S_V = \overline{f(\alpha_v \cap A)}.$$

Since $f(\alpha_v \cap A)$ is contained in either β_1 or β_2, we may assume $f(\alpha_v \cap A) \subset \beta_1$ Then $S_V \subset \bar{\beta}_1$ so that y_2 is not in S_V, and therefore y_2 is not in $F(x)$ Thus $F(x)$ is a single point.

If $x \in A$, then $f(x) \in S_V$ for all $V \in N(x)$ so that $f(x) \in \bigcap S_V$, and therefore $F(x) = f(x)$. Thus F is an extension of f It remains to be proved that F is continuous

Let U be any open set in Y containing $F(x)$ Choose a covering $\beta \in \text{Cov}^f(Y)$ consisting of two sets β_1 and β_2 such that $F(x) \in \beta_1 - \bar{\beta}_2$, and $\bar{\beta}_1 \subset U$. Let $\alpha \in \text{Cov}^f(X)$ be such that $f^{-1}\beta < \imath^{-1}\alpha$ and let $x \in \alpha_v$. Since $\alpha_v = V \in N(x)$ and $S_V = \overline{f(\alpha_v \cap A)}$, it follows that $F(x) \in \overline{f(\alpha_v \cap A)}$ Since $f(\alpha_v \cap A)$ is contained in either β_1 or β_2, since $F(x) \in \overline{f(\alpha_v \cap A)}$, and since $F(x) \in \beta_1 - \bar{\beta}_2$, it follows that $f(\alpha_v \cap A) \subset \beta_1$. For every $x' \in \alpha_v$, we have $\alpha_v = V \in N(x')$ so that $F(x') \in S_V = \overline{(f\alpha_v \cap A)} \subset \bar{\beta}_1 \subset U$. Thus $F(\alpha_v) \subset U$, which proves that F is continuous

LEMMA 9.7. *Let X,Y be completely regular, let Z be a normal space, and let f: $X \times Y \to Z$ An extension F. $X^\sim \times Y^\sim \to Z^\sim$ of f exists if and only if for every $\gamma \in \text{Cov}^f(Z)$ there exist $\alpha \in \text{Cov}^f(X)$ and $\beta \in \text{Cov}^f(Y)$ such that $f^{-1}\gamma < \alpha \times \beta$*

PROOF Consider the inclusion maps

$$\imath_1\colon X \to X^\sim, \qquad \imath_2\colon Y \to Y^\sim, \qquad \imath\colon Z \to Z^\sim.$$

Assume that an extension F $X^\sim \times Y^\sim \to Z^\sim$ is given If $\gamma \in \text{Cov}^f(Z)$, then by 9.3 there is a $\gamma' \in \text{Cov}^f(Z^\sim)$ such that $\gamma = \imath^{-1}\gamma'$ By 9.5 there exist coverings $\alpha' \in \text{Cov}^f(X^\sim)$, $\beta' \in \text{Cov}^f(Y^\sim)$ such that $F^{-1}\gamma' < \alpha' \times \beta'$. Let $\alpha = \imath_1^{-1}\alpha'$, $\beta = \imath_2^{-1}\beta'$. Then $\alpha \in \text{Cov}^f(X)$, $\beta \in \text{Cov}^f(Y)$, and since $\imath f = F(\imath_1 \times \imath_2)$, we have

$$\begin{aligned} f^{-1}\gamma &= f^{-1}\imath^{-1}\gamma' = (\imath_1^{-1} \times \imath_2^{-1})F^{-1}\gamma' < (\imath_1^{-1} \times \imath_2^{-1})(\alpha' \times \beta') \\ &= \imath_1^{-1}\alpha' \times \imath_2^{-1}\beta' = \alpha \times \beta. \end{aligned}$$

Conversely, assume that the condition of the lemma is satisfied. Consider the map $g = \imath f$ $X \times Y \to Z^\sim$, and let $\gamma \in \text{Cov}^f(Z^\sim)$, then $\imath^{-1}\gamma \in \text{Cov}^f(Z)$, and there exist coverings $\alpha \in \text{Cov}^f(X)$ and $\beta \in \text{Cov}^f(Y)$ such that $f^{-1}\imath^{-1}\gamma < \alpha \times \beta$. Then $g^{-1}\gamma = f^{-1}\imath^{-1}\gamma < \alpha \times \beta$, and by 9 6, the map g admits an extension F: $X^\sim \times Y^\sim \to Z^\sim$. Clearly F is also an extension of f, which concludes the proof.

DEFINITION 9 8. Let (X,A) and (Y,B) be normal pairs. A homotopy F: $(X \times I, A \times I) \to (Y,B)$ is called *uniform* if, for every finite open covering $\gamma \in \mathrm{Cov}^f(Y)$, there exist coverings $\alpha \in \mathrm{Cov}^f(X)$ and $\beta \in \mathrm{Cov}^f(I)$ such that $F^{-1}\gamma < \alpha \times \beta$.

As an immediate consequence of 9.7 we obtain

THEOREM 9.9 *Let f_0,f_1: $(X,A) \to (Y,B)$ be maps of normal pairs. A homotopy F: $(X \times I, A \times I) \to (Y,B)$ between f_0 and f_1 is uniform if and only if F can be extended to a homotopy G: $(X^\sim \times I, A^\sim \times I) \to (Y^\sim,B^\sim)$ between the maps $f_0^\sim,f_1^\sim$: $(X^\sim,A^\sim) \to (Y^\sim,B^\sim)$.*

Theorems 9 2 and 9 9 indicate clearly how to convert the category $\mathfrak{A}_N$ into an h-category The *couples* in $\mathfrak{A}_N$ are inclusions i $A \subset X$, j: $X \subset (X,A)$ where (X,A) is a normal couple *Excisions* are defined as usual, namely, type (E) of §5. *Homotopies* are defined to be uniform homotopies. *Points* are defined in the ordinary way. We now can state

THEOREM 9.10. *The Tychonoff compactification is a covariant h-functor $\sim$: $\mathfrak{A}_N \to \mathfrak{A}_C$*

DEFINITION 9.11. Let H be a partially exact homology [cohomology] theory on the category $\mathfrak{A}_C$. Regard H as a theory on $\mathfrak{A}_C$ and denote by $H^\sim$ the composition of H with the h-functor $\sim$: $\mathfrak{A}_N \to \mathfrak{A}_C$ The result is a partially exact homology [cohomology] theory on $\mathfrak{A}_N$, called the *normal space theory associated with H*. If H is exact, so also is $H^\sim$

Since $\mathfrak{A}_C$ is clearly a subcategory of $\mathfrak{A}_N$ and since $(X^\sim,A^\sim) = (X,A)$ for compact pairs, it is clear that $H^\sim$ is an extension of H from $\mathfrak{A}_C$ to $\mathfrak{A}_N$.

If H is the Čech theory on $\mathfrak{A}_C$, then the associated theory $H^\sim$ on $\mathfrak{A}_N$ may be given a direct description in terms of nerves of coverings. We recall that in IX,3, in addition to the Čech groups $H_q(X,A)$, we defined also the groups $H_q^f(X,A)$ using the finite coverings $\mathrm{Cov}^f(X,A)$ for the passage to the limit We will show that, if the pair (X,A) is normal, this group is essentially identical with $H_q^\sim(X,A) = H_q(X^\sim,A^\sim)$.

THEOREM 9.12. *Let (X,A) be a normal pair and let $H_q^f(X,A)$ [$H_f^q(X,A)$] be the homology [cohomology] groups defined using the Čech method and finite open coverings Then the inclusion map i: $(X,A) \subset (X^\sim,A^\sim)$ induces isomorphisms*

$$i_*: H_q^f(X,A) \approx H_q(X^\sim,A^\sim) \qquad [i^*: H^q(X^\sim,A^\sim) \approx H_f^q(X,A)].$$

PROOF. Since the subset A is closed, and in view of the discussion in IX,8, we may replace the directed set $\mathrm{Cov}^f(X,A)$ by the directed set $\mathrm{Cov}^f(X)$. By 9 3, the map i^{-1}: $\mathrm{Cov}^f(X^\sim) \to \mathrm{Cov}^f(X)$ is onto If $\alpha \in \mathrm{Cov}^f(X^\sim)$ and $\beta = i^{-1}\alpha \in \mathrm{Cov}^f(X)$, then, since X is dense in $X^\sim$ and A is dense in $A^\sim$, it follows readily that $(X^\sim_\alpha,A^\sim_\alpha) = (X_\beta,A_\beta)$ so that $H_q(X^\sim_\alpha,A^\sim_\alpha) = H_q(X_\beta,A_\beta)$. Thus the fact that i_* [i^*] is an isomorphism is a consequence of VIII,3 15 and VIII,4.13.

If we combine the above result with the obvious fact that the isomorphism i_* commutes with f_* and ∂ we obtain an isomorphism $H' \approx H^{\sim}$

THEOREM 9 13 *Let* f: $(X,A) \to (Y,B)$ *be a relative homeomorphism of normal pairs Then* $f^{\sim}$: $(X^{\sim},A^{\sim}) \to (Y^{\sim},B^{\sim})$ *is a relative homeomorphism if and only if* $f(C)$ *is closed in* Y *for each closed set* C *of* X *such that* $C \subset X - A$

PROOF: Suppose $f^{\sim}$ is a relative homeomorphism, and $C \subset X - A$ is closed in X By 9.1, $C^{\sim} \cap A^{\sim} = 0$ Hence $f^{\sim}$ maps $C^{\sim}$ topologically into $Y^{\sim} - B^{\sim}$. Since

$$C = X \cap C^{\sim} = (X - A) \cap C^{\sim},$$

and $f^{\sim}$ is 1-1 on $X - A$ and $C^{\sim}$, we have

$$f(C) = (Y - B) \cap f^{\sim}(C^{\sim}) = Y \cap f^{\sim}(C^{\sim}).$$

As $C^{\sim}$ is compact, $f^{\sim}(C^{\sim})$ is closed in $Y^{\sim}$ Hence $f(C)$ is closed in Y This proves the necessity of the condition

Assuming the condition fulfilled, we will show that $f^{\sim}$ is a relative homeomorphism Since Y is dense in $Y^{\sim}$ and $B \subset B^{\sim}$, it follows that $Y - B$ is dense in $Y^{\sim} - B^{\sim}$ Therefore $Y^{\sim} - B^{\sim} \subset f^{\sim}(X^{\sim})$. As $f^{\sim}(A^{\sim}) \subset B^{\sim}$, we have

$$Y^{\sim} - B^{\sim} \subset f^{\sim}(X^{\sim} - A^{\sim}).$$

Suppose $x_0,x_1 \in X^{\sim} - A^{\sim}$, and $x_0 \neq x_1$ Let N_0,N_1 be closures of neighborhoods in $X^{\sim}$ of x_0,x_1, respectively, such that A,N_0,N_1 are pairwise disjoint Let $C_i = N_i \cap X$ and $D_i = f(C_i)$ for $i = 0,1$ Since X is dense in $X^{\sim}$, it follows that C_i is dense in N_i, and therefore $C_i^{\sim} = N_i$. Since N_i is closed, C_i is closed in X Then, by hypothesis, D_i is closed in Y. Thus B,D_0,D_1 are closed disjoint sets of Y By 9 1, $B^{\sim},D_0^{\sim},D_1^{\sim}$ are disjoint Since

$$f^{\sim}(x_i) \in f^{\sim}(N_i) = f^{\sim}(C_i^{\sim}) \subset D_i^{\sim}, \qquad i = 0,1,$$

it follows that $f^{\sim}(x_0)$ and $f^{\sim}(x_1)$ are distinct points of $Y^{\sim} - B^{\sim}$. This combined with the result of the preceding paragraph shows that $f^{\sim}$ defines a 1-1 map of $X^{\sim} - A^{\sim}$ onto $Y^{\sim} - B^{\sim}$ Then 5 2 states that $f^{\sim}$ is a relative homeomorphism.

REMARK If the homology theory H on $\mathfrak{a}_C$ is invariant under relative homeomorphisms, e g, the Čech theory, then we may use in $\mathfrak{a}_N$ the larger class of "excisions" described in 9 13 Also, if H is invariant under excisions of type (E_1), the same holds for $H^{\sim}$.

10. COMPACT PAIRS AS LIMITS OF TRIANGULABLE PAIRS

The objective of this section is to prove the following theorem:

THEOREM 10.1. *Every compact pair is homeomorphic with the inverse limit of an inverse system of triangulable pairs.*

The proof will be preceded by two lemmas.

LEMMA 10.2. *If Q is a finite cube, R is a subcube of Q, then the pair (Q,R) is triangulable.*

PROOF. Since Q is a finite product of intervals an iterated application of II,8 9 yields a triangulation of Q in which R is a subcomplex.

LEMMA 10.3. *Let Q be a finite cube, R a subcube, X a closed subset of Q, and U an open set such that $X \subset U \subset Q$ Then there exists a set $Y \subset U$ such that $X \subset \text{Int } Y$ and the pair $(Y, Y \cap R)$ is triangulable.*

PROOF. Let T be a triangulation of (Q,R) and let ${}^iT$ denote the ith barycentric subdivision of T. Select a closed set X' such that

$$X \subset \text{Int } X' \subset X' \subset U,$$

and choose i large enough so that every closed simplex of ${}^iT$ intersecting X' is in U Such an i exists by II,6 5 Let Y be the union of all those simplexes of ${}^iT$ which meet X' Then Y and $Y \cap R$ are subcomplexes in the triangulation ${}^iT$, hence $(Y, Y \cap R)$ is triangulable. Since $X' \subset Y$, it follows that $X \subset \text{Int } Y$, and the lemma is proved.

PROOF OF 10 1. Let (X,A) be a compact pair. By 8.6 and 8 5, we may assume that X is a closed subset of a cube I^ξ, and that $A = X \cap R$ where $R = I^\alpha$ is a subcube of I^ξ, $\alpha \subset \xi$

For each subset a of ξ we shall consider the subcube I^a and the projection p_a: $I^\xi \to I^a$.

Let M denote the set of all pairs $m = [a,N]$ where

(1) a is a finite subset of ξ,

(2) $N \subset I^a$,

(3) $p_aX \subset \text{Int } N$ (relative to I^a),

(4) the pair $(N, N \cap R)$ is triangulable

A quasi-order is defined in M as follows: If $m_1 = [a_1,N_1]$ and $m_2 = [a_2,N_2]$, then $m_1 < m_2$ provided

(5) $a_1 \subset a_2$,

(6) $p_{a_1}N_2 \subset N_1$.

To prove that M is directed, let $m_1 = [a_1,N_1]$, $m_2 = [a_2,N_2]$, and define

$$a = a_1 \cup a_2, \qquad B = I^a \cap p_{a_1}^{-1}(N_1) \cap p_{a_2}^{-1}(N_2).$$

Then

$$p_aX \subset \text{Int } B \subset I^a,$$

and by 10.3 there is a set $N \subset I^a$ such that $m = [a,N]$ is an element of M. Then $m_1 < m$ and $m_2 < m$; hence M is directed.

We now define an inverse system of triangulable pairs $\{(X_m,A_m),\pi^{m_2}_{m_1}\}$ indexed by the set M as follows: If $m = [a,N]$, then $(X_m,A_m) = (N,N \cap R)$. If $m_1 = [a_1,N_1]$, $m_2 = [a_2,N_2]$ and $m_1 < m_2$, then $\pi^{m_2}_{m_1}$ is the map of $(N_2,N_2 \cap R)$ into $(N_1,N_1 \cap R)$ defined by the projection p_{a_1}: $I^{\xi} \to I^{a_1}$ The verification that this defines an inverse system of spaces is immediate Let (X_∞,A_∞) be the limit of this inverse system For each $m = [a,N]$, consider the map f_m $(X,A) \to (X_m,A_m)$ defined by the projection p_a: $I^{\xi} \to I^a$ which carries X into $N = X_m$, and $A = X \cap R$ into $N \cap R = A_m$ The system of maps $\{f_m\}$ constitutes a map of the pair (X,A) into the system (X_m,A_m), and defines a limit map f_∞ $(X,A) \to (X_\infty,A_\infty)$ (see VIII,2 and VIII,3)

We shall prove that f_∞ is a homeomorphism Let a be any finite subset of ξ Since, by 10 1, the pair $(I^a,I^{a \cap a})$ is triangulable, it follows that $m = [a,I^a]$ is an element of M Let $x \in X_\infty$ and set $x(a) = \pi_m(x)$ where a is any finite subset of ξ, and $m = [a,I^a]$. We prove the following properties of $x(a)$

(7) $x(a) \in p_aX$.

(8) If $x \in A_\infty$, then $x(a) \in R$.

(9) If $a_1 \subset a_2$, then $p_{a_1}x(a_2) = x(a_1)$.

(10) If $m' = [a,N]$, then $\pi_{m'}(x) = x(a)$.

To prove (7), assume that $x(a)$ is not in p_aX Then, by 10 2, there is an $m' = [a,N]$ with $x(a)$ not in N; but $x(a) = \pi^{m'}_m x_{m'} = x_{m'} \in N$ which provides a contradiction Propositions (8), (9), and (10) are obvious. Condition (9) implies the existence of an element $x' \in I^{\xi}$ such that $p_a(x') = x(a)$ for all finite subsets a of ξ. It is easy to see that $x' \in X$, $f_\infty(x') = x$, and that $x' \in A$ if $x \in A_\infty$ This shows that f_∞ maps (X,A) onto (X_∞,A_∞)

Suppose that $x_1,x_2 \in X$ and $x_1 \neq x_2$ Then, for some finite subset a of ξ, $p_ax_1 \neq p_ax_2$ Hence $\pi_mx_1 \neq \pi_mx_2$ where $m = [a,I^a]$, and therefore $f_\infty(x_1) \neq f_\infty(x_2)$. This concludes the proof.

11. CANONICAL MAPPINGS OF SPACES INTO NERVES

DEFINITION 11.1. Let α be a covering of X with nerve X_α. For every $x \in X$, the vertices v of X_α such that $x \in \alpha$, form a simplex of X_α that is denoted by $\Delta_\alpha(x)$ and is called *the simplex determined by x in X_α*. Clearly $\Delta_\alpha(x)$ is the largest simplex s of X_α such that x is in the carrier of s

LEMMA 11 2 *Let f. $X \to Y$, let β be a covering of Y, and let $\alpha = f^{-1}\beta$. For every $x \in X$,*

$$\Delta_\alpha(x) = \Delta_\beta(f(x)).$$

This follows from the fact that, for every vertex v of Y_β, the relations $f(x) \in \beta_v$ and $x \in \alpha_v = f^{-1}(\beta_v)$ are equivalent.

DEFINITION 11 3 Let α be a covering of X with nerve X_α. A mapping

$$\phi: \ X \rightarrow |X_\alpha|$$

s called *canonical relative to* α if, for every $x \in X$, the point $\phi(x)$ is in the closed simplex $\Delta_\alpha(x)$.

LEMMA 11 4 *If* $f\cdot \ X \rightarrow Y$ *and* $\phi\cdot \ Y \rightarrow |Y_\beta|$ *is a canonical map and* $\alpha = f^{-1}(\beta)$, *then* ϕf *maps* X *into* X_α *(recall that* $X_\alpha \subset Y_\beta$*), and* ϕf *is a canonical map relative to* α

This follows directly from 11 2 As a corollary of 11 4, we have

LEMMA 11 5. *If* ϕ: $X \rightarrow |X_\alpha|$ *is canonical and* $A \subset X$, *then* ϕ *defines a canonical map of* A *into* A_α *We then speak of the canonical map* ϕ. $(X,A) \rightarrow (|X_\alpha|,|A_\alpha|)$

LEMMA 11 6 *Let* α,β *be coverings of* X, *and let* β *be a refinement of* α. *If* ϕ. $(X,A) \rightarrow (|X_\beta|,|A_\beta|)$ *is canonical (relative to* β*) and* p: $(X_\beta,A_\beta) \rightarrow (X_\alpha,A_\alpha)$ *is a projection, then* $p\phi$ $(X,A) \rightarrow (|X_\alpha|,|A_\alpha|)$ *is canonical (relative to* α*)*

PROOF. Let $x \in X$ and let v be a vertex of $\Delta_\beta(x)$ Since $x \in \beta_v \subset \alpha_{p(v)}$, it follows that $p(v)$ is a vertex of $\Delta_\alpha(x)$ Hence p maps the (closed) simplex $\Delta_\beta(x)$ into the (closed) simplex $\Delta_\alpha(x)$, so that $p\phi(x) \in \Delta_\alpha(x)$, and $p\phi$ is canonical.

LEMMA 11.7 *Any two canonical maps* ϕ_0,ϕ_1 $(\dot{X},A) \rightarrow (|X_\alpha|,|A_\alpha|)$ *are homotopic*

This follows at once from the observation that, for every $x \in X$ (or $x \in A$), the points $\phi_0(x)$ and $\phi_1(x)$ are in the same simplex of X_α(or A_α)

THEOREM 11 8. *For every finite open covering* α *of a normal space* X *there exists a canonical map* ϕ $X \rightarrow |X_\alpha|$.

PROOF. By 3 3, there exists a closed covering β of X which is a reduction of α. By Urysohn's Lemma there exists, for each $v \in V_\alpha = V_\beta$, a continuous function f_v. $X \rightarrow I$ (I = closed unit interval [0,1]) such that

$$f_v(x) = 0 \qquad \text{for} \quad x \in X - \alpha_v,$$
$$f_v(x) = 1 \qquad \text{for} \quad x \in \beta_v.$$

Since $\bigcup \beta_v = X$, the sum $f(x) = \sum_v f_v(x)$ is positive. Define $\phi_v = f_v/f$. Then $\phi_v(x) \geqq 0$, $\sum \phi_v(x) = 1$, and $\phi_v(x) = 0$ for x not in α_v. Therefore

$$\phi(x) = \sum_v \phi_v(x)v$$

is a canonical map ϕ. $X \rightarrow |X_\alpha|$

Using the concept of a canonical map we prove the following approximation theorem that will be used in the next section.

THEOREM 11.9 *Let (X,A) be the inverse limit of the inverse system of compact pairs $\{(X_m,A_m),\pi_m^{m_1}\}$ For every mapping f: $(X,A) \to (Y,B)$ where (Y,B) is a triangulable pair, there exists an index m and a map f_m: $(X_m,A_m) \to (Y,B)$ such that the maps f and $f_m\pi_m$ of (X,A) into (Y,B) are homotopic*

PROOF. Let $T = \{t,(K,L)\}$ be a triangulation of (Y,B), and let τ be the associated covering of Y as defined in IX,9 1 Then $(Y_\tau,B_\tau) = (K,L)$, and the map t^{-1} $(Y,B) \to (|K|,|L|)$ is canonical Let $\alpha = f^{-1}\tau$; then (X_α,A_α) is contained in (K,L), and, by 11.4, the map $t^{-1}f$ is canonical. We now apply 3 8, and find an index m and a covering β of X_m such that the covering $\gamma = \pi_m^{-1}\beta$ of X is a refinement of α. and $(X_\gamma,A_\gamma) = (X_{m\beta},A_{m\beta})$.
Let

$$p: \quad (X_\gamma,A_\gamma) \to (X_\alpha,A_\alpha)$$

be a projection, and let

$$\phi: \quad (X_m,A_m) \to (|X_{m\beta}|,|A_{m\beta}|)$$

be a canonical map which exists by virtue of 11 8 and 11 5 It follows from 11 4 and 11 5 that the map

$$\phi\pi_m \quad (X,A) \to (|X_\gamma|,|A_\gamma|)$$

is canonical (relative to γ), and therefore, by 11.6, the map

$$p\phi\pi_m. \quad (X,A) \to (|X_\alpha|,|A_\alpha|)$$

is canonical (relative to α). Since $t^{-1}f$ is also canonical (relative to α), 11 7 implies that the maps $p\phi\pi_m$ and $t^{-1}f$ of (X,A) into $(|K|,|L|)$ are homotopic Since t is a homeomorphism, the maps $tp\phi\pi_m$ and f of (X,A) into (Y,B) are homotopic Thus the map $f_m = tp\phi$ satisfies the conditions of the theorem.

12. THE CONTINUITY UNIQUENESS THEOREM

It will be proved in this section that the Čech theory is essentially the only one satisfying the continuity requirement.

THEOREM 12.1 *Let H and $\overline{H}$ be two partially exact homology [cohomology] theories defined on the category $\mathcal{a}_C$ of compact pairs, and let $\overline{H}$ be continuous Let $G,\overline{G}$ be the coefficient groups of $H,\overline{H}$ respectively. Then any homomorphism*

$$h_0: \; G \to \overline{G} \qquad [h_0; \; \overline{G} \to G]$$

can be extended in just one way to a homomorphism

$$h: H \to \overline{H} \qquad [h: \overline{H} \to H] \qquad \text{on} \qquad \mathfrak{A}_C.$$

If H is also continuous and h_0: $G \approx \overline{G}$ [h_0: $\overline{G} \approx G$] *is an isomorphism, then h is also an isomorphism*

$$h: H \approx \overline{H} \qquad [h: \overline{H} \approx H] \qquad \text{on} \qquad \mathfrak{A}_C.$$

Taking $\overline{H}$ to be the Čech theory, one obtains

THEOREM 12.2. *Any continuous partially exact homology [cohomology] theory on the category* $\mathfrak{A}_C$ *of compact pairs is isomorphic with the Čech theory over the same coefficient group.*

In this theorem the values of the homology theory are either in $\mathcal{G}_R$ or $\mathcal{G}_C$. The cohomology theory has values in $\mathcal{G}_R$. The Čech cohomology theory with values in $\mathcal{G}_C$ are in general not defined, and the concept of continuity for a cohomology theory with values in $\mathcal{G}_C$ was not defined either (see §2).

PROOF OF 12.1. The proof will be restricted to homology. The proof for cohomology is dual.

On the category $\mathfrak{I}$ of triangulable pairs, both H and $\overline{H}$ are exact theories. By the Uniqueness theorem IV,10.2, we may assume that the homomorphism

$$h: H \to \overline{H} \qquad \text{on} \qquad \mathfrak{I}$$

is already given.

Let (X,A) be a compact pair. By 10.1 there exists an inverse system $(\mathbf{X},\mathbf{A})$ in Inv$\mathfrak{I}$ and a homeomorphism ϕ: lim $(\mathbf{X},\mathbf{A}) \to (X,A)$. The pair consisting of $(\mathbf{X},\mathbf{A})$ and of ϕ will be called a *development* of (X,A) and will be denoted by the letter D.

If M is the indexing set of $(\mathbf{X},\mathbf{A})$, then, for every $m \in M$, we have a homomorphism

$$h(q,X_m,A_m): \quad H_q(X_m,A_m) \to \overline{H}_q(X_m,A_m).$$

Their totality is a homomorphism of inverse systems

$$(1) \qquad h(q,\mathbf{X},\mathbf{A}): \quad H_q(\mathbf{X},\mathbf{A}) \to \overline{H}_q(\mathbf{X},\mathbf{A})$$

as defined in §2. Using the homomorphisms of 2.1, we obtain the diagram

$$\begin{array}{ccccc} H_q(X,A) & \xleftarrow{\phi_*} & H_q(\lim (\mathbf{X},\mathbf{A})) & \xrightarrow{l(q,\mathbf{X},\mathbf{A})} & \lim H_q(\mathbf{X},\mathbf{A}) \\ & & & & \downarrow h_\infty(q,\mathbf{X},\mathbf{A}) \\ \overline{H}_q(X,A) & \xleftarrow{\bar{\phi}_*} & \overline{H}_q(\lim (\mathbf{X},\mathbf{A})) & \xrightarrow{\bar{l}(q,\mathbf{X},\mathbf{A})} & \lim \overline{H}_q(\mathbf{X},\mathbf{A}) \end{array}$$

where $h_\infty(q,\mathbf{X},\mathbf{A})$ is the inverse limit of the map (1) Since ϕ is a homeomorphism, ϕ_* and $\bar{\phi}_*$ are isomorphisms Finally $\bar{l}$ is an isomorphism by virtue of the assumption that $\bar{H}$ is a continuous homology theory Define the homomorphism

$$h(q,X,A,D)\colon\ H_q(X,A) \to \bar{H}_q(X,A)$$

by setting

$$h(q,X,A,D) = \bar{\phi}_*\bar{l}^{-1}h_\infty l\phi_*^{-1}.$$

We shall establish the following properties of $h(q,X,A,D)$.

(2) If $(X,A) \in \mathfrak{J}$, then $h(q,X,A,D) = h(q,X,A)$

(3) If D' is another development of (X,A), then $h(q,X,A,D) = h(q,X,A,D')$

(4) If $f\colon (X,A) \to (Y,B)$ and D and D' are developments of (X,A) and (Y,B) respectively, then commutativity holds in the diagram

$$\begin{array}{ccc} H_q(X,A) & \xrightarrow{h(q,X,A,D)} & \bar{H}_q(X,A) \\ \downarrow f_* & & \downarrow \bar{f}_* \\ H_q(Y,B) & \xrightarrow{h(q,Y,B,D')} & \bar{H}_q(Y,B) \end{array}$$

(5) Let D' denote the development of A induced by the development D of (X,A), then commutativity holds in the diagram

$$\begin{array}{ccc} H_q(X,A) & \xrightarrow{h(q,X,A,D)} & \bar{H}_q(X,A) \\ \downarrow \partial & & \downarrow \bar{\partial} \\ H_{q-1}(A) & \xrightarrow{h(q-1,A,D')} & \bar{H}_{q-1}(A) \end{array}$$

Propositions (2)-(5) imply that setting $h(q,X,A) = h(q,X,A,D)$ does yield a homomorphism $h.\ H \to \bar{H}$ extending h_0

Without any loss of generality we may assume that (X,A) and (Y,B) actually are the inverse limits of $(\mathbf{X},\mathbf{A})$ and $(\mathbf{Y},\mathbf{B})$, and therefore that the homeomorphisms ϕ and ϕ_1 are identity maps We begin by proving two auxiliary propositions

(6) If π_m $(X,A) \to (X_m,A_m)$ is a projection, then commutativity holds in the diagram

$$\begin{array}{ccc}
H_q(X,A) & \xrightarrow{h(q,X,A,D)} & \bar{H}_q(X,A) \\
\downarrow{\scriptstyle \pi_{m*}} & & \downarrow{\scriptstyle \bar{\pi}_{m*}} \\
H_q(X_m,A_m) & \xrightarrow{h(q,X_m,A_m)} & \bar{H}_q(X_m,A_m)
\end{array}$$

Consider the diagram

$$\begin{array}{ccccc}
 & \lim H_q(\mathbf{X},\mathbf{A}) & \xrightarrow{h_\infty(q,\mathbf{X},\mathbf{A})} & \lim \bar{H}_q(\mathbf{X},\mathbf{A}) & \\
{\scriptstyle l}\nearrow & & & & \nwarrow{\scriptstyle \bar{l}} \\
H_q(X,A) & \downarrow{\scriptstyle \tau_m} & & \downarrow{\scriptstyle \bar{\tau}_m} & \bar{H}_q(X,A) \\
{\scriptstyle \pi_{m*}}\searrow & & & & \swarrow{\scriptstyle \bar{\pi}_{m*}} \\
 & H_q(X_m,A_m) & \xrightarrow{h(q,X_m,A_m)} & \bar{H}_q(X_m,A_m) &
\end{array}$$

in which τ_m and $\bar{\tau}_m$ are the projections of the limit groups $\lim H_q$ and $\lim \bar{H}_q$ into their m^{th} coordinates Commutativity in the square is a consequence of the definition of h_∞, and commutativity in the triangles follows from the definition of l and of $\bar{l}$ Hence

$$\begin{aligned}
\bar{\pi}_{m*}h(q,X,A,D) &= \bar{\pi}_{m*}\bar{l}^{-1}h_\infty l = \bar{\tau}_m h_\infty l \\
&= h(q,X_m,A_m)\tau_m l = h(q,X_m,A_m)\pi_{m*}.
\end{aligned}$$

This proves (6).

(7) If $f.$ $(X,A) \to (Y,B)$ and (Y,B) is in $\mathfrak{J}$, then commutativity holds in the diagram

$$\begin{array}{ccc}
H_q(X,A) & \xrightarrow{h(q,X,A,D)} & \bar{H}_q(X,A) \\
\downarrow{\scriptstyle f_*} & & \downarrow{\scriptstyle \bar{f}_*} \\
H_q(Y,B) & \xrightarrow{h(q,Y,B)} & \bar{H}_q(Y,B)
\end{array}$$

By 11 9 we may select an element m of the indexing set M of the system $(\mathbf{X},\mathbf{A})$, and a map

$$f_m\colon\ (X_m,A_m) \to (Y,B)$$

such that

$$f_m \pi_m \simeq f$$

where $\pi_m: (X,A) \to (X_m,A_m)$ is the projection. Consider the diagram

$$\begin{array}{ccccc}
H_q(X,A) & \xrightarrow{\pi_{m*}} & H_q(X_m,A_m) & \xrightarrow{f_{m*}} & H_q(Y,B) \\
\downarrow{\scriptstyle h(q,X,A,D)} & & \downarrow{\scriptstyle h(q,X_m,A_m)} & & \downarrow{\scriptstyle h(q,Y,B)} \\
\overline{H}_q(X,A) & \xrightarrow{\overline{\pi}_{m*}} & \overline{H}_q(X_m,A_m) & \xrightarrow{\overline{f}_{m*}} & \overline{H}_q(Y,B)
\end{array}$$

Commutativity in the left square follows from (6), and in the right square, commutativity is a consequence of the fact that $h: H \to \overline{H}$ is a homomorphism on $\mathfrak{I}$ and both (X_m,A_m) and (Y,B) are in $\mathfrak{I}$. Since, by the Homotopy axiom,

$$f_{m*}\pi_{m*} = (f_m\pi_m)_* = f_*, \qquad \overline{f}_{m*}\overline{\pi}_{m*} = (\overline{f_m\pi_m})_* = \overline{f}_*,$$

proposition (7) follows.

Proofs of (2)–(5). Proposition (2) follows from (7) by taking $(X,A) = (Y,B)$ and f = identity. Then f_* and $\overline{f}_*$ are identities so that (7) implies $h(q,X,A,D) = h(q,X,A)$.

Next we prove (4). Let n be an element of the indexing set N of D', and let $\rho_n: (Y,B) \to (Y_n,B_n)$ be the projection. Since (Y_n, B_n) is in $\mathfrak{I}$, proposition (7) applied to the maps ρ_n and $\rho_n f$ yields

$$\overline{\rho}_{n*}h(q,Y,B,D') = h(q,Y_n,B_n)\rho_{n*},$$
$$\overline{\rho}_{n*}\overline{f}_* h(q,X,A,D) = h(q,Y_n,B_n)\rho_{n*}f_*$$

Consequently

$$\overline{\rho}_{n*}\overline{f}_*h(q,X,A,D) = \overline{\rho}_{n*}h(q,Y,B,D')f_*.$$

Since this holds for every $n \in N$, the above relation holds with $\overline{\rho}_{n*}$ replaced by l. Since, by assumption, $\overline{H}$ is a continuous homology theory, l is an isomorphism, and therefore

$$\overline{f}_*h(q,X,A,D) = h(q,Y,B,D')f_*$$

which proves (4).

To prove (3) apply (4) with $(X,A) = (Y,B)$ and f = identity. Then f_* and $\overline{f}_*$ are both identities and (4) yields $h(q,X,A,D) = h(q,X,A,D')$.

Proposition (5) is a consequence of the commutativity relations in the diagram

$$\begin{array}{ccccccc} H_q(X,A) & \xrightarrow{l} & \lim H_q(\mathbf{X},\mathbf{A}) & \xrightarrow{h_\infty} & \lim \overline{H}_q(\mathbf{X},\mathbf{A}) & \xleftarrow{\bar{l}} & \overline{H}_q(X,A) \\ \downarrow \partial & & \downarrow \partial_\infty & & \downarrow \bar{\partial}_\infty & & \downarrow \bar{\partial} \\ H_{q-1}(A) & \xrightarrow{l} & \lim H_{q-1}(\mathbf{A}) & \xrightarrow{h_\infty} & \lim \overline{H}_{q-1}(\mathbf{A}) & \xleftarrow{\bar{l}} & \overline{H}_{q-1}(A) \end{array}$$

where ∂_∞ is the limit of ∂. $H_q(X_m,A_m) \to H_{q-1}(A_m)$ and similarly for $\bar{\partial}_\infty$ This concludes the proof (2)-(5)

We now show that the homomorphism $h\colon H \to \overline{H}$ extending h_0 is unique. Let then h': $H \to \overline{H}$ be another such homomorphism Then, by IV,10.2, we have $h = h'$ on $\mathfrak{J}$ Let (X,A) be a compact pair. Without loss of generality we may assume that (X,A) is the inverse limit of an inverse system $(\mathbf{X},\mathbf{A}) = \{(X_m,A_m),\pi_m^{m'}\}$ of triangulable pairs Let π_m: $(X,A) \to (X_m,A_m)$ be the projections Since (X_m,A_m) is triangulable and h and h' are homomorphisms of homology theories, we have

$$\bar{\pi}_{m*}h(q,X,A) = h(q,X_m,A_m)\pi_{m*} = h'(q,X_m,A_m)\pi_{m*} = \bar{\pi}_{m*}h'(q,X,A).$$

Since this holds for every element m of the indexing set, it follows from the definition of the map $\bar{l}(q,\mathbf{X},\mathbf{A})$ that

$$\bar{l}(q,\mathbf{X},\mathbf{A})h(q,X,A) = \bar{l}(q,\mathbf{X},\mathbf{A})h'(q,X,A).$$

Since $\overline{H}$ is continuous, $\bar{l}$ is an isomorphism, and therefore $h(q,X,A) = h'(q,X,A)$.

Finally assume that $\overline{H}$ is also a continuous theory and that h_0: $G \approx \overline{G}$ is an isomorphism. Let $\bar{h}_0$: $\overline{G} = G$ be the inverse of h_0 and let $\bar{h}$. $\overline{H} \to H$ be the homomorphism on $\mathfrak{a}_C$ extending $\bar{h}_0$. Then $\bar{h}h$: $H \to H$ is a homomorphism extending the identity map $G \to G$ Thus by the uniqueness condition, $\bar{h}h$ = identity Similarly $h\bar{h}$ = identity. Thus h is an isomorphism and $\bar{h}$ is its inverse.

NOTES

The Čech process. The method used in defining the Čech homology and cohomology groups is not restricted to these two instances. The general situation may be described somewhat as follows:

Suppose that a functor H (covariant or contravariant) on the category of simplicial pairs and simplicial maps with values in the category of groups and homomorphisms is given. Suppose further that, if f_0,f_1:

$(K,L) \to (K_1,L_1)$ are contiguous, then $H(f_0) = H(f_1)$ Then the definitions of §§2-4 of Chapter IX can be applied to yield a similar functor $\check{H}$ on the category of pairs (X,A) For compact pairs the functor $\check{H}$ satisfies the homotopy axiom (i e $\check{H}(f_0) = \check{H}(f_1)$ if f_0 and f_1 are homotopic) and the continuity axiom This procedure was used by Spanier to show that the cohomotopy groups satisfy the continuity axiom (see note to Chapter I).

Compactifications. In defining the compactification $\dot{X}$ of a locally compact space X, we required in §6 that the compactification be made by adding a point ω which does not belong to any of the spaces in the category $\mathfrak{A}_{LC}$. This actually amounts to replacing the category by a subcategory This highly artificial convention was caused by the requirement that the compactification $(\dot{X},A)$ of a pair (X,A) again be a pair; this forces us to compactify A using the same point as for X.

The proper solution in this situation calls for the generalization of the concept of a *pair* Let a *generalized pair* be defined as a triple (X,A,ϕ) where X and A are topological spaces and ϕ: $A \to X$ is a homeomorphism of A onto a subset of X. If X,A and ϕ are in $\mathfrak{A}_{LC}$, then we say that (X,A,ϕ) is in $\mathfrak{A}_{LC}$ Now the compactification $(\dot{X},\dot{A},\dot{\phi})$ is a generalized couple even if distinct points are used to compactify X and A.

There remains the question of a suitable natural choice of a point ω for each locally compact space X. To this end we remark that, in the von Neumann-Bernays-Godel axiomatics, a set may be an element of another set but never is an element of itself. Thus if $\{X\}$ denotes the set whose only element is X we have $\{X\} \cap X = 0$. Thus we may set $\dot{X} = X \cup \{X\}$ with the topology of 6 2

An analogous discussion can be carried out for the Tychonoff compactification.

EXERCISES

A. WEAK CONTINUITY.

1. Show that every compact pair may be represented as a nested intersection of pairs in the category $\mathfrak{I}_H$, i e pairs having the homotopy type of triangulable pairs (Hint. Use method employed to prove 10.1)

2. The conclusion of 2 6 will be referred to as the *weak continuity* property of a homology [cohomology] theory Prove 12 1 assuming that H is weakly continuous. (Hint: use preceding exercise and the fact that $\mathfrak{I}_H$ is a uniqueness category (IX, Exer. E4))

3. Use the preceding exercise to prove that continuity and weak continuity are equivalent properties.

4. Show that, in the category $\mathfrak{A}_C$, the Homotopy axiom is a consequence of Axioms 1, 2 and continuity. (See J. W. Keesee, *Annals of Math.* **54** (1951), 247–249.)

B. DIRECT SUM THEOREMS.

1 Show that the direct sum theorem I,13.2 remains valid in the Čech theory (arbitrary pairs)

2. Let $(X;X_1,X_2)$ be a compact triad with $X = X_1 \cup X_2$, $A = X_1 \cap X_2$ Show that the maps

$$H_q(X_i,A) \to H_q(X,A) \qquad i = 1,2$$

yield an injective representation of $H_q(X,A)$ as a direct sum, while the maps

$$H_q(X,A) \to H_q(X,X_i) \qquad i = 1,2$$

yield a projective representation. (All maps are induced by inclusions, and Čech groups are used) Establish similar results for cohomology.

3 Let X be a compact space, and let $\{U_\alpha\}$, $\alpha \in M$, be a family of pairwise disjoint open subsets of X with union U. Show that the maps

$$H_q(X,X - U) \to H_q(X,X - U_\alpha)$$

induced by inclusion yield a projective representation of the Čech homology group $H_q(X,X - U)$ as a direct product. Similarly show that

$$H^q(X,X - U_\alpha) \to H^q(X,X - U)$$

yield an injective representation of $H^q(X,X - U)$ as a direct sum

4 Transcribe 3 into the "single space notation" of §7.

C. EXACTNESS.

1. Let A be an open and closed subset of a space X. Show that the Čech homology sequence of the pair (X,A) is exact and is split (see VIII, Exer. D) by the homomorphisms ψ $H_q(X) \to H_q(A)$ where $\psi = k_*^{-1}l_*$, l: $X \subset (X,X - A)$, k $A \subset (X,X - A)$

2 Let $(X;X_1,X_2)$ be a compact triad with $X = X_1 \cup X_2$, $A = X_1 \cap X_2$. Show that the Čech homology sequence of the triple (X,X_2,A) is exact and is split by the homomorphisms ψ: $H_q(X,A) \to H_q(X_2,A)$ where $\psi = k_*^{-1}l_*$, l: $(X,A) \subset (X,X_1)$, k: $(X_2,A) \subset (X,X_1)$

3. Formulate similar results for the Čech cohomology sequences.

D. LC-GROUPS

1. Show that, in the definition of $H_q^\spadesuit(X,A)$, the directed set $\mathrm{Cov}_1(X)$

may be replaced by any of the directed sets listed below and that the resulting limit group is isomorphic with $H^{\blacktriangle}_q(X,A)$.

(a) $\mathrm{Cov}_2(X)$ = the set of all finite open coverings α of X such that at least one of the sets α_v is countercompact.

(b) $\mathrm{Cov}_0(X)$ = the set of all finite open coverings α of X such that all the sets α_v except perhaps one are bounded

Hint· Show $\mathrm{Cov}_0 \subset \mathrm{Cov}_1 \subset \mathrm{Cov}_2$, and Cov_0 is cofinal in Cov_2.

(c) $\mathrm{Cov}_3(X)$, the set of all finite open coverings of X but with the ordering $\ll$ defined as follows $\alpha \ll \beta$ means $\alpha < \beta$ and the union of the bounded sets of β contains the union of the bounded sets of α (Hint· set up a map $\mathrm{Cov}_3(X) \rightarrow \mathrm{Cov}_0(X)$).

(d) $\mathrm{Cov}'(X)$

(e) The sets $\mathrm{Cov}_0(x)$, $\mathrm{Cov}_1(X)$, $\mathrm{Cov}_2(X)$ with the ordering $\ll$

2 Given $\alpha \in \mathrm{Cov}'(X,A)$ define, in addition to the subcomplex Ω_α, the subcomplex Ω'_α which is the full subcomplex of X_α determined by the vertices v for which α_v is not bounded Thus Ω'_α is the least full subcomplex of X_α containing Ω_α Show that, for any of the families of coverings named above, one may replace the groups $H_q(X_\alpha,A_\alpha \cup \Omega_\alpha)$, used in the limiting process, by the groups $H_q(X_\alpha,A_\alpha \cup \Omega'_\alpha)$ without altering (up to an isomorphism) the limit groups

REMARK The limit groups of $H_q(X_\alpha,A_\alpha \cup \Omega'_\alpha)$ over the directed set $\mathrm{Cov}'(X)$ are those treated in detail by Alexandroff [Trans Amer Math Soc 49(1941), 41-105] Thus the Alexandroff groups are isomorphic with the $^{\blacktriangle}$-groups

3 Let (X,A) be a locally compact pair Let $\{X_\alpha\}$ be the family of all countercompact subsets of X Show that the Čech groups $H_q(X,A \cup X_\alpha,G)$ $[H^q(X,A \cup X_\alpha;G)]$ together with the homomorphisms induced by inclusions form an inverse [direct] system of groups with a limit group isomorphic with $H^{\blacktriangle}_q(X,A,G)$ $[H^q_{\blacktriangle}(X,A,G)]$

4 Let K be a locally finite simplicial complex and L a subcomplex of K Show that the pair $(|K|,|L|)$ is locally compact and establish the natural isomorphisms

$$H^{\blacktriangle}_q(|K|,|L|,G) \approx \overleftarrow{\mathfrak{K}}_q(K,L;G), \qquad H^q_{\blacktriangle}(|K|,|L|;G) \approx \overrightarrow{\mathfrak{K}}^q(K,L;G)$$

with $\overleftarrow{\mathfrak{K}}_q$ and $\overrightarrow{\mathfrak{K}}^q$ as defined in VIII, Exer F3.

5 Transcribe the relative Mayer-Victoris sequence of a compact triad $(X;A_1,A_2)$ into the single space language of §7.

E. TYCHONOFF COMPACTIFICATION

1 Show that, for any topological space X, any map $f.$ $X \rightarrow I$ can be factored uniquely into maps $X \xrightarrow{T} T(X) \rightarrow I$ Show that this property describes $T(X)$ among compact spaces up to a homeomorphism.

2. Replace the interval I in exercise 1 by any completely regular space.

3. Let X be a normal subset of a compact space Y such that $\overline{X} = Y$ and for any two subsets A,B of X the relation $\overline{A} \cap \overline{B} \cap X = 0$ implies $\overline{A} \cap \overline{B} = 0$. Show that Y is a Tychonoff compactification of X.

4. Let X be normal and Y compact metric with a metric ρ. Show that a homotopy F, $X \times I \to Y$ is uniform if and only if for each $\epsilon > 0$ there is a $\delta > 0$ such that $t_1,t_2 \in I$, $|t_1 - t_2| < \delta$ implies $\rho(F(x,t_1),F(x,t_2)) < \epsilon$ for all $x \in X$.

5. Let X be a normal space and let $\omega(X)$ denote the set of those points $x \in X$ such that x has a countable set N of neighborhoods with the property that any neighborhood of x contains an element of N. Show that $\omega(X) = \omega(X^\frown)$.

6 Show that the Čech homology and cohomology theories of normal pairs obtained using finite open coverings is isomorphic with the theory obtained using finite closed coverings.

F. Solenoids.

In these exercises use the fact (established in XI,4 5) that the map f_k: $S^1 \to S^1$ defined by $f_k(z) = z^k$ satisfies $f_{k*}u = ku$ for $u \in H_1(S^1;G)$ and $f_k^*(u) = ku$ for $u \in H^1(S^1;G)$.

1. Let Σ_p be the p-adic solenoid (see VIII, Exer. E). Show that for the Čech groups

$$\begin{aligned} \tilde{H}_0(\Sigma_p;G) &= 0, & \tilde{H}^0(\Sigma_p;G) &= 0 \\ H_q(\Sigma_p;G) &= 0, & H^q(\Sigma_p;G) &= 0 \qquad \text{for} \qquad q > 1. \end{aligned}$$

$H_1(\Sigma_p,G)$ is isomorphic with the limit of the inverse system

$$G \overset{\phi}{\leftarrow} G \overset{\phi}{\leftarrow} \cdots \overset{\phi}{\leftarrow} G \overset{\phi}{\leftarrow} \cdots \qquad \text{where} \qquad \phi(g) = pg.$$

$H^1(\Sigma_p;G)$ is isomorphic with the limit of the direct system

$$G \overset{\phi}{\to} G \overset{\phi}{\to} \cdots \overset{\phi}{\to} G \overset{\phi}{\to} \cdots .$$

Using the above results, prove that $H_1(\Sigma_p;J) = 0$, $H^1(\Sigma_p;J)$ is isomorphic with the group of all rational numbers of the form m/p^n where m and n are integers, and that $H_1(\Sigma_p;S^1) \approx \Sigma_p$ where S^1 is the multiplicative group of complex numbers of absolute value 1 (or equivalently, the addive group of real numbers reduced modulo integers).

2. Let p and q be two integers. Consider the exact sequence

$$S\colon\ 0 \to J \overset{\tau}{\to} J \overset{\eta}{\to} J/qJ \to 0$$

where $\tau(n) = qn$ and η is the natural map Consider the mapping γ: $S \to S$ defined by $\gamma(x) = px$. Analyze the inverse limit S_∞ of the inverse sequence

$$S \overset{\gamma}{\leftarrow} S \overset{\gamma}{\leftarrow} \cdots \overset{\gamma}{\leftarrow} S \leftarrow \cdots$$

For what values of p and q is S_∞ an exact sequence? (see Example VIII,5 5)

3. Let (E,S) be a 2-cell and its boundary defined in the complex plane by the conditions $|z| \leqq 1$ and $|z| = 1$ respectively. Consider the map ϕ: $S \to S$ defined by $\phi(z) = ze^{2\pi i/q}$. Let (P,C) be the pair obtained from (E,S) by identifying each $z \in S$ with $\phi(z)$. Show that the homology sequence (integer coefficients) of the pair (P,C) is isomorphic with the exact sequence S of the preceding exercise Show that the map f: $(E,S) \to (E,S)$, $f(z) = z^p$ induces a map $\tilde{f}$: $(P,C) \to (P,C)$ such that $\tilde{f}_*(u) = pu$. Let (P_∞,C_∞) be the limit of the inverse sequence

$$(P,C) \overset{\tilde{f}}{\leftarrow} (P,C) \overset{\tilde{f}}{\leftarrow} \cdots \overset{\tilde{f}}{\leftarrow} (P,C) \overset{\tilde{f}}{\leftarrow} \cdots$$

Show that the homology sequence of (P_∞,C_∞) (Čech group with integer coefficients) is isomorphic with the sequence S_∞ of the preceding exercise (see Example X,4 1) Show that C_∞ and Σ_p are homeomorphic.

G. Miscellaneous.

1 Let (X,A) be a compact pair, and let u be in the kernel of the natural homomorphism $H^q(X) \to H^q(A)$ (Čech cohomology groups). Show that there exists a closed neighborhood V of A such that u is in the kernel of $H^q(X) \to H^q(V)$

2 Let α be a covering of X Show that a map ϕ. $X \to |X_\alpha|$ is canonical relative to α if and only if, for each $v \in V_\alpha$, we have $\phi^{-1}(\mathrm{st}(v)) \subset$

CHAPTER XI

Applications to Euclidean Spaces

1. INTRODUCTION

The objective of this chapter is twofold First we derive a number of theorems concerning Euclidean space among which are some of the most classical and widely used ones such as the Brouwer fixed-point theorem and the invariance of domain Secondly we show how such theorems can be derived using the axioms without appeal to any concretely defined homology or cohomology theory

The first five articles depend only on Chapters I and II The last section (6) uses, in addition, the continuity axiom of Chapter X

The notations of I,16 are used throughout the chapter R^n = cartesian n-space, E^n = the n-cell $||x|| \leq 1$, S^{n-1} = the $(n-1)$-sphere $||x|| = 1$, etc These spaces appear in most of the theorems *The theorems are still valid if R^n, E^n, S^{n-1} are replaced by homeomorphs* For example, the Brouwer fixed point theorem 3 3 holds for any set X homeomorphic with E^n For let f' be a map $X \to X$, and h a homeomorphism $X \to E^n$ Then $f = hf'h^{-1}$ maps E^n into itself By 3 3, there is a fixed point, say x, of f Then $h^{-1}(x)$ is a fixed point of f' The proofs of invariance of the other theorems of this chapter are equally trivial.

2. MAPPINGS INTO SPHERES

LEMMA 2 1 *If (X,A) is a normal pair (i e X is normal, and A is closed in X) and f $A \to S^n$, then there is an open set U of X containing A and an extension f': $U \to S^n$ of f*

PROOF Regard f as a map of A into R^{n+1}. Then, by Tietze's extension theorem, there exists an extension g $X \to R^{n+1}$ of f Define $U = X - g^{-1}(0)$ where 0 is the origin of R^{n+1} and set

$$f'(x) = \frac{g(x)}{||g(x)||} \qquad x \in U.$$

This yields the desired extension of f

LEMMA 2 2 *Let (X,A) be a normal pair such that $X \times I$ is normal Every map f. $X \times 0 \cup A \times I \to S^n$ admits an extension F. $X \times I \to S^n$.*

PROOF Let $B = X \times 0 \cup A \times I$ Since the pair $(X \times I, B)$ is normal, there is, by 2 1, a neighborhood U of B and an extension ϕ: $U \to S^n$ of f. Since $A \times I \subset U$ and I is compact, we may construct a neighborhood V of A in X such that $V \times I \subset U$ Select a Urysohn function θ $X \to I$ such that θ is 0 on $X - V$ and is 1 on A Then for each $x \varepsilon X$ and $t \varepsilon I$, we have $(x,\theta(x)t) \varepsilon U$ Define $F(x,t) = \phi(x,\theta(x)t)$ Then $F(x,0) = \phi(x,0) = f(x,0)$, and for $x \varepsilon A$, we have $\theta(x) = 1$ so that $F(x,t) = \phi(x,t) = f(x,t)$ Thus F is the desired extension

LEMMA 2 3 *Let (X,A) be a normal pair such that $X \times I$ is normal, and let f $X \to S^n$ Every homotopy of $f|A$ can be extended to a homotopy of f*

PROOF Let h $A \times I \to S^n$ be a homotopy of $f|A$ Extend h to a map h' $X \times 0 \cup A \times I$ by setting $h(x,0) = f(x)$ for $x \varepsilon X$ Then, by 2 2, h' admits an extension H $X \times I \to S^n$, which is the desired homotopy

DEFINITION 2 4 A map f $X \to S^n$ is called *inessential* if f is homotopic to a map of X into a single point of S^n Otherwise f is called *essential*

LEMMA 2 5 *If f $X \to S^n$ is essential then $f(X) = S^n$*

PROOF Suppose $y \varepsilon S^n - f(X)$. Since $S^n - y$ is contractible to a point and $f(X) \subset S^n - y$, it follows that f is inessential

LEMMA 2 6 *Let (X,A) be a normal pair such that $X \times I$ is normal. Every inessential map f $A \to S^n$ admits an inessential extension f' $X \to S^n$.*

PROOF. Let h. $X \to S^n$ be a map of X into a single point of S^n. Then $h|A$ is homotopic with f The existence of f' follows now from 2 3.

LEMMA 2 7 *Let $T = \{t,K\}$ be a triangulation of X such that* dim *$K < n$ Then every map f $X \to S^n$ is inessential*

PROOF Select a triangulation $T' = \{t',K'\}$ of S^n with dim $K' = n$ By the theorem on simplicial approximation there exists a map g. $X \to S^n$ homotopic with f and which is simplicial relative to the triangulations ${}^kT,T'$ where ${}^kT = \{{}^kt,{}^kK\}$ is a sufficiently high barycentric subdivision of T Since the dimension of a simplicial complex does not change under barycentric subdivision, we have dim ${}^kK < n$. Since g is simplicial, it follows that g does not map X onto S^n Thus, by 2 5, g is inessential, and therefore f is inessential.

COROLLARY 2.8. *If $m < n$, every map f. $S^m \to S^n$ is inessential and admits an extension f' $E^{m+1} \to S^n$*

PROOF The fact that f is inessential follows from 2 7. The existence of an extension f' follows from 2 6.

LEMMA 2 9. *Let $T = \{t,K\}$ be a triangulation of X with* dim $K \leq n$, *let A be a closed subset of X, and let f: $A \to S^n$. Then f admits an extension f': $X \to S^n$.*

LEMMA 2.10. *Let $T = \{t,K\}$ be a triangulation of X with* dim $K \leqq n + 1$, *let A be a closed subset of X, and let f: $A \to S^n$. There exists a finite set $F \subset X - A$ such that f admits an extension f': $X - F \to S^n$.*

PROOF OF 2.9 AND 2.10. By 2.1, there is an open set U in X containing A, and an extension f': $U \to S^n$ of f. If we replace T by a sufficiently high barycentric subdivision ${}^kT = \{{}^kt,{}^kK\}$, we can find a subcomplex A' of X (relative to ${}^kT$) such that $A \subset A' \subset U$. Since dim K = dim ${}^kK$, it follows that we may assume in 2.9 and 2.10 that A is a subcomplex of X relative to T.

Let X^q denote the q-dimensional skeleton of X relative to T. Clearly f: $A \to S^n$ admits an extension f^0: $X^0 \cup A \to S^n$. Suppose inductively that an extension f^q: $X^q \cup A \to S^n$ is already given for some $q < n$. Let $s_1, \cdots, s_r$ be the $(q + 1)$-dimensional simplexes of $X - A$. Then f^q is defined on $s_1, \cdots, s_r$, and, by 2.8, f^q may be extended over every one of the simplexes $s_1, \cdots, s_r$, thus yielding f^{q+1}: $X^{q+1} \cup A \to S^n$ extending f. Therefore there exists an extension f^n: $X^n \cup A \to S^n$ of f. If dim $K \leqq n$, then $X^n = X$, which proves 2.9. If dim $K \leqq n + 1$, let $s_1, \cdots, s_r$ be the $(n + 1)$-simplexes of $X - A$, and let $F = \{b_1, \cdots, b_r\}$ be the set of their barycenters. A radial projection in each simplex $s_1, \cdots, s_r$ yields a retraction h: $X - F \to X^n \cup A$. The map f^qh: $X - F \to S^n$ is then an extension of f: $A \to S^n$. Thus 2.10 is proved.

LEMMA 2.11. *Let A be a closed subset of S^n and let B be a set containing exactly one point out of each component of $S^n - A$. Then, for every f: $A \to S^{n-1}$, there exists a finite subset F of B and an extension f': $S^n - F \to S^{n-1}$ of f.*

PROOF. By 2.10, there exists a finite subset $(x_1, \cdots, x_k)$ of $S^n - A$ and an extension $S^n - (x_1, \cdots, x_k) \to S^{n-1}$ of f. For each x_i, let b_i denote the point of B in the component of $S - A$ containing x_i, and let F be the set $(b_1, \cdots, b_k)$. To prove that f admits an extension $S^n - F \to S^{n-1}$, we shall prove inductively that, if f admits an extension $S^n - (b_1, \cdots, b_{i-1}, x_i, \cdots, x_k) \to S^{n-1}$, then f also admits an extension $S^n - (b_1, \cdots, b_i, x_{i+1}, \cdots, x_k) \to S^{n-1}$. Since x_i and b_i lie in the same component of $S^n - A$, there is a finite sequence

$$x_i = y_0, \cdots, y_l = b_i$$

of points and a sequence of convex n-cells

$$E_1, \cdots, E_l$$

in $S^n - A$ such that $y_{j-1}, y_j \in E_i$ for $j = 1, \cdots, l$, and the boundary S_i of E_i contains none of the b's, x's, and y's

It clearly suffices to show that, if f admits an extension

$$f_{i,j-1}: \quad S^n - (b_1, \cdots, b_{i-1}, y_{j-1}, x_{i+1}, \cdots, x_k) \to S^{n-1},$$

then f also admits an extension

$$f_{i,j}: \quad S^n - (b_1, \cdots, b_{i-1}, y_j, x_{i+1}, \cdots, x_k) \to S^{n-1}.$$

Let $r \quad S^n - y_j \to \overline{S^n - E_i}$ be a retraction. Then setting

$$f_{i,j}(x) = f_{i,j-1}r(x) \text{ for } x \in S^n - (b_1, \cdots, b_{i-1}, y_j, x_{i+1}, \cdots, x_k)$$

yields the desired extension $f_{i,j}$. This concludes the proof.

3. THEOREMS OF BROUWER AND BORSUK

THEOREM 3 1. *The n-sphere S^n is not contractible (over itself) to a point, i e the identity map of S^n is essential.*

PROOF. Let H be any homology (or cohomology) theory defined on the category $\mathfrak{I}$ of triangulable pairs and with a coefficient group $G \neq 0$. Then, by I,16 6, S^n is not homologically trivial Thus, by I,11 5, S^n is not contractible to a point

REMARK. The above statement is the only one of this section that requires the use of homology theory in the proof All the other propositions follow from 3.1 using the results of §2 and other simple geometric arguments

THEOREM 3 2 S^{n-1} *is not a retract of* E^n.

PROOF. Let r: $E^n \to S^{n-1}$ be a retraction. Then

$$F(x,t) = r((1 - t)x) \qquad x \in S^{n-1}, \quad t \in I$$

yields a homotopy contracting S^{n-1} to a point, contrary to 3 1.

THEOREM 3.3 (*Brouwer's Fixed Point Theorem*) *Each map f. $E^n \to E^n$ has at least one fixed point, i e. a point $x \in E^n$ such that $f(x) = x$:*

PROOF. Assume $f(x) \neq x$ for all $x \in E^n$ Let $r(x)$ be the point of S^{n-1} such that x lies on the line segment from $f(x)$ to $r(x)$. The continuity of r is proved using that of f and some elementary geometry. Since $r(x) = x$ when $x \in S^n$, r is a retraction $E^n \to S^{n-1}$ contradicting 3.2.

THEOREM 3 4 *Let $T = \{t,K\}$ be a triangulation of a space X. In order that* dim $K \leq n$, *it is necessary and sufficient that for every closed subset A of X, and every map f: $A \to S^n$, there exists an extension f': $X \to S^n$*

PROOF The necessity of the condition follows from 2.9. Suppose that dim $K > n$, and let s be a simplex of X of dimension $n + 1$. Let

$A = s$ and let $f.\ A \to S^n$ be a homeomorphism Suppose that f admits an extension $f'\ X \to S^n$ Then $f^{-1}f'\ X \to A$ is a retraction Therefore s is a retract of the closed simplex s contrary to 3 2

THEOREM 3 5 (*Invariance of Dimension*). *If* $T_1 = \{t_1,K_1\}$ *and* $T_2 = \{t_2,K_2\}$ *are triangulations of the same space* X, *then* $\dim K_1 = \dim K_2$.

This is an immediate consequence of 3 4.

THEOREM 3 6 (*Borsuk's Separation Criterion*) *Let* X *be a compact subset of cartesian space* R^n *and let* $x_0 \varepsilon R^n - X$ *In order that* x_0 *lie in the unbounded component of* $R^n - X$, *it is necessary and sufficient that the map* $p\ X \to S^{n-1}$ *given by*

$$p(x) = \frac{x - x_0}{||x - x_0||} \qquad x \varepsilon X$$

be inessential

PROOF. By a translation and a similarity transformation of R^n, it can be arranged that X lies in the interior of E^n, and x_0 is the origin. The map p is then defined by $p(x) = x/||x||$

Suppose x_0 lies in the unbounded component C of $R^n - X$ Since C is arcwise connected, there exists a mapping $f.\ I \to C$ such that $f(0) = x_0, f(1) = x_1 \varepsilon R^n - E^n$. Consider the mapping $F\ X \times I \to S^{n-1}$ defined by

$$F(x,t) = \frac{x - f(t)}{||x - f(t)||} \qquad x \varepsilon X,\ t \varepsilon I.$$

Then $F(x,0) = p(x)$ and $F(x,1) = (x - x_1)/||x - x_1||$ Since $x \varepsilon E^n$ and x_1 is not in E^n, it is easy to see that $F(x,1) \neq x_1/||x_1||$ It therefore follows from 2.5 that p is inessential

Assume now that the component C of $R^n - X$ containing x_0 is bounded Then $\bar{C} \subset E^n$ and $C \cup X$ is closed Suppose p is inessential Then, by 2 6, p admits an extension $p'.\ C \cup X \to S^{n-1}$ Define a map $r\ E^n \to S^{n-1}$ by setting

$$r(x) = p'(x) \qquad \text{for } x \varepsilon C \cup X,$$

$$r(x) = \frac{x}{||x||} \qquad \text{for } x \varepsilon E^n - C.$$

The two definitions agree on X, hence r is continuous When $x \varepsilon S^{n-1}$, we have $r(x) = x$, so that r is a retraction of E^n onto S^{n-1} This contradicts 3 2.

THEOREM 3.7 (*Borsuk's Theorem*) *Let* X *be a closed subset of* S^n. *The set* $S^n - X$ *is connected if and only if every map* $f\ X \to S^{n-1}$ *is inessential.*

PROOF Suppose $S^n - X$ is not connected and let x_0,x_1 be two points of $S^n - X$ lying in distinct components of $S^n - X$ If we regard

$S^n - x_1$ as the cartesian space R^n, it follows that x_0 lies in a bounded component of $R^n - X$ Thus, by 3 6, there exists an essential map $p.\ X \to S^{n-1}$

Suppose now that $S^n - X$ is connected and let $f.\ X \to S^{n-1}$ Let $x_0 \varepsilon S^n - X$. By 2 11, f admits an extension $f'\cdot\ S^n - x_0 \to S^{n-1}$ Since $S^n - x_0$ is contractible, it follows that f' is inessential Thus f is also inessential

LEMMA 3 8. *Let (X,A) be a pair in S^n homemomorphic with (E^n,S^{n-1}). Then $S^n - A$ decomposes into two components, namely $S^n - X$ and $X - A$. In particular, $X - A$ is an open subset of S^n*

PROOF Since E^n does not admit essential maps into S^{n-1}, it follows from 3 7 that $S^n - X$ is connected On the other hand S^{n-1} does admit an essential map into S^{n-1} (e g the identity map), thus $S^n - A$ is not connected The set $X - A$ is homeomorphic with $E^n - S^{n-1}$, and is therefore connected Since

$$S^n - A = (S^n - X) \cup (X - A),$$

and $S^n - A$ is not connected, and $S^n - X$, $X - A$ are connected, it follows that the latter two sets are the components of $S^n - A$

THEOREM 3 9 (*Invariance of Domain*) *If U_1,U_2 are homeomorphic subsets of S^n, and U_1 is open, then U_2 is also open*

PROOF Let f: $U_1 \to U_2$ be a homeomorphism Let $x_2 \varepsilon U_2$ and $x_1 = f^{-1}(x_2)$ Select a spherical neighborhood V_1 of x_1 such that $\bar{V}_1 \subset U_1$ Let $\dot{V}_1 = \bar{V}_1 - V_1$, then $(\bar{V}_1,\dot{V}_1)$ and (E^n,S^{n-1}) are homeomorphic pairs Let $(X,A) = (f\bar{V}_1,f\dot{V}_1)$ Then (X,A) and (E^n,S^{n-1}) are homeomorphic, and, by 3 8, $X - A$ is open Since $x_2 \varepsilon X - A \subset U_2$, it follows that U_2 is open

DEFINITION 3 10 A space M will be called *locally euclidean of dimension n* if every point $x \varepsilon M$ possesses a neighborhood homeomorphic with R^n

THEOREM 3.11 *Let U_1 and U_2 be homeomorphic subsets of spaces M_1 and M_2, respectively, both of which are locally euclidean of dimension n If U_1 is open, then U_2 is also open*

PROOF Let f: $U_1 \to U_2$ be a homeomorphism. Let $x_2 \varepsilon U_2$, and let $x_1 = f^{-1}(x_2)$ Select neighborhoods V_1,V_2 of x_1,x_2 respectively such that

$$x_1 \varepsilon V_1 \subset U_1, \quad x_2 \varepsilon V_2 \subset M_2, \quad f(V_1) \subset V_2,$$
$$V_1,V_2 \text{ are homeomorphic with } R^n$$

Since R^n is homeomorphic with an open subset of S^n, we may select a map $\phi_1\ V_1 \to S^n$, $\phi_2.\ V_2 \to S^n$ mapping V_1 and V_2 homeomorphically onto open subsets W_1 and W_2 of S^n Then $\phi_2 f \phi_1^{-1}$ maps W_1 homeomorphically onto a subset of W_2 Thus, by 3 9, the set $\phi_2 f \phi_1^{-1}(W_1)$ is open in S^n and therefore also in W_2. Therefore the set $f\phi^{-1}(W_1) = f(V_1)$ is open in V_2 and therefore also in M_2. Since $x_2 \in f(V_1) \subset U_2$, it follows that U_2 is open.

COROLLARY 3.12. *Two locally euclidean spaces M_1 and M_2 of different dimensions are not homeomorphic.*

PROOF. Let m_1, m_2 be the dimensions of M_1 and M_2 respectively and let $m_1 = m_2 + n$, $n > 0$. Then $M_2 \times R^n$ is locally euclidean of dimension m_1, and contains a nonopen subset homeomorphic with M_2. Thus, by 3.11, M_1 and M_2 cannot be homeomorphic.

4. DEGREES OF MAPS

In this section and the following one, we assume that a homology theory H is given with a coefficient group G isomorphic to the group of integers J. Analogous results also hold for a cohomology theory with a coefficient group $G = J$.

DEFINITION 4.1. Consider a map f: $S^n \to S^n$, $n > 0$. Since $H_n(S^n) \approx J$ (see I,16.6) there is an integer d such that $f_*u = du$ for all $u \in H_n(S^n)$. This integer d is called the *degree* of f, written $d = \text{degree}(f)$.

Clearly two homotopic mappings of $S^n \to S^n$ have the same degree. The converse is also true but will not be proved until Chapter XV. It is easy to see that the degree of the composition of two maps is the product of their degrees.

LEMMA 4.2. *A mapping f: $S^n \to S^n$ of degree $\neq 0$ is essential, and therefore maps S^n onto S^n.*

PROOF. Suppose f is inessential. Then we may replace f by a homotopic map f' such that $f'(S^n)$ is a single point $x_0 \in S^n$. Since $H_n(x_0) = 0$, it follows that $f'_* = 0$. Thus the degree of f is zero.

LEMMA 4.3. *Let $n > 1$, and let f be a map of the triad (S^n, E^n_+, E^n_-) into itself:*

$$f: (S^n; E^n_+, E^n_-) \to (S^n; E^n_+, E^n_-).$$

Let f_1: $S^n \to S^n$, f_2: $S^{n-1} \to S^{n-1}$ be the maps defined by f. Then

$$\text{degree}(f_1) = \text{degree}(f_2).$$

PROOF. By I,16.2, the triad $(S^n; E^n_+, E^n_-)$ is proper. Consider the diagram

$$\begin{array}{ccc} H_n(S^n) & \xrightarrow{\Delta} & H_{n-1}(S^{n-1}) \\ \downarrow{\scriptstyle f_{1*}} & & \downarrow{\scriptstyle f_{2*}} \\ H_n(S^n) & \xrightarrow{\Delta} & H_{n-1}(S^{n-1}) \end{array}$$

where Δ is the boundary operator in the Mayer-Vietoris sequence of the triad By I,15 4, commutativity holds in this diagram Further, by I,16.5, Δ is an isomorphism Let $u \in H_n(S^n)$, $u \neq 0$ Then

$$\text{degree}(f_2)\Delta u = f_{2*}\Delta u = \Delta f_{1*}u = \text{degree}(f_1)\Delta u.$$

Since $\Delta u \neq 0$, it follows that degree (f_2) = degree(f_1)

LEMMA 4 4 *Suppose that a function* Γ: $S^n \times S^n \to S^n$, $n > 0$, *is given such that there exists an element* $e \in S^n$ *(unit element) satisfying*

(i) $$\Gamma(e,x) = x = \Gamma(x,e)$$

for all $x \in S^n$ *(see note at end of this chapter). Given two maps* f_1,f_2: $S^n \to S^n$, *define* f: $S^n \to S^n$ *by setting*

$$f(x) = \Gamma(f_1(x),f_2(x)).$$

Then

$$\text{degree}(f) = \text{degree}(f_1) + \text{degree}(f_2).$$

PROOF. We first replace f_1,f_2 by homotopic maps g_1,g_2 such that

(ii) $$g_1(E_-^n) = e, \qquad g_2(E_+^n) = e.$$

To show that such maps exist, consider $f_1|E_-^n$ Since E_-^n is contractible to a point, there is a homotopy F $E_-^n \times I \to S^n$ of $f_1|E_-^n$ such that $F(x,1) = e$ for $x \in E_-^n$ By 2 3, this homotopy may be extended to a homotopy F $S^n \times I \to S^n$ of f_1 Set $g_1(x) = F(x,1)$. The existence of g_2 is proved similarly The map f is now homotopic with the map g: $S^n \to S^n$ defined by

(iii) $$g(x) = \Gamma(g_1(x),g_2(x))$$

so that it suffices to prove that

$$g_* = g_{1*} + g_{2*}$$

in the dimension n.

From (i)-(iii) it follows that

$$g(x) = \begin{cases} g_1(x) & \text{for } x \in E_+^n, \\ g_2(x) & \text{for } x \in E_-^n. \end{cases}$$

If we denote by g',g_1',g_2' the maps $(S^n,S^{n-1}) \to (S^n,e)$ defined by g,g_1,g_2 respectively, then the conditions of I,14.6 are satisfied; and therefore

$$g'_* = g'_{1*} + g'_{2*}.$$

Consider the inclusion maps $i: S^n \subset (S^n, S^{n-1})$ and $j: S^n \subset (S^n, e)$. Since $jg = g'i$ and similarly with subscripts 1 and 2, it follows that

$$j_* g_* = j_* g_{1*} + j_* g_{2*}.$$

Since $n > 0$, we have $H_n(e) = 0$. Hence, by exactness, j_* has kernel zero in the dimension n. This implies $g_* = g_{1*} + g_{2*}$ in the dimension n

THEOREM 4.5. *Regard S^1 as the set of all complex numbers z with $|z| = 1$. For any integer k, the map $f_k: S^1 \to S^1$ defined by $f_k(z) = z^k$ has degree k*

PROOF. Define $\Gamma: S^1 \times S^1 \to S^1$ by setting $\Gamma(x,y) = xy$ For any two integers k,l, we have

$$f_{k+l}(z) = f_k(z)f_l(z) = \Gamma(f_k(z), f_l(z))$$

Thus 4.4 yields

$$\text{degree}(f_{k+l}) = \text{degree}(f_k) + \text{degree}(f_l).$$

Since f_1 is the identity map, it has degree 1. This implies that f_k has degree k.

REMARK The proof above actually yields a stronger result, namely: $f_{k*}u = ku$ for each $u \in H_1(S^1)$ in any homology theory. Similarly $f_k^* u = ku$ for $u \in H^1(S^1)$

THEOREM 4.6 *For any $n > 0$ and any integer k there exist maps $f: S^n \to S^n$ of degree k*

PROOF. The proposition has been established for $n = 1$ Assume inductively that $g: S^{n-1} \to S^{n-1}$ $(n > 1)$ is a map of degree k Extend g to a map $f: S^n \to S^n$ such that $f(E_+^n) \subset E_+^n$ and $f(E_-^n) \subset E_-^n$(such extensions obviously exist). Then 4.3 implies that f and g have the same degree. Thus f has degree k.

5. THE FUNDAMENTAL THEOREM OF ALGEBRA

Consider the compactification $\dot{R}^n$ of the cartesian space R^n obtained by adjoining a point ω at infinity (see X,6 2) A mapping $f: R^n \to R^n$ is admissible (in the category $\mathfrak{A}_{LC}$ of locally compact pairs) if its extension $\dot{f}: \dot{R}^n \to \dot{R}^n$ defined by $\dot{f}(\omega) = \omega$ is continuous.

DEFINITION 5.1. The *degree* of an admissible map $f: R^n \to R^n$ is defined as the degree of the map $\dot{f}: \dot{R}^n \to \dot{R}^n$. The latter degree is defined since $\dot{R}^n$ is homeomorphic with the n-sphere S^n.

LEMMA 5.2. *If $f: R^n \to R^n$ has degree $\neq 0$, then f maps R^n onto R^n*

PROOF. 4.2 implies that $\dot{f}$ maps $\dot{R}^n$ onto $\dot{R}^n$. Thus $f(R^n) = R^n$.

LEMMA 5.3 *Let $D^n = \overline{R^n - E^n}$. Thus D^n is defined by the condition $||x|| \geq 1$ If $f: R^n \to R^n$, $n > 1$, is an admissible map such that $f(E^n) \subset$*

E^n, $f(D^n) \subset D^n$, then the degree of f coincides with the degree of the map g: $S^{n-1} \to S^{n-1}$ defined by f

PROOF. The result follows from 4.3 since the triads $(\dot{R}^n;E^n,D^n)$ and $(S^n;E^n_+,E^n_-)$ are homeomorphic.

LEMMA 5.4. *Let f_0,f_1: $R^n \to R^n$ and let F: $R^n \times I \to R^n$ be a homotopy between f_0 and f_1 such that, for every real number A, there is a real number B such that*

$$\|F(x,t)\| > A \qquad \text{if} \qquad \|x\| > B, \qquad\qquad t \in I.$$

Then the maps f_0 and f_1 are both admissible and have the same degree.

PROOF. The condition of the lemma implies that setting $\dot{F}(\omega,t) = \omega$ yields an extension $\dot{F}$: $\dot{R}^n \times I \to \dot{R}^n$ of F. This implies that f_0 and f_1 are admissible and that $\dot{F}$ is a homotopy between $\dot{f}_0$ and $\dot{f}_1$. Therefore f_0 and f_1 have the same degree.

THEOREM 5.5 *A polynomial of degree $k > 0$*

$$f(z) = a_kz^k + a_{k-1}z^{k-1} + \cdots + a_1z + a_0, \qquad a_k \neq 0$$

with complex coefficients is an admissible map f: $R^2 \to R^2$ which has topological degree k.

COROLLARY 5.6 (*Fundamental Theorem of Algebra*). *The equation $f(z) = 0$ has at least one solution*

PROOF. Let $g(z) = a_{k-1}z^{k-1} + \cdots + a_1z + a_0$ and $a = |a_{k-1}| + \cdots + |a_0|$ Define F: $R^2 \times I \to R^2$ by setting

$$F(z,t) = a_kz^k + (1 - t)g(z), \qquad z \in R^2,\ t \in I.$$

Let A be any real number, and let

$$|z| > \frac{1}{|a_k|}(A + a), \qquad |z| > 1.$$

Since $k > 0$, we have

$$|F(z,t)| \geqq |a_kz^k| - |g(z)| \geqq |a_kz^k| - a|z^{k-1}| > |a_k||z| - a > A.$$

Thus the homotopy F satisfies the condition of 5.4 Consequently f is admissible and has the same degree as the map f': $R^2 \to R^2$ where $f'(z) = F(z,1) = a_kz^k$.

Let θ: $I \to R^2 - (0)$ be such that $\theta(0) = a_k$, $\theta(1) = 1$. The homotopy G: $R^2 \times I \to R^2$ defined by $G(z,t) = \theta(t)z^k$ again satisfies the condition of 5.4 Therefore f' has the same degree as the map $f''(z) = G(z,1) = z^k$. Since $|f''(z)| = |z|^k$, the conditions of 5.3 are satisfied and f'' has the same degree as the map f_k. $S^1 \to S^1$ defined by f''. Since, by 4.5, f_k has degree k, the proof is complete.

We now proceed to establish an analog of the fundamental theorem of algebra in a setting more general than that of complex numbers

We shall assume that a mapping Γ: $R^n \times R^n \to R^n$ is given. For simplicity we shall denote $\Gamma(x,y)$ by the product xy. This product is subject to the following conditions:

(i) $x \neq 0$, $y \neq 0$ imply $xy \neq 0$.

(ii) There exists an element $e \in R^n$ such that $ex = x = xe$ for all $x \in R^n$.

(iii) The relation $(tx)y = t(xy) = x(ty)$ holds of each real positive number t. The last condition implies by continuity: $0x = 0 = x0$.

These properties are satisfied if R^n carries the structure of a real division algebra. Typical examples are real numbers ($n = 1$), complex numbers ($n = 2$), quaternions ($n = 4$), Cayley numbers ($n = 8$). It is not known for what values of n a product such as described above exists (see note at end of this chapter).

We define inductively a *monomial* m (in one variable) of degree k as follows: A constant in R^n different from zero is a monomial of degree zero. The identity function x is a monomial of degree 1. If m_1 and m_2 are monomials of degree k_1 and k_2, then m_1m_2 is a monomial of degree $k_1 + k_2$.

It follows that, if m is a monomial and $x \in R^n$, then $m(x) \in R^n$ and m: $R^n \to R^n$ is continuous.

THEOREM 5.7. *Assume $n > 1$ Let f be a polynomial of algebraic degree $k > 0$ of the form $m + g$ where m is a monomial of degree k and g is a finite sum of monomials of degree $< k$. Then the map f: $R^n \to R^n$ is admissible and has topological degree k.*

COROLLARY 5.8. *The equation $f(x) = 0$ has at least one solution.*

Applying this to the equations $ax - e = 0$ and $xa - e = 0$, we find

COROLLARY 5.9. *Each $a \neq 0$ has at least one left inverse and at least one right inverse.*

Applying 5.8 to the polynomial $xx + e$ yields

COROLLARY 5.10 *There exists an element whose square is $-e$.*

The proof of 5.7 will be preceded by some lemmas.

LEMMA 5.11. *There exist real numbers $0 < A < B$ such that*

$$A\|x\|\,\|y\| \leqq \|xy\| \leqq B\|x\|\,\|y\|.$$

PROOF. Let γ: $S^{n-1} \times S^{n-1} \to R^n$ be defined by $\gamma(x,y) = xy$. Then $\gamma(S^{n-1} \times S^{n-1})$ is a compact subset of R^n not containing zero. Thus there exist real numbers $0 < A < B$ such that $A \leqq \|xy\| \leqq B$ for $x,y \in S^{n-1}$. Then (iii) yields the conclusion of the lemma

LEMMA 5.12. *If m is a monomial of degree k, then there exist real numbers $0 < A_m < B_m$ such that*

$$A_m\|x\|^k \leqq \|m(x)\| \leqq B_m\|x\|^k.$$

PROOF. The conclusion is obviously valid for a monomial of degree 0, and for the monomial x Suppose now that $m = m_1 \cdot m_2$ is a product of monomials of degrees k_1, k_2 respectively with $k = k_1 + k_2$, and that the conclusion of the lemma is valid for m_1 and m_2 with the constants $A_{m_1}, B_{m_1}, A_{m_2}, B_{m_2}$ respectively. Then, by 5.11,

$$||m(x)|| \leqq B||m_1(x)||\ ||m_2(x)|| \leqq BB_{m_1}B_{m_2}||x||^{k_1}||x||^{k_2} = B_m||x||^k$$

where $B_m = BB_{m_1}B_{m_2}$. Similarly $A_m||x||^k \leqq ||m(x)||$ where $A_m = AA_{m_1}A_{m_2}$.

PROOF OF 5.7. Consider the homotopy F: $R^n \times I \to I$ defined by

$$F(x,t) = m(x) + (1 - t)g(x), \qquad x \in R^n,\ t \in I.$$

Let $g = m_1 + \cdots + m_l$ be a decomposition of g into a sum of monomials of degrees $< k$, and let

$$B' = B_{m_1} + \cdots + B_{m_l}.$$

If A' is a real number, and

$$||x|| > \frac{1}{A_m}(A' + B'), \qquad ||x|| > 1,$$

then $k > 0$ implies

$$\begin{aligned} ||F(x,t)|| &\geqq ||m(x)|| - ||g(x)|| \geqq A_m||x||^k - \sum B_{m_i}||x||^{k-1} \\ &> A_m||x|| - B' > A'. \end{aligned}$$

Thus the homotopy F satisfies the condition of 5 4. Consequently f is admissible and has the same topological degree as the map $m(x) = F(x,1)$ This reduces the proof to the case when $f = m$ is a monomial of degree k.

With each monomial m of degree k we associate a function g: $S^{n-1} \to S^{n-1}$ defined by

$$g(x) = \frac{m(x)}{||m(x)||}, \qquad x \in S^{n-1}.$$

It follows from 5 11 that $m(x) \neq 0$ for $x \neq 0$, thus g is a well-defined continuous function. We shall prove that m and g have the same topological degree. If $k = 0$, both m and g are constant and have topological degree 0. Let $k > 0$. Define a homotopy F: $R^n \times I \to R^n$ by setting

$$F(x,t) = \frac{m(x)}{||m(x)||}[(1 - t)||m(x)|| + t||x||^k], \qquad x \in R^n - (0),\ t \in I,$$

$$F(0,t) = 0.$$

By 5.12

$$[(1 - t)A_m + t]\,||x||^k \leqq ||F(x,t)|| \leqq [(1 - t)B_m + t]\,||x||^k,$$

and therefore

$$A_F\,||x||^k \leqq ||F(x,t)|| \leqq B_F\,||x||^k$$

where $A_F = \min\,(1,A_m)$, $B_F = \max\,(1,B_m)$. This inequality shows that F is continuous at $x = 0$, and that F satisfies the condition of 5.4. Since $F(x,0) = m(x)$, it follows from 5.4 that the map

$$M(x) = F(x,1) = m(x)\frac{||x||^k}{||m(x)||}, \qquad x \neq 0,$$
$$M(0) = 0$$

is admissible and has the same topological degree as m The function M satisfies the condition of 5.3 and therefore has the same degree as the map $S^{n-1} \to S^{n-1}$ defined by M. Since $M(x) = g(x)$ for $x \in S^{n-1}$, it follows that m and g have the same degree

Consider the mapping

$$\gamma\colon\; S^{n-1} \times S^{n-1} \to S^{n-1}$$

defined by

$$\gamma(x,y) = \frac{xy}{||xy||}, \qquad x,y \in S^{n-1}.$$

Setting $e' = e/||e||$ we have

$$\gamma(e',x) = x = \gamma(x,e'), \qquad x \in S^{n-1}.$$

Now consider the product $m = m_1m_2$ of two monomials and let g,g_1,g_2 be the associated maps of S^{n-1} into itself Then

$$\gamma(g_1(x),g_2(x)) = \frac{g_1(x)g_2(x)}{||g_1(x)g_2(x)||} = \frac{m_1(x)m_2(x)}{||m_1(x)||\,||m_2(x)||}\,\frac{||m_1(x)||\,||m_2(x)||}{||m_1(x)m_2(x)||}$$
$$= \frac{m(x)}{||m(x)||} = g(x).$$

Since $n - 1 > 0$, 4.4 gives

$$\text{degree}(g) = \text{degree}(g_1) + \text{degree}(g_2).$$

This implies that the topological degree of m is the sum of the topological degrees of m_1 and m_2. Thus the topological degree of a monomial obeys the same additivity rule as the algebraic degree. Since for a monomial of algebraic degree zero (i.e a constant $\neq 0$) the topological degree is also zero, and the monomial x has both algebraic and topological degree 1, it follows that the algebraic and topological degrees coincide for all monomials. This concludes the proof.

6. MANIFOLDS

DEFINITION 6 1. If X is a compact space, A is closed in X, and $X - A$ is locally euclidean of dimension n (see 3 10), then the pair (X,A) is called a *relative n-manifold* If there exists a relative homeomorphism f $(E,S) \to (X,A)$ (see x,5 1), where (E,S) is an n-cell and its boundary, then (X,A) is called a *relative n-cell*

The objective of this article is to show that the homology and cohomology groups of a relative n-manifold (X,A) are zero for dimensions greater than n, and to establish results concerning the structure of the n-dimensional groups

In case (X,A) is triangulable, the Uniqueness theorem of Chapter III shows that the results are independent of the choice of the homology theory It has been proved that any compact, differentiable, absolute manifold can be triangulated [S. Cairns, Bull. Amer. Math Soc. 41 (1935), 549-552]. It is an unsolved problem of long standing to determine if the result is valid without the assumption of differentiability. It is easy to construct *relative* n-manifolds which are not triangulable. For example, let X be an n-sphere, and let A be a closed subset of X which is not triangulable, then (X,A) is a relative n-manifold and is not triangulable A less trivial example is obtained as follows. Let Y be a locally-euclidean space which is not separable. Let X be the compact space obtained by adjoining a point at infinity to Y, and let A consist of this point Since Y is not separable, X is not triangulable.

We shall use the Čech homology-cohomology theory. Its continuity will play a vital role in the computations We repeat that the results are independent of this choice if (X,A) is triangulable If the singular theory were used, there would be no correspondingly simple results valid in the nontriangulable case

Since the results we will obtain are more significant in the case of cohomology, we restrict ourselves to that case The corresponding results for homology are stated at the end of this article

LEMMA 6 2. *Let Y be a space having a triangulation T of dimension n, and let (X,A) be a compact pair in Y. Then $H^q(X,A)$ is zero for each $q > n$.*

PROOF. Let ${}^iT$ denote the ith barycentric subdivision of T. Let X_i and A_i be the least subcomplexes of Y, in the triangulation ${}^iT$, containing X and A respectively Clearly $H^q(X_i,A_i)$ is zero for $q > n$, and $(X_{i+1},A_{i+1}) \subset (X_i,A_i)$ Since the mesh of ${}^iT$ tends to zero as i tends to infinity, it follows that $(X,A) = (\bigcap X_i, \bigcap A_i)$. By applying x,2.6, we obtain that $H^q(X,A)$ is zero, and the proof is complete.

Since a relative homeomorphism induces isomorphisms of the Čech groups (x,5.4), it follows that the Čech groups of a relative n-cell (X,A) are those of an n-cell, i.e. all groups are zero except $H^n(X,A) \approx G$.

LEMMA 6.3. *If (X,A) and (X,B) are relative n-cells such that $(X,A) \subset (X,B)$, then the inclusion map induces an isomorphism*

$$H^n(X,A) \approx H^n(X,B).$$

PROOF. Consider first the special case where X is the euclidean n-cell E: $\sum_1^n x_i^2 \leqq 1$, and A is the sphere S: $\sum_1^n x_i^2 = 1$. Also, let x_0 be a point of $E - S$, let U be the interior of a sphere with center x_0 such that $U \subset E - S$, and let $B = E - U$. Then (E,B) is a relative n-cell containing (E,S). Radial projection from x_0 defines a retraction r: $B \to S$. If we set

$$r(x,t) = (1 - t)x + tr(x),$$

we obtain a strong deformation retraction of B into S Then I,10 5c applies to give the desired result in this case.

For the next case, let (X,A) be a relative n-cell, let f: $(E,S) \to (X,A)$ be a relative homeomorphism, let U be the interior of a sphere contained in $E - S$, and let $B = X - f(U)$ Then (X,B) is a relative n-cell, and f defines a relative homeomorphism f_1: $(E,E - U) \to (X,B)$. Let j: $(E,S) \subset (E,E - U)$ and h: $(X,A) \subset (X,B)$ Since $f_1 j = hf$, we have $j^*f_1^* = f^*h^*$. As f_1,f are relative homeomorphisms, f_1^* and f^* are isomorphisms. By the preceding case, j^* is an isomorphism. It follows that h^* is an isomorphism.

Consider now the general case. Let f. $(E,S) \to (X,A)$ and g: $(E,S) \to (X,B)$ be relative homeomorphisms. Now $f^{-1}(X - B)$ is open in $E - S$, hence it contains a sphere with interior U. Let $A' = X - f(U)$. Then $g^{-1}(X - A')$ is open in $E - S$; hence it contains a sphere with interior V. Let $B' = X - f(V)$. Inclusion relations induce homomorphisms

$$H^n(X,B') \xrightarrow{j^*} H^n(X,A') \xrightarrow{k^*} H^n(X,B) \xrightarrow{l^*} H^n(X,A).$$

By construction, the preceding case applies to the inclusions jk and kl. Hence k^*j^* and l^*k^* are isomorphisms. It is easily seen that this implies that l^* is an isomorphism. This completes the proof.

DEFINITION 6.4. Let (X,A) be a relative n-manifold An ordered pair of relative n-cells $(X,B),(X,B')$ such that

$$(X,A) \subset (X,B) \subset (X,B'), \qquad \text{or} \qquad (X,A) \subset (X,B') \subset (X,B)$$

will be called a *step* in $X - A$ By 6 3, the inclusion map induces an isomorphism $H^n(X,B) \approx H^n(X,B')$ called the *isomorphism of the step* A finite sequence P of relative n-cells

$$(X,B_1),(X,B_2), \cdots, (X,B_k),$$

such that each pair of adjacent terms is a step in $X - A$, will be called a *path* in $X - A$ from (X,B_1) to (X,B_k) The composition of the isomorphisms of the successive steps is an isomorphism $H^n(X,B_1) \approx H^n(X,B_k)$ called the *isomorphism* P^* *of the path* In case $(X,B_1) = (X,B_k)$ the path is said to be *closed*, and the isomorphism of the path becomes an automorphism The manifold (X,A) is said to be *orientable* if the automorphism of each closed path in $X - A$ is the identity; otherwise (X,A) is called *nonorientable*. (Note that this concept depends on the coefficient group)

LEMMA 6 5 *Let $(X,A),(X',A')$ be relative n-manifolds such that $X' \subset X$, and $X' - A' \subset X - A$ If (X,A) is orientable, so also is (X',A')*

PROOF. Let $(X',B_1'), \cdots, (X',B_k')$ be a closed path in $X' - A'$. Set

$$B_i = B_i' \cup (X - X'), \qquad i = 1, \cdots, k.$$

Since $X' - B_i' = X - B_i$ is open in $X' - A'$, by 3 11, it is also open in $X - A$, hence B_i is closed It follows that g_i: $(X',B_i') \subset (X,B_i)$ is a relative homeomorphism, (X,B_i) is a relative n-cell, and (X,B_1), $\cdots$, (X,B_k) is a closed path in $X - A$ We obtain the diagram

$$\begin{array}{ccccccc} H^n(X,B_1) & \approx & H^n(X,B_2) & \approx & \cdots & \approx & H^n(X,B_k) \\ \downarrow g_1^* & & \downarrow g_2^* & & & & \downarrow g_k^* \\ H^n(X',B_1') & \approx & H^n(X',B_2') & \approx & \cdots & \approx & H^n(X',B_k') \end{array}$$

The isomorphisms of the upper [lower] line compose to give the automorphism ϕ [ϕ'] of the path in $X - A$ [$X' - A'$] Since g_i is a relative homeomorphism, g_i^* is an isomorphism Since all isomorphisms are induced by inclusions, commutativity holds in each square It follows that $g_k^*\phi = \phi' g_1^*$ Since the original path is closed, $B_1' = B_k'$; hence

$g_1 = g_k$. Since (X,A) is orientable, ϕ is the identity. It follows that ϕ' is the identity; therefore (X',A') is orientable.

LEMMA 6.6. *If P is a path in $X - A$ from (X,B) to (X,B'), and P^* is the isomorphism of P, then commutativity holds in the diagram*

$$\begin{array}{ccc} H^n(X,B) & \overset{P^*}{\approx} & H^n(X,B') \\ & {\scriptstyle j^*}\searrow \quad \swarrow{\scriptstyle j'^*} & \\ & H^n(X,A) & \end{array}$$

where j,j' are the indicated inclusions

PROOF. If P consists of a single step, commutativity is obvious since all homomorphisms are induced by inclusions. In the general case, we enlarge the diagram by inserting the individual steps and the homomorphisms $H^n(X,B_i) \to H^n(X,A)$ induced by inclusions Commutativity in each small triangle implies the desired result.

LEMMA 6.7. *Let (X,A) be a relative n-manifold with $X - A$ connected. Let $(X,B),(X,B')$ be relative n-cells such that $A \subset B$ and $A \subset B'$. Then there exists a path in $X - A$ from (X,B) to (X,B')*

PROOF Let F be the family of those relative n-cells (X,C) which are the end terms of paths in $X - A$ starting from (X,B) Let W be the union of the sets $X - C$ for $(X,C) \in F$. Clearly W is an open connected set in $X - A$. Let $x \in X - A$ be a limit point of W. As $X - A$ is locally euclidean, there is a relative n-cell (X,D) such that $D \supset A$ and $x \in X - D$ As $X - D$ is open, it contains a point $y \in W$. Then $y \in X - C$ for some $(X,C) \in F$. As $X - A$ is locally euclidean, there is a relative n-cell (X,C') such that $C' \supset C \cup D$ and $y \in X - C'$. Then $(X,C),(X,C'),(X,D)$ is a path in $X - A$ from (X,C) to (X,D). Combining this with a path from (X,B) to (X,C), we obtain a path from (X,B) to (X,D). Thus $x \in W$, and W is also closed in $X - A$. As $X - A$ is connected, we must have $W = X - A$. Therefore $X - B'$ contains a point $z \in W$. Treating $(X,B'),z$ in the same manner as we treated $(X,D),y$ above, we obtain the desired path from (X,B) to (X,B').

We are now prepared to state and prove the main results of this section.

THEOREM 6.8. *Let (X,A) be a relative n-manifold with $X - A$ connected; let (X,B) be a relative n-cell containing (X,A), and let j: $(X,A) \subset (X,B)$. Then we have the following five propositions:*

(i) $H^q(X,A) = 0$ *for* $q > n$.

(ii) j^* *maps* $H^n(X,B)$ *onto* $H^n(X,A)$.

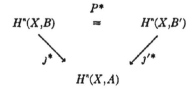

(iii) *The kernel of* j^* *contains all elements of the form* $u - P^*u$ *where* $u \in H^n(X,B)$ *and* P *is any closed path in* $X - A$ *starting with* (X,B)

(iv) *If* (X,A) *is orientable, then* j^* $H^n(X,B) \approx H^n(X,A)$

(v) *If the coefficient group* G *is the group of integers, and* (X,A) *is nonorientable, then* $H^n(X,A)$ *is cyclic of order* 2.

PROOF. To prove (iii) we apply 6.6 with $j = j'$, to obtain $j^*P^* = j^*$; hence $j^*(u - P^*u) = 0$.

We shall say that (X,A) is *finite* if there exists a finite set (X,B_1), $\cdots$, (X,B_k) of relative n-cells such that $X - A = \bigcup_{i=1}^{k} X - B_i$. The propositions (i), (ii), (iv), (v), restricted to the case (X,A) finite, will be denoted by (i)', (ii)', (iv)', (v)' respectively We will show first that the finite case implies the general case.

Let $\{(X,B_\alpha)\},\alpha \in M$, be an indexed collection of relative n-cells such that $A = \bigcap_{\alpha \in M} B_\alpha$. The existence of such a collection is implicit in the assumption that $X - A$ is locally euclidean Let $V_\alpha = X - B_\alpha$. For each finite set $\xi \subset M$, define $W_\xi = \bigcup_{\alpha \in \xi} V_\alpha$ and $A_\xi = X - W_\xi$. Let N be the set of those ξ's such that W_ξ is connected Inclusion relations in M provide a partial order in N. We will show that N is a directed set with respect to this order. If $x \in X - A$, let $C(x)$ be the union of all W_ξ such that $\xi \in N$ and $x \in W_\xi$. Then $C(x)$ is open If y is a limit point of $C(x)$ in $X - A$, choose a V_α containing y Then $V_\alpha \cap C(x)$ contains a point of some W_ξ which contains x Hence $V_\alpha \cup W_\xi$ is connected, contains x and y, and is a W_η for $\eta \in N$. Therefore $y \in C(x)$. As $C(x)$ is open and closed in $X - A$, and $X - A$ is connected, we have $C(x) = X - A$. If $\xi,\eta \in N$, choose $x \in W_\xi$, $y \in W_\eta$. Since $C(x) = X - A$, there is a $\zeta \in N$ such that $x,y \in W_\zeta$. Let $\omega = \xi \cup \eta \cup \zeta$ Then $W_\omega = W_\xi \cup W_\eta \cup W_\zeta$ is a connected set. Hence $\omega \in N$, and ω contains ξ and η This proves that N is directed.

It follows that (X,A) is the intersection of the nested system $\{(X,A_\xi)\},\xi \in N$. By x,2 6, and the continuity of the Čech groups, the group $H^q(X,A)$ is the limit group of the groups $\{H^q(X,A_\xi)\}$

Now (i)' asserts that each $H^q(X,A_\xi) = 0$ for $q > n$ and each $\xi \in N$. Then the limit group $H^q(X,A)$ is also zero; hence (i)' implies (i)

If $u \in H^n(X,A)$, then by definition of direct limit, there is an element $v \in H^n(X,A_\xi)$ such that $g^*v = u$ where g: $(X,A) \subset (X,A_\xi)$ Let (X,B') be a relative n-cell such that $(X,A_\xi) \subset (X,B')$, and let j' denote the inclusion map. Then (ii)' asserts the existence of $w \in H^n(X,B')$ such that $j'^*w = v$. Thus the image of $(j'g)^*$. $H^n(X,B') \to H^n(X,A)$ contains u By 6 7 and 6.6, the image of $(j'g)^*$ coincides with the image of j^* As u is arbitrary, it follows that (ii)' implies (ii)

Choose $\alpha_0 \in M$, let $\xi_0 \in N$ consist of the single element α_0, and let N' be the subset of those $\xi \in N$ such that $\xi \supset \xi_0$. Clearly N' is cofinal in N, hence $H^n(X,A)$ is the limit group of $\{H^n(X,A_\xi)\}$ for $\xi \in N'$. Let $\xi \subset \eta$ in N' Then we have the diagram

$$\text{(1)} \qquad H^n(X,B_{\alpha_0}) \xrightarrow{f^*} H^n(X,A_\xi) \xrightarrow{g^*} H^n(X,A_\eta) \xrightarrow{h^*} H^n(X,A)$$

where f,g,h are inclusions Suppose now that (X,A) is orientable By 6 5, (X,A_ξ) and (X,A_η) are orientable Then (iv)$'$ asserts that f^* and g^*f^* are isomorphisms It follows that g^* is an isomorphism Thus all projections of the direct system $\{H^n(X,A_\xi)\},\xi \in N'$, are isomorphisms It follows from VIII,4 8 that $(fgh)^*$ is an isomorphism If we apply 6 6 with $B_{\alpha_0} = B'$, it follows that j^* is an isomorphism. Thus (iv)$'$ implies (iv)

Suppose (X,A) is nonorientable relative to integer coefficients. Let P be a path such that P^* is not the identity We may suppose that the collection $\{(X,B_\alpha)\}$ includes the relative n-cells of P (adjoin them otherwise), and let $\xi_0 \subset M$ denote the set of indices of these cells Clearly W_{ξ_0} is connected, so $\xi_0 \in N$. Let N' be the set of those $\xi \in N$ such that $\xi \supset \xi_0$ Proceed now as in the preceding paragraph, and obtain the diagram (1) where $\alpha_0 \in \xi_0$ Now P is a path in $X - A_\xi$ for each $\xi \in N'$, hence (X,A_ξ) is nonorientable Then (v)$'$ asserts that $H^n(X,A_\xi)$ and $H^n(X,A_\eta)$ are cyclic of order 2; and (ii)$'$ asserts that f^* and g^*f^* are onto It follows that g^* is an isomorphism. By VIII,4 8, h^* is an isomorphism Hence $H^n(X,A)$ is cyclic of order 2 Thus (ii)$'$ and (v)$'$ imply (v)

It remains to establish the propositions in the finite case, i e $X - A = \bigcup_{i=1}^k (X - B_i)$ If $k = 1$, then (X,A) is a relative n-cell, so (i) holds, and 6 3 implies that j^* is an isomorphism Hence (ii) holds Since (iii) has been proved, and the kernel of j^* is zero, it follows that (X,A) is orientable, and then (iv) holds. Proposition (v) does not apply if $k = 1$.

Suppose inductively that (i), (ii), (iv), (v) have been proved for finite (X,A) such that $k \leq m$. Consider now the case $k = m + 1$. Since $X - A$ is connected, the union of some m of the sets $X - B_i$ is connected Let the indexing be such that $\bigcup_1^m (X - B_i)$ is connected. Set $A' = \bigcap_1^m B_i$, and $B' = B_{m+1}$. Then $X - A'$ is connected, and the inductive hypotheses apply to (X,A') The relative Mayer-Vietoris sequence (I,15 6) of the triad $(X;A',B')$ contains the portion

$$\text{(2)} \qquad H^q(X,A' \cup B') \xrightarrow{\phi} H^q(X,A') + H^q(X,B')$$
$$\xrightarrow{\psi} H^q(X,A) \xrightarrow{\Delta} H^{q+1}(X,A' \cup B').$$

To prove (i) for (X,A), choose a relative homeomorphism f: $(E,S) \rightarrow (X,B')$. Let $C = f^{-1}(A' \cup B')$ Then C is closed, and f defines a relative homeomorphism g $(E,C) \rightarrow (X,A' \cup B')$ By 6 2, $H^{q+1}(E,C) = 0$ for $q \geqq n$ By x,5 4, g^* is an isomorphism Hence $H^{q+1}(X,A' \cup B') = 0$ for $q \geqq n$ Then the exactness of (2) implies that $H^q(X,A)$ is the image of ψ. By the inductive hypothesis, $H^q(X,A')$ and $H^q(X,B')$ are zero for $q > n$ Therefore $H^q(X,A) = \psi(0) = 0$; and (i) has been proved.

$$\begin{array}{ccccc} & & H^n(X,A') & & \\ & \nearrow f^*_\alpha & \uparrow f'^*_\alpha & \searrow f^* & \\ H^n(X,D_\alpha) & \xrightarrow{h^*_\alpha} & H^n(X,C_\alpha) & \longrightarrow & H^n(X,A) \\ & \searrow g^*_\alpha & \downarrow g'^*_\alpha & \nearrow g^* & \\ & & H^n(X,B') & & \end{array} \tag{3}$$

For each component α of $X - A' \cup B'$, let $C_\alpha = X - \alpha$; and let (X,D_α) be a relative n-cell such that $D_\alpha \supset C_\alpha$ We obtain the diagram (3) in which all homomorphisms are induced by inclusions. By 6.3,

(4) g^* is an isomorphism.

By the inductive hypothesis (ii) on (X,A'),

(5) $H^n(X,A') = \text{image } f^*_\alpha$.

As shown in the preceding paragraph,

(6) $H^n(X,A) = \text{image } \psi$

Then, if $w \in H^n(X,A)$, we can choose $u \in H^n(X,A')$ and $v \in H^n(X,B')$ such that

$$w = \psi(u,v) = f^*u - g^*v, \qquad \text{see I,15 6}$$

By (4) and (5) we can choose $u',v' \in H^n(X,D_\alpha)$ such that $f^*_\alpha u' = u$ and $g^*_\alpha v' = v$ Since $f_\alpha f = g_\alpha g$, we have

$$g^*g^*_\alpha(u' - v') = f^*f^*_\alpha u' - g^*g^*_\alpha v' = f^*u - g^*v = w.$$

This proves

(7) $H^n(X,A) = \text{image } g^*$.

By 6.7 and 6.6, $H^n(X,B')$ and $H^n(X,B)$ have the same image in $H^n(X,A)$. Thus (7) implies (ii)

Having proved (ii), it follows that h_α^* is onto. Since $g_\alpha'^* h_\alpha^* = g_\alpha^*$, (4) yields

(8) $$h_\alpha^* \text{ is an isomorphism.}$$

According to x, Exer. B3, $H^n(X, A' \cup B')$ is the injective direct sum

$$H^n(X, A' \cup B') = \sum_\alpha H^n(X, C_\alpha) \approx \sum_\alpha H^n(X, D_\alpha).$$

Referring to the definition of ϕ, it follows that $\{(f_\alpha^*, g_\alpha^*)\}$ is the collection of components of a homomorphism

$$\phi'\colon \quad \sum_\alpha H^n(X, D_\alpha) \to H^n(X, A') + H^n(X, B')$$

having the same image as ϕ. Since the sequence (2) is exact,

$$\text{kernel } \psi = \text{image } \phi = \text{image } \phi'.$$

Hence by (6)

$$H^n(X, A) \approx [H^n(X, A') + H^n(X, B')]/\text{image } \phi'.$$

Let us regard $H^n(X, B')$ as imbedded in the direct sum. Using (7) and the fact that $\psi(0, v) = -g^* v$, we obtain

(9) $$H^n(X, A) \approx H^n(X, B')/K$$

where

$$K = H^n(X, B') \cap \text{image } \phi'.$$

Now an element of $\sum_\alpha H^n(X, D_\alpha)$ is a sum $\sum u_\alpha$ where $u_\alpha \in H^n(X, D_\alpha)$ is nonzero for only a finite number of α's. By definition

$$\phi'(\sum u_\alpha) = (\sum f_\alpha^* u_\alpha, \sum g_\alpha^* u_\alpha).$$

Hence $u \in K$ if and only if u has the form

(10) $$u = \sum g_\alpha^* u_\alpha \qquad \text{where} \qquad \sum f_\alpha^* u_\alpha = 0.$$

Choose a fixed component γ of $X - A' \cup B'$; and, for any component α, let Q_α be a path in $X - A'$ from (X, D_α) to (X, D_γ). The path $(X, D_\gamma), (X, B'), (X, D_\alpha)$ followed by Q_α is a closed path P_α in $X - A$ beginning and ending with (X, D_γ). Set

$$v_\alpha = g_\gamma^{*-1} g_\alpha^* u_\alpha \in H^n(X, D_\gamma).$$

Using 6.6 we have

$$\sum f_\alpha^* u_\alpha = \sum f_\gamma^* Q_\alpha^* g_\alpha^{*-1} g_\gamma^* v_\alpha = f_\gamma^* \sum P_\alpha^* v_\alpha.$$

Then (10) becomes

(11) $$u = g_\gamma^* \sum v_\alpha \qquad \text{where} \qquad f_\gamma^* \sum P_\alpha^* v_\alpha = 0.$$

To prove (iv), let (X,A) be orientable. Then

$$\sum P_\alpha^* v_\alpha = \sum v_\alpha.$$

By 6.5, (X,A') is orientable. By the inductive hypothesis (iv) for (X,A'), f_γ^* is an isomorphism. Hence (11) reduces to $u = 0$. By (10), we have $K = 0$, so $g^*\colon H^n(X,B') \approx H^n(X,A)$. Using 6.6 and a path P from (X,B) to (X,B'), it follows that j^* is an isomorphism. This proves (iv) for (X,A).

To prove (v), let G = integers, and (X,A) nonorientable. The only nontrivial automorphism of G is the reversal of sign, hence $P_\alpha^* v_\alpha = \pm v_\alpha$. Let x be the sum of those v_α such $P_\alpha^* v_\alpha = v_\alpha$, and y the sum of those v_α such that $P_\alpha^* v_\alpha = -v_\alpha$. Then (11) becomes

$$u = g_\gamma^*(x + y) \qquad \text{where} \qquad f_\gamma^*(x - y) = 0. \tag{12}$$

If (X,A') is orientable, the inductive hypothesis (iv) for (X,A') gives $x - y = 0$; hence $u = 2g_\gamma^* x$. If (X,A') is nonorientable, the inductive hypothesis (v) for (X,A') states that $x - y$ has the form $2z$. Hence $u = 2g_\gamma^*(x - z)$. In either case $u \equiv 0 \bmod 2$. Thus K consists of the even elements of $H^n(X,B')$. It follows that $H^n(X,A)$ is cyclic of order 2. This proves (v) for (X,A), and thereby completes the proof of the theorem.

Remark. The restriction to integer coefficients in proposition (v) is not necessary. If (X,A) is nonorientable, then, for any G, $H^n(X,A)$ is isomorphic to G reduced mod 2. The proof just given holds in general if we know, for any closed path P, that $P^* = \pm I$ where I = identity. This is true by virtue of the universality of integer coefficients which we have not proved (see v, Exer G3). Specifically, if $P^* = I$, respectively $-I$, for integer coefficients, then $P^* = I$, respectively $-I$, for any G. It is therefore customary to use the terms orientable and nonorientable always in the sense of integer coefficients.

Theorem 6.9. *The n-sphere S^n is orientable relative to any coefficient group.*

Proof. If $x_0 \in S^n$, a sterographic projection shows that (S^n,x_0) is a relative n-cell. Since $H^n(S^n,x_0) \approx H^n(S^n)$ under the inclusion, it follows from (6.8) (iii) that S^n is orientable.

Theorem 6.10. *Let (X,A) be a relative n-manifold such that X is homeomorphic to a subset of the n-sphere. For each connected component α of $X - A$, let $A_\alpha = X - \alpha$. Then each (X,A_α) is orientable, so that $H^n(X,A_\alpha) \approx G$; and the homomorphisms $H^n(X,A_\alpha) \to H^n(X,A)$ induced by inclusions provide an injective direct sum representation*

$$H^n(X,A) \approx \sum_\alpha H^n(X,A_\alpha).$$

The orientability of (X,A_α) follows from 6 9 and 6 5. The direct sum decomposition follows from x, Exer B3

We turn now to the corresponding results for homology. Up through 6 7, the exactness property was not used Hence all statements and proofs through 6 7 hold with cohomology replaced by homology.

The proof of 6 8 uses exactness. The Čech homology theory on compact pairs is not generally exact To obtain exactness we add to the hypotheses of 6 8 that G is compact or a vector space over a field. Then the duals of (i)-(iv) of 6 8 are valid. These are

(i) $H_q(X,A) = 0$ for $q > n$.

(ii) *The kernel of* j_*: $H_n(X,A) \to H_n(X,B)$ *is zero.*

(iii) *If* u *is in the image of* j_*, *then* $u = P^*u$ *where* P *is any closed path in* $X - A$ *starting with* (X,B)

(iv) *If* (X,A) *is orientable, then* j_* $H_n(X,A) \approx H_n(X,B)$.

The duals of 6 9 and 6 10 hold with the same restriction on G The dual of the conclusion of 6 10 reads The homomorphisms $H_n(X,A) \to H_n(X,A_\alpha)$ provide a projective representation of $H_n(X,A)$ as a direct product

$$H_n(X,A) \approx \prod_\alpha H_n(X,A_\alpha)$$

Without restricting G, the Čech homology theory is exact for triangulable pairs. This suggests an alternative procedure, namely, add to 6 8 the hypothesis that (X,A) is triangulable, then show that the proof carries through using only triangulable pairs at each stage One would use triangulable relative n-cells having triangulable unions and intersections For this it would suffice to use a single triangulation of (X,A) and its barycentric subdivision Then the duals of (i) to (iv), as stated above, would hold; and, in addition, the dual of (v), which is

(v) *If* $G =$ *integers, and* (X,A) *is nonorientable, then* $H_n(X,A) = 0$

NOTE

The existence of multiplications In §5 we assumed the proposition:

(A) There exists a multiplication Γ. $R^n \times R^n \to R^n$ satisfying conditions (i), (ii), and (iii) of §5

We also remarked that examples of such a multiplication, for $n = 1,2,4$, and 8, were provided by the real, complex, quaternionic, and Cayley number systems respectively. No other examples are known If we require in addition that $\Gamma(x,y)$ be bilinear and $|\Gamma(x,y)| = |x|\,|y|$, then Hurwitz has proved [Nachr Ges d Wiss Gottingen (1898), 309-316] that $n = 1,2,4$, or 8, and Γ is isomorphic to the corresponding abovementioned number system. If we drop the norm condition, but retain bilinearity, then Hopf has shown [Comm Math Helv. 13 (1941), 219-239] that n must be a power of 2.

It can be shown that (A) is equivalent to each of the propositions (B), (C), and (D) listed below. For the equivalence of (C) and (D) as well as the definition of the terms involved see S Eilenberg, Ann. of Math 41 (1940), 662-673.

(B) There exists a multiplication $S^{n-1} \times S^{n-1} \to S^{n-1}$ with a two-sided unit e: $ex = x = xe$ for all $x \in S^{n-1}$.

(C) There exists a map $S^{n-1} \times S^{n-1} \to S^{n-1}$ of type (1,1).

(D) There exists a map $S^{2n-1} \to S^n$ having Hopf invariant 1.

Proposition (D) is of considerable importance in the study of homotopy groups of spheres, and much effort has been expended in trying to determine the values of n for which it is valid. Hopf has shown [Fund Math 25 (1935), 427-440] that, if $n > 1$, then (D) implies that n is even G. W Whitehead has proved [Ann of Math 51 (1950), 192-237] that, if $n > 2$, then (D) implies that n is divisible by 4 J. Adem has announced [Proc Nat. Acad Sci. 38 (1952)] that (D) implies that n is a power of 2 H Toda has announced [C R Acad Sci Paris 241(1955), 849-850] that D is false for $n = 16$

EXERCISES

A DEGREES OF MAPS.

1 Define the degree of a map $f.\ S^0 \to S^0$ by using the reduced group $\tilde{H}_0(S^0)$. Show that this degree is always 0, 1, or -1 Show that 4 3 remains valid for $n = 1$

2. Define the degree of a map $f\ (E^n,S^{n-1}) \to (E^n,S^{n-1})$, and show that it is equal to the degree of the map $g.\ S^{n-1} \to S^{n-1}$ defined by f.

3 Show that a map $f.\ (E^1,S^0) \to (E^1,S^0)$ has degree 0, 1, or -1.

4. Prove that, for each $n > 1$ and each integer k, there exists a map f: $(E^n,S^{n-1}) \to (E^n,S^{n-1})$ of degree k

B. INVARIANCE OF DOMAIN.

1 Let X be a closed subset of S^n and let $x_0 \in X$. Show that, if $x_0 \in \text{Int}\ X$, then $H^n(X,X - U) \neq 0$ for all sufficiently small neighborhoods U of x_0

2 Prove the converse of 1 assuming both continuity and exactness.

C. CARTAN'S MATCHING PROCESS.

Let X be a compact space and $\{A_\alpha\}$ a family of closed subsets of X. Let $A = \subset A_\alpha$ and i_α: $(X,A) \subset (X,A_\alpha)$. Assume that an exact and continuous homology theory (on the category $\mathfrak{A}_C$ of compact pairs) is given and that, for a fixed integer q,

$$H_{q+1}(X,B \cup A_\alpha) = 0 \qquad \text{for all } \alpha$$

where B is any finite intersections of the sets $\{A_\alpha\}$.

1. Prove that, if $u \in H_q(X,A)$ is such that $i_{\alpha*}u = 0$ for all α, then $u = 0$

2. Each $u \in H_q(X,A)$ determines a family $\{u_\alpha\}$ where $u_\alpha = i_{\alpha*}u \in H_q(X,A_\alpha)$ satisfying $k_{\alpha\beta}u_\alpha = k_{\beta\alpha}u_\beta$ where $k_{\alpha,\beta}$: $H_q(X,A_\alpha) \to H_q(X,A_\alpha \cup A_\beta)$ is induced by inclusion. Show that each family $\{u_\alpha\}$ satisfying $k_{\alpha\beta}u_\alpha = k_{\beta\alpha}u_\beta$ is obtained in this fashion from some $u \in H_q(X,A_\alpha)$.

3. Assume that the open sets $X - A_\alpha$ form a base for $X - A$ and that $u_\alpha \in H_q(X,A_\alpha)$ are such that $k_{\alpha\beta}u_\alpha = u_\beta$ whenever $A_\alpha \subset A_\beta$. Show that there is some $u \in H_q(X,A)$ with $i_{\alpha*}u = u_\alpha$ for all α.

4 Transcribe the results of §6 and of the preceding exercises into the "single space notation" of x,7.

5 Assume that the coefficient group G is a domain of integrity, and deduce from §6 results concerning co-Betti numbers. In particular show that if A is a proper closed subset of S^n then $S^n - A$ has $R^{n-1}(A;G) + 1$ components

Index

Index

Ingram Content Group UK Ltd.
Milton Keynes UK
UKHW051119100523
421436UK00024B/161

9 781015 432451